DICTIONARY OF ZOOLOGY

DICTIONARY OF ZOOLOGY

KIRAN NARANG
M.Sc., Ph.D.

ANMOL PUBLICATIONS PVT. LTD.
NEW DELHI - 110 002 (INDIA)

ANMOL PUBLICATIONS PVT. LTD.
Regd. Office: 4360/4, Ansari Road, Daryaganj,
New Delhi-110002 (India)
Tel.: 23278000, 23261597, 23286875, 23255577
Fax: 91-11-23280289
Email: anmolpub@gmail.com
Visit us at: www.anmolpublications.com

Branch Office: No. 1015, Ist Main Road, BSK IIIrd Stage
IIIrd Phase, IIIrd Block, Bangalore-560 085 (India)
Tel.: 080-41723429 • Fax: 080-26723604
Email: anmolpublicationsbangalore@gmail.com

Dictionary of Zoology

First Edition, 1988

Reprint, 1989, 1990, 1991, 1992, 1993, 1994

Second Revised Edition, 1995

Reprint, 2002, 2003, 2004, 2010

PRINTED IN INDIA

Printed at Mehra Offset Press, Delhi.

Preface

It is a great satisfaction that the first edition of this dictionary got exhausted within a short span of time. Now the editor is presenting, revised and uptodate edition of this dictionary.

This dictionary has been written to provide students of zoology at all levels with a handy and reliable source for the meanings of terms in the field of zoology. These terms have been carefully selected and presented in brief analytical or functional phrases, thus avoiding the more comprehensive type of treatment appropriate to larger reference works. An attempt has been made to include new terms which have been introduced and have come into usage with the developments of modern zoology.

Line drawings have been included whenever the meaning of a word can be best understood by means of a diagram. The network of cross-references has been kept throughout this book.

The dictionary will be of value to students of zoology and to naturalists, geographers and others studying or working in related fields.

The editor is happy to give credit here to all those who had a hand in getting the dictionary into its present form and specially to put on record great debt to Ms Timmy Katyal, University of Saskatechowan, Canada, for the care and interest she took in clearing up complex points so that they might be put in simple English.

In compiling a dictionary of this kind, it becomes necessary to draw upon the work of many authorities and seek the advice of colleagues to all of whom the editor is deeply indebted.

All comments from users on omissions or short comings will be most welcome.

Kiran Narang

Preface

A

Abactinal. The surface of the body opposite to the mouth in an animal that is having radial symmetry. Also, *see aboral.*

A-Bands. A-Discs. Refer to dark bands which are running transversely across the fibres of striped muscle. They are formed by filaments of the protein called *myosin* partly interlocking with thinner filaments of another protein called *actin,* the latter forming the so-called I-discs.

Abaxial. Not on, pointing away from or on the side. Abaxial is synonymous with dorsal when the latter term is being applied to lateral organs. Compare *adaxial.*

Abdomen. (1) Refers to the section of the body cavity of vertebrates that is having the stomach, intestines, and other viscera.

(2) Refers to the posterior section of the body in arthopods. Primitively it is made up of a series of similar segments.

Abducens Nerve (Cranial nerve VI). One of the pair of nerves arising from the anterior end of the medulla oblongata in the vertebrate brain to supply the posterior rectus muscle of each eyeball. It consists mainly of motor nerve fibres. It serves to rotate the eyeball. See cranial nerves.

Abductor. A muscle which is able to pull a limb away from the main axis of the body (*cf. Adductor*).

Abiogenesis. The development of living from non-living matter, as in the origin of life.

Abomasum. The fourth and last region of the specialized stomach of ruminants (e.g. the cow), sometimes called the *rennet stomach*. It is lined with normal mucosa having gastric pits, producing hydrochloric acid, pepsin, and rennin, and is the true digestive stomach.

Aboral. Refers to the side of an animal which is away from the mouth in those animals that do not have definite dorsal and ventral surfaces, or whose dorsal and ventral surfaces change position in the course of development (e.g., in a starfish).

Accessory Nerve (Cranial nerve XI). One of the pair of nerves arising from the posterior region of the medulla oblongata in the brain of higher vertebrates. It carries motor nerve fibres and merges with the adjacent vagus nerve close to its root. In fishes and amphibians it is regarded as an integral part of the vagus. (See *cranial nerves*).

Accessory Sex Organs. Organs assisting in the process of sexual reproduction, e.g., the nuptial pads of male frogs, claspers of insects etc.

Acclimatisation. Used for describing adjustment of an animal organism with some particular condition of its environment by making necessary physical, physiological, biochemical and behavioural changes e.g., as frog being cold blooded animal it cannot withstand excessive hot or cold weather and, therefore, during winters it undergoes hibernation which involves lowering of BMR, consequent suspended animation thereby minimising the vital needs of the body to cope with the conditions of cold while during summers it undergoes aestivation.

Accommodation. Changing the focus of the eye by altering the curvature of the position of the lens.

Accretion. Process of growth in which material is added to the outside in non-living matter.

Acellular. Having no cells or compartments e.g. Protozoa. The term is introduced by considering the protozoan cell as a complete organism. As a single cell of these animals, as compared to metazoan cells, is capable of carrying out all the functions of life and therefore regarded equivalent to organism and not the cell.

Acephalous. Without a head. The term is also referred for such creatures whose heads are much reduced or are invaginated into the thorax, as in certain insect larvae.

Acetabulum. Refers to the socket for the articulation of the femur in the pelvic girdle of a tetrapod vertebrate. It consists of the ilium, ischium and pubis.

Acetylcholine. A chemical substance ($CH_3CO.O.CH_2CH_2N(CH_3)_3OH$) which is produced at the synapses and nerve endings of the parasympathetic nerves. Its effects are the exact opposite to those which are produced by *sympathin*, a substance similar to adrenalin, which is produced by the sympathetic nerves. Acetylcholine for example, slows down the heart while sympathin speeds it up.

Acetylcholinesterase. Refers to an enzyme hydrolysing acetylcholine into choline and acetate; present on the surface of post-synaptic membrane. Enzyme molecule is having two distinct sites: an anionic site which binds the quaternary nitrogen of acetylcholine and an esteratic site which can donate electron to the acetate portion of acetylcholine molecule separating it from the choline moiety. This permits the hydrolysis of acetylcholine and terminates the post-synaptic action of the transmitter.

Acetyl Coenzyme A (Acetyl Co-A). A compound of key importance in intermediary metabolism because it is having higher energy content than acetic acid. It reacts readily with substances in reactions which would otherwise need outside energy. From energetics point of view, formation of 1 mol of compound is equal to the formation of I mol of ATP. Formed by pyruvic acid in Kreb's cycle and also by β-oxidation of fatty acids.

Achromatic Figure. The metaphase plate chromosomal line up.

Achilles, Tendon. The tendon of the heel: a strong tendon connecting the gastrocnemius and solaeus muscles to the calcaneum in terapods.

Aciculum. Refers to an internal needle-like chaeta or bristle which gives support to the external chaeta on certain polychaete worms. Free-swimming polychaetes such as *Nereis* are having on each segment a pair of flap-like projections called parapodia on which there are a number of external chaetae.

Acne. A disease which is marked by pimples.

Acoelous. The first two centrum of a vertebra in which both the sides are convex e.g. IX vertebra of frog.

Acorn-Worm. A common name used for hemichordate *Balano-glossus.*

Acoustico-Lateralis System. Refers to a system in vertebrates (especially fish and aquatic amphibians) which is having neuromast cells in various tracts on or near the surface of the body (i.e. the *lateral-line* system) and the inner ear (*membranous labyrinth*). Neuromasts and receptor cells of the inner ear are similar. They consist of groups of cells having sensory processes which are innervated by nerves from the medulla oblongata of the brain. They are nearly helpful in detection of sounds in water.

Acquired Characteristics. Refers to the changes in the structure or function of an organ or system during the life of an organism. These changes occur by the use or disuse of that organ or system, or by environmental influences. Acquired characteristics are not inherited. See also *Lamarckism*.

Acrania (Protochordata). A subdivision of the Chordata including two subphyla called the Urochordata and the Cephalochordata, (i.e. all the invertebrate chordates). Brain, skull, heart, and kidneys are not present and the notochord is never replaced by cartilage or bone. Sometimes the term is used for the Cephalochordata only.

Acrocentric. A chromosome having one arm shorter than the other one because centremere is close to one end.

Acrodont. Refers to condition when teeth are fused to the outer side of jaw-bone e.g., teeth in some bony fishes and reptiles like *Uromastix*. Such a condition in an individual is known as *acrodonty*.

Acromegaly. A disease which is caused by hypersecretion of anterior pituitary after bones have altarned full growth; bones thickened especially in the limbs and face; characterised by an overgrowth of mandibles, phalangeal certilages and related areas, forehead, nose, hands, feet etc. Body is becoming disproportionate giant.

Acromion. A combination of the spine of the scapula which is forming a projection for articulation with the clavicle in many mammals and birds.

Acrosomal Cap. Covering over the head of a sperm which is derived from the golgi body.

Acrosauria. Extinct crocodile-like reptiles of the Upper Jurrasic period.

Acrosome. Refers to the pointed anterior projection of a spermatozoon by which it is able to penetrate the ovum.

ACTH (Adrenocorticotropic Hormone). A peptide which is secreted from anterior pituitary lobe (sdenohypophysis); acts on adrenal cortex for increasing the synthesis and secretion of steroids in adrenal cortex thereby resulting in conversation reaction during stress. (Also see *homeostasis*). The Corticotropie Releasing Hormone (CRH) from hypothalamus stimulates release of the hormone while the hormone itself is responsible for inhibiting its release.

Actin. Refers to a protein whose presence forms light bands which are running transversely across the fibrils of striated muscle. These are called I-bands (formerly known as J-discs). When the muscle is relaxed the actin molecule attached to an ATP molecule whose energy is liberated muscle is stimulated, (See also *Mysoin*).

Actinal Surface. The oral surface, i.e., that on which the mouth is situated in an animal that is showing radial symmetry. (*Cf. Abactional.*)

Action Potential. The transitory charge in electrical potential occurring across the membrane of a nerve or muscle fibre during the passage of a nervous impulse. The degree of change has been found to be independent of the strength of the impulse; it either takes place or does not (see all or-none). In the absence of an impulse there is a *resting potential* of about-70 mv across the membrane (inside negative) due to the intracellular and extracellular media (*see sodium pump*). However, during the passage of an impulse the resistance of the membrane decreases and permits an inward current of sodium ions, which makes the inside less negative and thereby change the membrane potential to about + 30 mv (inside positive). It is followed by an outward flow of potassium ions shortly afterwards and restores the resting potential. Local currents are following ahead of the action potential that in turn result in further action potentials. In this way a wave of activity is propagated along the membrane to give rize the impulse.

Activator. Refers to a substance which increases the activity of an enzyme. It may be an inorganic ion such as calcium, an organic substance or sometimes another enzyme, e.g. enterokinase.

Active Site. Refers to the region of an enzyme molecule that undergoes combination with and acts on the substrate. It consists of amino acids which are arranged in a configuration specific to a particular substrate or type of substrate. Binding of an inhibiting compound elsewhere on the enzyme molecule may alter this configuration and thereby the efficiency of the enzyme activity.

Active Transport. Refers to the transport of molecules or ions taking place across a cell membrane against a concentration gradient, with the expenditure of energy. Anything which is interferring with the provision of energy will interfere with active transport. The exact mechanism is not known with certainly but there is a carrier involved, probably a protein or lipoprotein.

Actomyosin. A protein complex of *actin* and *myosin*, two of the active substances involved in muscle contraction. When stimulation has been applied to the muscle the actomyosin probably helps in the liberation of energy from ATP.

Acupuncture. Old system of medical treatment which involves the piercing the body at specific points in specific numbers and pattern with metalic needles (particularly Ag, Au, Zn etc.) for treating diseases.

Acyl Carrier Protein (ACP). A protein to which 'acyl' residues get attached during fatty acid synthesis.

Adaptation (1) Refers to the extent to which an organism, or a physiological or structural characteristic of an organism, becomes suited to a particular environment.

(2) (*Accommodation*). Refers to the decrease with time of the frequency of response of sensory receptors which are subjected to a constant intensity of stimulus. A more intense stimulus is then needed to elicit a response.

Adaptive Convergence. Refers to the superficial similarity between different species of animals due to similarity of habits, or sometimes due to the fact that both are exhibiting protective resemblance to some other species, e.g. several flies, beetles, spiders etc., show a protective resemblance to ants.

Adaptive Orientation. A habit which is shown by many insects as well as some birds and other animals, of standing or resting in such an attitude as to make the best possible use of their protective colouration.

Adaptive Radiation. Refers to the tendency for a group of animals, the members of which are more or less similar, to produce local races which become suited to different environments, e.g., coloured for camouflage in different habitats.

Adductor. (1) A muscle is pulling a limb towards the main axis of the body.

(2) A muscle which is pulling two parts together, as for example those which close a bivalve shell.

Adenine. Constituent of DNA; organic base of purine group, on combining with dexyribose sugar and phosphoric acid it forms nucleotide; associated with complementary base thymine forming cross-bar of hydrogen bound in the DNA heli.

Adenohypoyhysis. The glandular lobe of the pituitary organ which is derived embryonically from an upgrowth of the stomodaeum. It consists of the *pars distalis, pars tuberalis pars intermedia.*

Adenosine. A phosphorylated organic compound which takes part in intracellular energy transfers.

Adenosine Triphosphate (ATP). Energy rich nucleotide which is used in celluear metabolism and activities requiring input of energy (endergonic reactios). When ATP is converted to ADP or AMP, one or two phosphate groups are released giving up one or two energy-rich bonds to the reaction. Hydrolysis or first bond liberates 3000 cals and that of second bond liberates 12,000 cals. (Fig. 1).

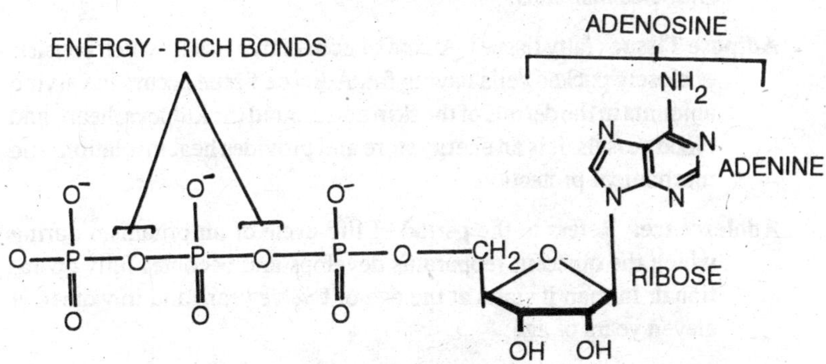

Fig. 1 A.T.P.

Adenotrophic Viviparity. Used for describing the phenomenon in certain insects (e.g. in *Glossina*) in which the larvae after hatching are retained in the 'uterus' of the parent, develop by feeding on uterine secretions, moult twice and are ultimately deposited as mature larvae are almost ready to pupate.

Adenylate Cyclase. Refers to an enzyme which is associated with hormone action at its target cell catalysing the hydrolysis of ATP (adenosine triphosphate) to *c*-AMP when it gets allosterically activated due to conformational changes taking place in receptor molecule present in target cell membrane which is brought about by interaction of receptor molecule. The compound has been closely associated with receptor molecules but faces towards cytoplasm.

ADH (Antidiuretic Hormone). Refers to hormone which is secreted by posterior pituitary lobe. It is involved in reabsorption of water from glomerular filtrate and thereby controls urine formation (diuresis). In kidney tubule Na^+ concentration around the proximal and distal convoluted tubules are equivalent to that of most of body (300-350) milliosmols but the concentration of Na^+ ions at the turn of tubule's loop of Henle and proximal to that becomes much higher (about 1500 milliosmols). This ionic concentration gradient pulls water out of collecting tubule and helps capillary water reabsorption. This whole mechanism has been controlled by hormone. In hyposecretion of the hormone the Na^+ ions get absorbed and thereby concentration gradient gets disturbed resulting in passing out large amounts of water in the urine thus resulting a pathological condition called *Diabetes insipidns.*

Adipose Tissue (fatty tissue). A type of connective tissue which consists of closely packed cells having fat. Adipose tissue occurs in varying amounts in the dermis of the skin and around the kidneys, heart, and blood vessels. It is an energy store and provides heat insulation, and mechanical protection.

Adolescence. Refers to the period of life-cycle of an organism during which the quiscent repparatus develops and becomes fully operational. In man it starts at the age of twelve years and in women at eleven years of age.

ADP (Adenosine Diphosphate). Produced in the cells during metabolism by involving a hydrolytic removal of terminal bond to ATP; also acts as a source if energy for metabolic process and releases energy by hydrolytic removal of one terminal phosphate bond and gets converted to AMP. It is converted to ATP as soon as free phosphate group becomes available.

Adrenal Cortex. Refers to the outer part of adrenal, gland; divided into three zones of variable distinctness—outer zona glomerulosa having whorls of cells which become continuous with columns of cells of middle zone called zona fasciculata (these columns are separated by venous sinuses), innermost zona reticularis where all the columns are interlaced in network. The cells have abundant lipids which are present especially more in outer portion of zone fasciculata. It secretes glucocorticoids which are having widespread effects on metabolism of carbohydrates and proteins, mineralocorticoids which play essential role in maintenance of sodium balance and extracellular fluid volume and sex hormones which tend to exert minor effects on the reproductive functions.

Adrenal Glands. Endocrine glands which are situated near the kidneys in all higher vertebrates; secreting a number of hormones, the chief of which is *adrenalin*. In lower vertebrates such as fish two sets of glands are known *inter-renals* between the kidneys and *supra-renals* above the kidneys. The adrenals of higher animals are formed by a fusion of these two.

Adrenaline (epinephrine). A hormone which is secreted by the adrenal glands. The middle part of these glands, the adrenal medulla, secretes the hormone. This hormone is chemically almost identical to the transmitter substance noradrenaline which is produced at the ends of sympathetic nerves. Adrenaline secretion into the bloodstream in stress brings about the acceleration of the heart, constriction of arterioles, and dilation of the pupils. In addition, adrenaline brings about a marked increase in metabolic rate thereby preparing the body for emergency.

Adrenal Medulla. Refers to the portion of adrenal gland; consists of interlacing cords of densely intervated granule-containing cells which about on venous sinuses. Morphologically there are tow types of cells-epinephrine secreting cells (having dense granules) and nor-epinephrine secreting cells (in which the smaller dense granules do not fill the vesicles in which they are present). Paraganglia—

small groups of cells resembling those of medulla, occur near the thoracic and abdominal sympathetic ganglia. Medulla secretes epinephrine (adrenaline) and nor-epinephrine (nor-adrenaline) which are collectively known as catecholamines.

Adrenergic. A term for describing nerves which liberate a substance similar to adrenalin at their ending, e.g., many of the sympathetic nerves. Nerves of this type usually possess an action antagonistic to the other type which secrete acetylcholine at their endings. These are termed as *cholinergic*. An example of this antagonism is, that adrenergic nerves increase the beating of the heart and cholinergic nerves show it down.

Adrenergic Receptor. Refers to the membrane receptor molecules which are specific to interact with catecholamines; type of nerve fibre which releases adrenaline like substance from axon terminal when nerve impulse transmission is taking place across the synapses. It is possible to distinguish two types of adrenergic receptors (α-adrenergic receptors and β-adrenergic receptors) by use of drugs that activate one or the other receptor types, eliciting responses which are characteristic of one or the other type of catecholamine.

α-Adrenergic Receptors. Refer to the adrenergic receptors to which adrenaline (epinephrine) binds most effectively.

β–Adrenergic Receptors. Refer to the adrenergic receptors to which adrenaline (epinephrine) binds most effectively.

Adrenotrophic Hormone. A hormone which is produced by the anterior pituitary lobe, stimulates activity of the adrenal glands.

Adventitia. Said of an outer layer which is surrounding some structure, e.g., dense fibrous connective tissue layer which is surrounding alimentary canal in neck and thorax region where serosa is not present.

Aedeagus. The male intromittent organ or 'penis' in certain insects.

Aequorin. A type of an animal protein extracted from the jellyfish *Aequorea*. When it is combined with calcium ions, it emits a photon of bluish-green colour. Discovered in 1963. The protein finds use in the measurement of the Ca^{2+} in various tissues and body regions photocolourimetrically.

Aerial. An animal/organism that is inhabiting or frequenting the air e.g. birds and insects.

(Aerobe) Aerobic. Refers to organisms able to live and grow only in presence of free oxygen; refers to processes that can only take place when oxygen is available e.g., conversion of pyruvic acid to acetic acid before Kreb's cycle.

Aerobic Respiration. Respiration in which free oxygen is consumed in oxidising organic substrates to carbon dioxide and water, evolving a high yield of energy. The reaction overall is: $C_6H_{12}O_6 + 6O_2 = 6CO_2 + 6H_2 +$ energy. It involves a number of stages, the first of which (glycolysis) also takes place in anaerobic respiration in the cell cytoplasm. With glucose as the substrate, a sequence of reactions gives rise to the formation of pyruvate. The remaining stages, which do not take part in anaerobic respiration, take place in the mitochondria. Pyruvate is converted to acetyl coenzyme A which enters a cyclic series of reactions, the Kreb's cycle, with the formation of carbon dioxide and hydrogen atoms. There is a net production of 36 ATPs per molecule of glucose during aerobic respiration.

Aestivation. Refers to a state of inactivity which some animals undergo during long periods of drought or heat. Superficially it resembles hibernation; feeding, movement, respiration and other bodily activities are very much reduced. In many fish aestivation occurs in response to the drying up of the water.

Afferent. Refers to conveying (impulses, blood, etc.) from the outer regions of an organ or body towards the center. For example, afferent nerves are conveying nerve impulses from sense organs to the central nervous system. *Compare* efferent.

Afterbirth. After the birth of the young the placenta, umbilical cord, and membranes discharge from the body.

Agglutination. Used for describing the clumping together of red blood cells or bacteria due to the action of antibodies. Agglutination may take place in transfusion if blood of the wrong group is given.

Aggression. The type of animal behaviour involving threats and actual attacks on other animals. Aggression is generally a response to opposition (e.g. defence of territory).

Agnatha. The class having the earliest and most primitive vertebrates that are characterized by the absence of jaws. The only living order is the Cyclostomata, which is including lampreys (e.g. *Petromyzon*) and . hagfish (e.g. *Myxine*).

Agnathostomata. Animal without jaws in mouth e.g. cyclostomes.

Albinism. Used for describing the absence of pigmentation in the eyes, skin, feathers, hair, etc., of some animals. Albinism in humans is due to a recessive gene and occurs in the absence of a dark brown pigment (melanin). The complete albino is having milk-white skin and hair and the irises of the eyes appear pink.

Albinism-Ocular. Refers to the X-linked trait in which melanin pigment is absent from eyes but not from other parts.

Albinism-Oculocutaneous. Refers to the absence of melanin pigment from eyes, skin and hair.

Albumen. See *albumin*.

Albumin. A group of simple proteins, which are normally deficient in glycine. They are water soluble and when heated coagulate. Examples are legumelin in peas and the serum albumin of blood. The protein in egg white (*albumen*) is a mixture of albumins. These also have carbohydrate groups and consequently have been classified as conjugated proteins.

Alchoholic Fermentation. A type of anaerobic respiration in which glucose is converted to ethyl alcohol and CO_2; carried out by yeasts and various other fungi and by certain bacteria; takes place outside the organism and catalysed by zymase secreted either by living cells or released on their death; used in brewing, wine making, backing etc.

Aldosterone. A hormone which is secreted by the cortex of the adrenal glands. It is a steroid, and plays an important role in the control of sodium and potassium ion concentrations in mammals.

Alimentry Canal. Used for describing the gut or passage through which food is passing and in which it is being digested and absorbed. It may be having a number of regions: oesophagus, stomach, intestines etc., but usually the actual digestion occurs only in the central part called the mid-gut or mesenteron which is lined with cells of endodermal origin (Fig. 2).

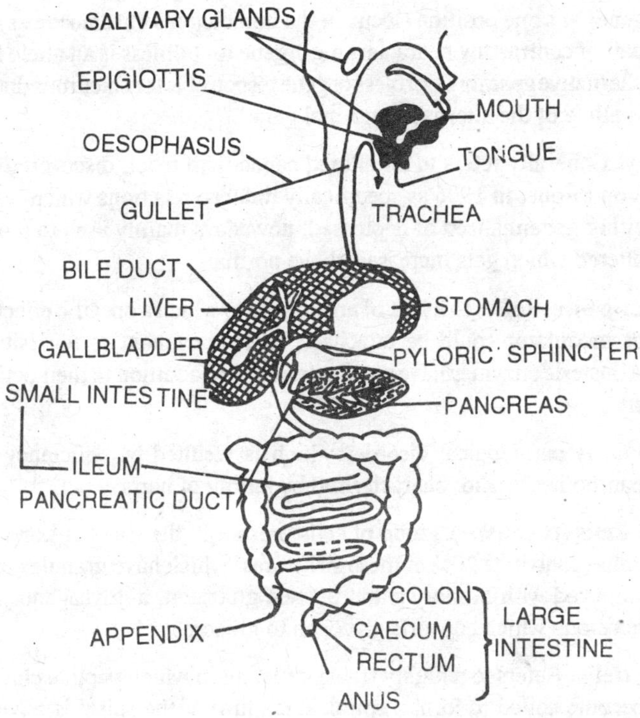

Fig. 2. *Alimentary Canal of Man.*

Allantochorion (Chorio-Allantois). A compound membrane which is obtained by fusion of the chorion and the allantois in eutherian mammals. It takes part in the formation of the placenta.

Allantoic Acid. An excretory product of many vertebrate embryos, which is formed by the oxidation of allantoin.

Allantoin. An excretory product of many vertebrate emoryos which is formed from breakdown of uric acid.

Allantois. Used for describing a sac or bladder which grows out from the hind-gut in the embryos of reptiles, birds and mammals. As it enlarges it gets very vascular, extending outside the embryo, and is the chief organ by which dissolved food and oxygen get absorbed.

Allele. Used for one of a pair or any one of series of alternate genes which possesses same locus in homologous chromosomes; refers to two genes at same position (locus in the homologous chromosomes); a pair of contrasting characters e.g., gene for tallness is an allele for alternative gene for dwarfness or gene responsible for colourblindness is allele of the alternative normal gene.

Allergy. Generally refers to an altered capacity to react; discovered by Von Pirquet in 1906 as specifically induced reactions which were either get enhanced or depressed; nowadays mainly used to mean altered which gets increased above normal.

Allosteric Site. Refers to a part of an enzyme to which a specific effector or modulator could be attached. This attachment is reversible. Allosteric enzymes have an allosteric site in addition to their active site.

Alopecia. A pathological disorder which is resulted by deficiency of panthothenic acid; characterised by falling of hairs.

Alpha-Cells (α-cells). A group of cells present in the islets of Langerhans (constitute 20% of these structures) which have granules that stain red with Mallory stain. Secrete glucagon, a glycogenolytic hormone which converts glycogen to glucose.

Alpha Helix. Refers to a highly stable structure in which peptide chains become coiled to form a spiral. Each turn of the spiral is having approximately 3.6 amino-acid residues. The R group of these amino-acids extends outward from the helix. Hydrogen bonding between successive coils is responsible for holding the helix together.

Alpha-Keto Glutaric Acid (α-Keto Glutaric Acid). An organic compound which is formed during Kerb's cycle by dehydrogenation of

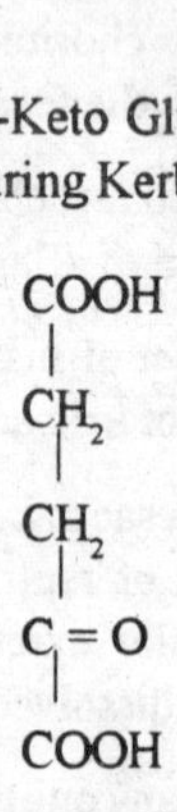

Alpha-Keto Glutaric Acid

isocitric acid and is changed into succinic acid by its own dehydrogenation. (*Also see Kerb's cycle*). Also formed in tissue by glutamic acid during transamination. (Glutamic acid+ Pyruvic acid = α - Keto glutaric acid+ Alanine).

Alpha-Motoneurons. Refer to the neurons which are exiting the spinal cord and synapsing with extrafusal fibres. These initiate the muscle contraction.

Alternation of Generations. Means alternate development of two types of individual during the life-cycle of an organism. Usually one type reproduces asexually while the other sexually. In many parasitic protozoa (e.g., the malarial parasite), the sexual phase is in are host and asexual phase in another.

Alveolar Ducts. The ducts which arise from each respiratory bronchiole in mammals and terminate in air sacs whose walls get outfolded to form alveoli (Also see *alveoli*).

Alveolar Glands. Refer to multicellular glands (acinous glands) which have flask shaped bodies and their necks serve as ducts e.g., mucous and poison gland of frogs and toads.

Alveolar Membranes. Tooth in mammals is kept in bony socket termed as alveolus and part of tooth inside the body socket is termed as root. The root of the tooth separates from bone of eveolus by collagenous periosteum of jaw bone which is as alveolar membrane (also called periodontal membrane).

Alveoli. (1) Minute air sacs making up the lungs of a vertebrate.

(2) Acini or secretory cavities of a gland.

(3) Any tiny pit or cavity.

Ambient. The surrounding/prevailing conditions e.g., ambient temperature, ambient pressure etc., to which the living organisms are put.

Ambulacrum. Refers to groove lined with tube feet in certain echiniderms.

Amennorrhea. Used for the absence of menstrual periods in human female; it is of two types—*primary,* when menstrual bleeding has never taken place and *secondary,* if there occurs cessation of cycles in woman with previously normal periods.

Amino Acids. Derivatives of carboxylic acids in which a hydrogen atom in an aliphatic acid is replaced by an amino group. All are optically active. In the body the various proteins are formed from the necessary amino acids. In man, all but ten of the twenty amino acids can be synthesized by the body itself. As these are not required in the diet they are called non-essential amino acids. The remaining ten cannot be synthesized by the body and are to be supplied in the diet. They are called essential amino acids.

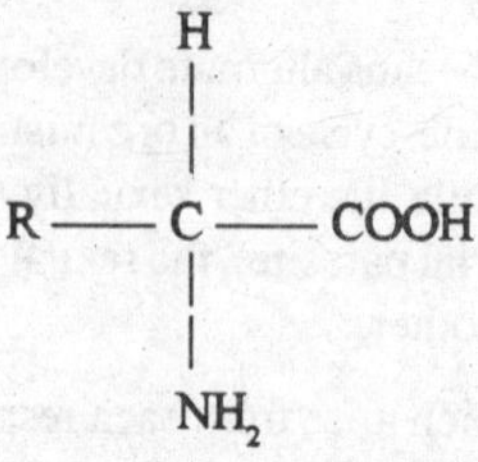

The amino acids most commonly found in proteins

alanine	glycine	prolinet[2]
arginine	histidine[1]	serine
asparagine	isoleucine[1]	threonine[1]
aspartic acid	leucine[1]	tryptopher[1]
cysteine	lysine[1]	tyrosine[1]
glutamic acid	methionine[1]	valine[1]
glutamine	phenylanine[1]	

[1] essential amino acids in animal diets

[2] an imino acid derived from pyrollidine

Aminopeptidase. An exopeptidase attacking terminal bond at the free amino end of the peptide chain liberating amino acids and dipeptides. They contain Mg^{2+} or Mn^{2+}.

Amitosis. Nuclear division characterized by the absence of a nuclear spindle and resulting in the formation of daughter nuclei with unequal sets of chromosomes. Cells formed amitotically inherit variable numbers of chromosomes.

Ammocoete. The larva of a cyclostome like the lamprey which exhibits many primitive features and in some respects, notably the ciliary method of feeding, resembles *Amyhioxus*.

Ammonotelic. Having the ability to excrete ammonia. This occurs in many small aquatic organisms where the ammonia, which is very soluble and very toxic, is immediately carried away by the water.

Amnio-Cardiac Vesicles. Two pouches of splanchnic mesoderm which get fused together to form the pericardium in the embryonic development of a vertebrate.

Amnion. Refers to one of the three embryonic membranes of reptiles, birds and mammals. It is the inner layer of membrane which encloses the embryo, first as a hood and later as a complete bubble. The amniotic cavity has *amniotic fluid*, which may be removed in pregnant women by amniocentesis to diagnose some types of genetic abnormality in the fetus.

Amniota. Reptiles, birds and mammals, i.e., land-living vertebrates whose embryos possess an amnion, chorion and allantois.

Amniotic Fluid. The fluid filling the amniotic cavity (see *Amnion*).

Amoeba. The member of the Protozoa. A microscopic organism occurring universally in fresh water, it is continually changing shape due to the formation of pseudopodia for locomotion and food capture. Food and water vacuoles undertake digestion and osmoregulation. Reproduction takes place by binary fission and spore formation occurs in adverse conditions.

Amoebocyte. Refers to a cell that can wander freely in animal tissues. Amoebocytes are found in the walls of sponges and in the blood and body fluids of mammals. Their general appearance particularly in exhibiting amoeboid movement is similar to amoeba.

Amoeboid. Having the qualities of an *Amoeba*: i.e., moving and feeding by the action of pseudopodia.

Amoeboid Movement. Refers to the movement such as that of *Amoeba* in which some of the protoplasm temporarily changes from a *gel* into a *sol* and flows forwards in the form of pseudopodia. Such movement takes place in many animal cells, e.g., the white corpuscles of the blood.

Amoebulae. Small amoebae that emerge after multiple fission within a cyst.

AMP (Adenosine Monophosphate). A nucleotide which consists of adenine, ribose and phosphate. See *ATP, cyclic AMP*.

Amphiaster. Used for describing the figure which is produced by two asters and spindle in cytoplasm in the division of nucleus during mitosis.

Amphibia. Refers to cold-blooded, smooth skinned vertebrates that are able to start life as an aquatic larva breathing by gills and later undergo a somewhat rapid metamorphosis into the adult lung-breathing form. Frogs, toads and newts are well-known examples.

Amphibious. An organism that is capable of leading life both in water and on land e.g. frog and small.

Amphiblastic. An alternative word for *telolecithal;* a term which is used to describe eggs which are having most of the yolk on one side (the vegetative pole), and the nucleus on the other side (animal pole).

Amphiblastula. The larva of certain type of sponge: it consists of a spherical mass of cells half of which get flagellated and the other half large and granular.

Amphichelydia. One of the most ancient groups of tortoises and turtles found as fossils of the Triassic period.

Amphicoela. Refers to a group of frogs in which the vertebrae are amphicoelous or biconcave, e.g. the New Zealand frog.

Amphicoelous. Biconcave: a term which is used to describe vertebrae in which the centrum is having a conical depression on both its anterior and its posterior face. Such vertebrae occur in most fish.

Amphimixis. True sexual reproduction which takes place by fusion of gametes. *Compare* apomixis.

Amphipathic. Refers to molecules which bear groups with different properties e.g., hydrophilic and hydrophobic groups.

Amphiplatyan. Refers to flat on both the sides e.g., vertebrae of mammals in which centra and pleurocentra are both having more or less flat surfaces.

Amphisytlic Jaw Suspension. An arrangement occurring in a few sharks and many fossil fish (e.g. the Cladoselachii) whereby the jawa are being suspended by ligaments from the skull and by the hyomandibular bone. *Compare* autostylic jaw suspension, hyostylic jaw suspension.

Amphioxus (Branchiostoma). Refers to a genus of small marine burrowing cephalochordates, the lancelets. *Amphioxus* is having a fish-shaped body with a dorsal and caudal fin and segmentally arranged muscle blocks (myotomes). The pharynx and gill slits have been modified for food collection as well as respiration. Excretion takes place by nephridia, which is unique among chordates. See *Cephalochordata*.

Amphoteric. A chemical substance having qualities of both an acid (capable of liberating H^+ ions) and a base (capable of accepting H^+ ions) e.g., aminoacids.

Amplexary Pads. Used for describing sexual dimorphic character present in hand of male frog which forms in the first digit grows and develops during breeding season under hormonal influence and serves to hold the female during amplexus.

Amplexus. Refers to copulatory embrace, e.g. of a frog in which the male climbs on the back of the female and presses tightly with the nuptial pads.

Ampulla. Refers to bladder-like enlargement; generally refers to water sacs which are attached to tube feet echinoderms; sac of ampulla of Lorenzini in sharks; also refers to canals of internal ear.

Ampullae of Ear. Refers to the dilatations of the three semi-circular canals of the vertebrate ear at the point where each is entering the sacculus.

Ampullae of Lorenzini. Refers to the small pits in skin which are situated close to the lateral line canals or in patterns on the heads of many fish.

Amylase. An enzyme which hydrolyses polysaccharides to maltose (starch/polysaccharides⟶ Dextrins, achrodcextrins, maltriose and maltose). Universally present in pancreatic juice of all the vertebrates and also in the digestive juices of other animals, in the blood and in the tissues other tan alimentary canal. A major enzyme of saliva of mammals and starts digestion right from buccal cavity.

Amylopsin. Refers to pancreatic amylase which acts on carbohydrates during digestion of food.

Anabolism. Refers to a constructive phase of metabolism in which cells synthesis protoplasm from food materials e.g., protein synthesis.

Anadromons. Refers to the fishes living in sea going to fresh water for spawning e.g. *Salmon*.

Anaemia. Refers to the condition in which there occurs the low percentage of haemoglobin in blood than normal (16 g/dl in male and 14 g/dl in female); may be occurring due to red blood cell volume deficiency (microcytic anaemia) or with respect to haemoglobin concentration (normochromic or hypochromic). Vitamin B_{10} plays vital role in curing perniceous or microcytic anaemia. The normochromic anaemia is cured by iron therapy.

Anaerobe. An organism that can live and grow in the absence of free oxygen, i.e., it respires anaerobically.

Anaerocic Respiration. A type of respiration found in yeasts, bacteria, and occasionally in muscle tissue, in which oxygen does not take part. The organic substrate is not completely oxidized and the energy evolved is low. In the absence of oxygen in animal muscle tissue, glucose gets degraded to pyruvate by glycolysis, with the evolving of a small amount of energy and also lactic acid, which may get oxidized later when oxygen becomes available. Fermentation is another example of anaerobic respiration, in which certain yeasts produce ethanol and carbon dioxide as end products. Only two molecules of ATP are formed in this process.

Analogous. Refers to the parts which are similar in function but not necessarily alike in genetic relationship, e.g., wings of birds and insects.

Analogous Organs. Organs which, although they may have quite different origins, carry out the same functions in different animals, e.g., the kidneys of an amphibian and those of a mammal (*cf. Homologous Organs*).

Anal Plate. A thin sheet of skin which is formed by fusion of the wall of the hindgut with the ectoderm in the vertebrate embryo. It marks the position where the future anus will be formed by perforation.

Anal Respiration. Refers to the absorption of dissolved oxygen which takes place through the wall of the hind-gut in certain arthropods, e.g. *Cyclops*. To facilitate the process the water is kept moving by rhythmic contractions of the intestine.

Anal Style. A pair of small structures that are attached to sternite of 9th abdominal segment in most insects; the characteristic feature and identification marks of male cockroach.

Anal Vesicles. Refers to the chief excretory organs of echinoids; a pair of caeca which are leading into the hind-gut, each caecum having numerous ciliated funnels opening into it from the coelom.

Anamnia (Anamniota). Vertebrates whose embryos lack an amnion, i.e. Agnatha, Fish and Amphibia (see *Amniota*).

Anamorpha. A subclass of Chilopoda (centipedes) which is having fifteen pairs of legs; comprising the two orders Lithobiomorpha and Scutigeromorpha (q.v.).

Anamorphosis. Refers to addition of extra abdominal segments to an insect after it has hatched from the egg. In the Protura, a primitive group of wingless insects, eight segments are present in the abdomen at the time of hatching and three more develop at a later stage.

Anaphase. Pertaining to the phase in cell division when the chromosomes, having arranged themselves round the equator of the cell, move away from one another towards opposite poles.

Anapophyses. A pair of small prominences below the post-zygapophyses on certain vertebrae, e.g., on the lumbar vertebrae of mammals.

Anapsid. A type of skull having a complete covering of dermal bones with no temporal fossae; such a skull is present in the primitive reptile *Seymouria* and also in the Chelonia (tortoises and turtles).

Anapsida. Reptiles in which the skull is having a complete covering of dermal bones with no temporal fossae, although the bones may be emarginated from the hind-end. Examples are the fossil Cotylosauria as well as tortoises and turtles.

Anaspida. Used for describing ostracoderm fish of the Silurian and Devonian periods that has a laterally compressed fusiform body, a heterocercal tail and long thin rectangular scales arranged in parallel rows on the body.

Anastomosis. Refers to a union of some body structures e.g., two or more blood vessels, nerves, other structures.

Anatomy. (1) Deals with the study of the structure of the body of an organism as learned by dissecting it.

(2) Refers to the organization of the parts of the body and the structural relationships among them.

Androgen. A male sex hormone (a steroid) which is able to control the development, function and maintenance of secondary male characteristics (e.g., facial hair and deepening of the voice), male accessory sex organs, and spermatogenesis. These are mainly produced by the testis (smaller amounts are produced by the ovary and adrenal cortex); the most important is testosterone.

Androsterone. A steroid hormone which is formed in the liver by the metabolism of testosterone. It has only weak androgenic activity. See *androgen*.

Anemone. Pertaining to the common name for anthozoan *Metridium dianthus*; a cylindrical animal having crown of tentacles which are arranged in a circle in a flower-like manner; sessile having usually brownish/yellowish colour. Body cavity consists of six radial chambers.

Aneurism. Refers to the localised dialation of an artery wall. The elastic walls tend to balloon and if collagen fibres sheath breaks down artery then it is having tendency to undergo ballooning.

Angina Pectorts. A substernal pain which is developed in chest when myocardium gets ischemic (Also see *myocardial ischemia*) during exertion due to the accumulation of a p-factor in muscles.

Angiotensin. A protein in the blood, produced from angiotensinogen by the action of renin produced by kidneys.

Angiotonin. The old name for hypertension or angiotensin-II.

Animal. An organism that is able to feed on other organisms or on organic matter, is generally motile, and reacts to stimuli quickly. Animal cells are surrounded by cell membranes. There is no chlorophyll and growth is usually limited.

Animal Pole. Refers to that part of the surface of animal eggs to which the nucleus (germinal vesicle) is closest. It is usually opposite the vegetal, or yolky, pole.

Animation. Pertaining to animal-like activities that are not totally natural but tend to simulate the essential features of normal activities e.g. living activities of frog during hibernation which resemble the normal life activities in essential features but are not exactly similar to them.

Anisogamy. Refers to the sexual fusion in which male and female gametes are of unequal size e.g., fusion in malarial parasite (*Plasmodium*).

Ankylosis. Refers to a union or knitting together of two or more bones or parts of bones.

Annelids. Pertaining to the segmented animals that are having ring or ring-like structure e.g., earthworms and leeches etc. They are triploblastic and coelomates.

Annovulatory Cycles. Menstrual cycles in a woman during which there occurs no ovulation.

Annulus. Any ring or ring-like structure.

Anoxia. Refers to condition of the lack of oxygen in tissue which diving animals e.g. mammals and birds suffer during submergence. Its effects in first on central nervous system so the diving animals are utilising oxygen stores in the body, oxygen is utilised only by brain and heart during this period and blood flow to other parts/organs get reduced and these tissues get adapted to anaerobic pathways.

Antagonist. A structure, process or reaction which is acting in opposition e.g., antagonist muscle (a muscle acting in opposite direction to the movement of another muscle).

Aetenna. Refers to the movable sense organ on the head of insects, myriapods and crustaceans. Insects and myriapods possess only one pair, crustaceans have two pairs.

Antennal Gland. Refers to a major organ of osmoregulation in crustaceans like cray-fish. Part of it, coelomosac has tubules which resemble in structure with vertebrate glomerulus; also exhibits structures which are physiologically analogous to proximal and distal tubules of the vertebrate kidney.

Antennule. Used for describing many jointed feelers which are found on some athropods and located near the antenna; often known as first antenna.

Anterior. The front or head end of an animal.

Anterior Choroid Plexus. The non-nervous tissue together with pia mater covering it forms villi-like processes which is extending into diocoel (Cavity/ventricle of diencephalon) and is forming anterior choroid plexus; secretes cerebrospinal fluid.

Anterior Vena Cava. A large vein which is collecting blood from anterior regions of the body; bilaterally symmetrical and opens into sinus venosus (in frogs, reptiles etc.) and sinuauricular opening in mammals.

Anthropomorphic. Pertaining to human characteristics of other organism.

Antibiotic. One of a group of organic compounds, possessing variation in structure, that are produced by microorganisms and are able to kill or inhibit the activities of other micro-organisms. One of the best known examples is penicillin, which was discovered by Sir Alexander Fleming. Another example is streptomycin.

Antibody. Refers to a substance which is usually a soluble protein and is produced in the body of an animals as a protection against invasion by parasites or other foreign bodies like bacteria and viruses. The foreign body has substances called *astigens* which bring about the production of specific antibodies. The interaction of the two usually results in the death of the invading parasites.

Antibrachium. Anterior part of fore limb that is joined anteriorly with upper arm and posteriorly with hand, supported with radius and ulna bones. Also known as forearm.

Anticoagulant. A chemical substance which is able to inhibit the tendency of thrombosis or blood clotting by antagonising the blood clotting factor, by destruction of key substance fibrinogen or by inhibiting the synthesis of factors II, VII, IX and X, e.g. heparin (secreted by mast cells).

Anticodon. Refers to a sequence of 3 nucleotides in t-RNA which pair with specific codon in m-RNA. It is part of t-RNA which imparts identity to an amino acid in the cell for positioning it at the proper place in polypeptide chain.

Antidiuretic Hormone. A substance which is secreted by the nerve endings passing from the hypothalamus into the posterior pituitary lobe in a vertebrate. It is able to control the absorption or resorption of water by the kidney tubules. Insufficiency of this hormone brings about an increase in water intake resulting in *diabetes inspidus.*

Antienzyme. A substance inhibiting the normal action of an enzyme without actually destroying it. Such substances are given out by the epithelium of the stomach, and prevent its being digested by its own enzymes.

Antigen. A substance that is able to induce the production of antibodies.

Antilymphocyte Serum (ALS). A serum which is used to suppress the immune reaction in patients receiving tissue or organ transplants. It is produced by injecting human lymphocytes into a horse, which then produces antibodies against them. ALS is then collected from the horse and the antibodies are extracted and purifed. When this preparation is injected into the transplant patient, it destroys his lymphocytes, which would otherwise have produced antibodies resulting rejection of the graft.

Antimycin. An antibiotic which is isolated from *Streptomyces* strain. It blocks electron transport system by inhibiting transport of electrons from cytochrome-b to cytochrome-c.

Antitoxin. A chemical substance which is capable of neutralizing other chemical substance that is a toxin e.g., injection of atropin on exposure to insecticides.

Antrum. A cavity or chamber pertaining specially to one within a bone or a follicle etc.

Antrum Folliculi. A space/cavity which is formed in between cumulus cophorus (cell surrounding immature ovum) and stratum grauulosum (outer Cells) of Graffian follicic; filled with a liquid which is known as liquor folliculi (Also see *Graffian follicle.*)

Anura (Batrachia). Frogs and toad; tailless Amphiqia having four limbs and with a smooth skin; without gills in the adult but developing from a larva of tadpole which is having external gills at first and internal gills later. The hindlimbs are generally enlarged for jumping and possess webbed feet for swimming.

Anus. Refers to the posterior opening to the alimentary canal , which is found in nearly all animals. Through it, faeces and sometimes semisolid wastes get expelled from the body, often under muscular control.

Aorta. In mammals, it refers to the large artery which is arising from the left ventricle of the heart. It is carrying oxygenated blood, via various branches, to all parts of the body. It is divided into an ascending portion, an arch, and a descending portion. The aorta forms the left branch of the systemic arch.

Aortic Arches. Refer to six pairs of blood vessels which are present in all vertebrate embryos, and link the ventral aorta leaving the heart with the dorsal aorta.

Aortic Body. Refers to a nodule on the aortic arch which is having chemoreceptors that sense the composition of blood; it monitors changes in O_2 and CO_2 in arterial blood e.g., carotied body of mammals and birds, carotid labyrinth of amphibia. The chemoreceptors have been innervated by branches of IX (glossopharyngeal) and X (vagus) cranial nerves.

Apatetic Coloration. Coloration which helps to protect an animal by misleading its enemies. In some cases the outline of the animal has been broken up (see *Disruptive Coloration*); in others the makings reveal the presence of an eye or other organ where in fact there is none.

Aperture. Refers to orfice or opening.

Aphaniptera (Siphonaptera). Fleas; parasitic blood-sucking insects without having wings; with a laterally compressed body and having the hind legs very much enlarged for jumping. The larvae do not have legs and have biting mouth-parts. Pupae are enclosed in silken cocoons.

Aphasmidia. A subclass of Nematoda incorporating the whipworm, kidney worm and trichina. They are characterized by the absence of *phasmids* or precaudal sensorp organs, but they are having *amphids* or anterior chemoreceptors.

Apical. The end or outermost part as of a conical structure.

Apne. Refers to the suspension or absence of breathing, brief periods of respiratory arrest. May be generated by injecting varatridine and nicotine into coronary and pulmonary circulation thereby producing bradycardia and hypotension and due to stimulation of chemoreceptors.

Apneumaic. Not filled with air.

Apneusis. If all cranial nerves are cut and brain stem gets transected above pons, regular breathing continues. However, an additional transection in the inferior portion of pons causes inspiratory neuron discharge continuously resulting in a sustained contraction of inspiratory muscles and such arrest of respiration during inspiration is termed as apneusis.

Apneustic Centre. When vagi are cut and animal develops a peculiar type of breathing in which prolonged respiratory cycle is spent in inspiration and after a brief expiration there occurs a sustained inspiration. Such a breathing occurs due to apneustic centres that are existing in pontine are a of vagus. The centre discharges tonically upon medullary respiratory centre thereby promoting more sustained respiratory activity.

Apocrine. Refers to a gland in which the products are collected at the outer part of cells and squeezed out with a little cytoplasm thereby a part of cells gets damaged. For example, the sweat glands of mammals.

Apodemes. Refer to ingrowths of the cuticle of insects, crustaceans and other arthropods which serve as points of attachment for the muscles. They may unite to form an internal framework called the endophragmal skeleton.

Apoenzyme. An enzyme whose cofactor has been removed (e.g. via dialysis) making it catalytically inactive. It is the protein part of a conjugate enzyme. When it is combined with its prosthetic group (coenzyme) it yields a complete enzyme (holoenzyme).

Apophyses. Refer to various kinds of projections that are arising from a vertebra i.e., zygapophyses, tranverse processes etc.

Apopyle. Refers to the pore which is leading from flagellated chamber into central cavity or spongocoel of sponges; associated with sycon type canal systems.

Appendage. Refers to a portion of body that is projecting and is having a free end e.g., fore and hind limbs.

Appendicular Skeleton. Constituting collectively the skeletal elements of girdles and limbs; includes pectoral and pelvic girdles.

Appendix. Refers to the slender tubular pouch which is projecting from caecum of large intestine in some mammals; vestigeal organ in human beings. Enlargement results in appendicitis.

Appetite. Used for a psychological condition which gets manifested as a relish for something used usually in relation to food; controls amount of food intake in animals and is related with the energy requirements of body; depends upon colour, smells, taste, cooking medium, place of serving food etc. Controlled by staiety centres present in hypothalamus.

Apposition Image. When light is intense, the rays of light given out by some objects in front of ommatidia of compound eye (in insects etc.) penetrate through cornea. The oblique rays of eye will strike iris pigment and gets terminated there while straight rays of light reach upto focusing part of ommatidium and the image of only that part of the object gets formed from which the straight rays are passed thus separate images will be formed by each ommatidium. This is termed as apposition image and gets magnified and more clear (also known as mosaic vision).

Apterus. Wingless e.g., all apterygote insects like silver-fish.

Apterylae. Area of bird skin in which feathers do not grow (Compare with pterylae).

Aquatic. Pertaining to water, living in water e.g., fishes.

Aqueduct. A canal/channel which is used for conduction of fluid; very narrow mesocoel (cavity of mid brain) which connects diocoel (cavity of diencephalon) with metacoel (cavity of cerebellum), also known as iter/aqueduct of Sylvius.

Aqeuous Humor. Jelly-like liquid which is present within the eye in the area between cornea and lens.

Arachnida. The class of the Arthropoda having mostly terrestrial and carnivorous scorpions and spiders. They typically are having spinnerets on the abdomen for web spinning, and the parasitic ticks and mites. The body is divided into two parts, the anterior cephalothorax (*prosoma*), and the posterior abdomen (*opisthosoma*), and they have four pairs of walking legs. They lack antennae and the eyes are simple.

Arachnid Membrane. Refers to the middle one of the three membranes (meaning) which surround and protect the brain and spinal cord in vertebrates.

Arboreal. Pertaining to trees e.g., tree living animals such as monkeys.

Arhorvitae. Refers to internal white matter present within cerebellum of brain; resembles a multibranched tree.

Archaezoic. A geological era that started after formation of earth crest about 3700 million years ago and had duration/age of 2100 million years. Protista (prokaryotic animals) were evolved during this period.

Archenephros. Used for describing a type of kidney which is present in most primitive vertebrates; having one nephric tubule of each side per body segment and all the tubules joining into a longitudinal duct which is opening into cloaca. Also known as holonephros.

Archenteron. Refers to the earliest gut cavity of most animal embryos which is produced by an infolding of part of the outer surface of a blastula to form an internal cavity that is in coutinuity with the outside via the blastopore.

Archeocytes. Amoeboid cells in sponges which receive, digest and transport food; also take part in regeneration and can form any type of body cell.

Archeptrygium. A type of fish fin having a single segmented axial rod from sides of which numerous secondary segmented radials arise; present in Neocerarodus (a lung-fish).

Area Centralis. Refers to a small depression present towards the bosterior side of retina (where there are more cones than rods) and its centre is exclusive region of cones termed as fovea centralis and in primates area centralis possesses yellow-pigment thus termed as macula lutea.

Area Pellucida. Refers to the central area/circle in the blastodise of reptiles and birds over blastocoel which is less opaque and also called area opeca. Constituent of opiblast and the region where entire chick gets developed.

Areola. Refers to the distal extent of iris; closed cells of forewing of some lepidopterans (group of butterflies etc.); longitudinal strip of crayfish. Carapace posterior to cervical groove; pigmented area around nipples of breast.

Areolar Tissue. Refers to a type of loose connective tissue which is present around visceral organs, among muscles, around blood vessels, around nerve fibres and structural element of atorma of various organs. It is also forming basement membrane of epithelia. It has jelly-like matrix with elastic, collagen and reticular fibrils. There are fibroblasts (help in healing), histeocytes (phagocytic), plasma cells (antibody forming), mast cells (heparin producing) and lymphocytes (antibody forming) in this tissue.

Aristotle's Lantern. Refers to a specialised calcareous structure surrounding mouth of sea-urchins having teeth like structures and helping in ingestion and chewing of food.

Arousal. A general level of alertness in an animal, which results from activity of a particular part of the brain.

Arrector Pilli. A small bundle of unstriped muscle fibres which are attached to hair whose contraction makes hair almost vertically erect to the surface of skin.

Arterioles. Branches of an artery having lesser diameter and divisible into capillaries.

Artery. A large thick-walled blood vessel carrying blood from the heart to the limbs and organs. All arteries except the pulmonary artery are carrying oxygenated blood. The artery wall is having an inner layer of endothelium (*tunica intima*); a thick middle layer (*tunica media*), consisting of smooth muscle and elastic tissue; and a thin outer layer (*tunica externa*) of collagen fibres. It is thus able to with stand the pressures which are resulting from the pumping of the heart.

Arthropoda. The largest phylum in the animal kingdom and the only invertebrate phylum including aquatic, terrestrial, and aerial members. Arthropods are bilaterally symmetrical segmented animals having a characteristic tough chitinous protective exoskeleton flexible only at the joints; growth takes place by ecdysis.

Articular. Used for small bone of the lower jaw in bony fish (Osteichthyes), amphibians, and reptiles which is forming a hinge joint with the quadrate bone of the upper jaw. The articular has been derived from the ossification of Meckel's cartilage.

Articulation. Refers to the surface where two skeletal elements meet, thereby forming a movable joint.

Artificial Insemination. Refers to the process of the artificial introduction of semen into the reproductive tract of a female animal. It finds extensive use in breeding animals (e.g. sheep, cattle). Semen obtained from a male animal having desirable hereditary characters can be frozen and transported long distances to fertilize a large number of females. The method is also applicable to human females who wish to conceive where the partner fails to copulate successfully or is sterile.

Artiodactyla. The order of mammals having the even-toed ungulates, in which the weight of the body is supported on the third and fourth digits only. This order includes herbivorous mammals such as sheep, goats, deer, domestic cattle, antelopes, pigs, camel, and giraffes.

Arytenoid. Referring to a pair of laryngeal cartilages which are lying posterior to thyroid cartilages in mammals; anterior to cricoid cartilages in other vertebrates.

Asbestosis. A lung disease in which there occurs the formation of asbestos fibres.

Aschelminthes. Animals of organ grade of organisation, triploblastic having pseudocoel and alimentary canal e.g., nematodes like *Ascaris, Enterobius* etc.

Ascon Type Canal System. Refers to a simple type of canal system. It is common in sponge *Leucosolenia* whose diploblastic body wall is perforated at regular intervals and flagellate cells (choanocytes) are lining the entire spongocoel.

Ascorbi Acid. Also known as vitamin-C; water soluble substance which is necessary for bringing about hydroxylation of lysine in collagen synthesis. Its chief sources are citrus fruits, leafy green vegetables etc.; daily requirement for infants is 35 mg, children 35 mg and adults 45 mg (both for men and women); its deficiency causes scurvy.

Asexual Reproduction. The reproduction from only one parent; as, for example, the budding of a hydra or the simple fission of Amoeba.

Asphyxia. Death-like condition caused by need for oxygen in blood, as when breathing is stopped by unnatural causes.

Assimilation. Refers to the process of incorporation of simple molecules of food that has been digested and absorbed into living cells of an animal and converted into the complex molecules making up the organism.

Assortative Mating. Refers to the type of reproduction of animals in which the males and females appear not to pair at random but tend to select partners of a similar phenotype.

Astacus. A genus of crayfish. Species are widely found in fresh water and are omnivorous. The carapace protecting the cephalothorax get prolonged in front to form a pointed *rostrum*, on each side of which there is a stalked mobile compound eye. The thorax bears greatly enlarged pincers (chelae) used in feeding and defence, and four pairs of walking legs.

Aster. Refers to a spherical array of fibres (microtubules) found around the centriole at each end of the spindle at nuclear division. It is so-called from its appearance in the light microscope. Its association with the centrioles means that it is absent from most plant cells. Its function is obscure.

Asteroidea. The class of the Echinodermata which is including the starfish (*e.g. Asterias*). They are usually found just below the low tide mark. A starfish typically has five arms radiating from a central disc, which contains the main body organs and has the mouth on its ventral surface. Chalky plates in the skin form a skeletal test. Suckered tube feed on the underside of the arms take part in locomotion and holding prey.

Astgmatism. Defocussing phenomenon in eye which is caused by deformation of cornea. In this defect oblique lines are not focused correctly because impinging of light rays are deflected from what we regard a normal course.

Asymmetrical. An animal whose two sides of body are not alike: lacking any symmetry.

Asynchronous Muscle. A type of muscle which is found in thorax of some insects. Undergoes Contraction without any one to one relation to motor impulses (also called fibrillar muscle).

Atavism. Refers to the resemblance of an off spring to a remote ancestor.

Ataxia. Refers to the incoordination due to errors in rate, range, force and direction of movement; occurred due to leisions in cerebellum. There occurs no abnormality at rest.

Atlas. Used for a ringlike bone in tetrapods which forms the first neck (cervical) vertebra of the vertebral column, on which the skull rests. In reptiles, birds, and mammals, it articulates with the skull to permit nodding of the head and articulates with the axis to permit rotatory movement.

ATP (adenosine triphosphate). Refers to the universal energy carrier of living cells. Energy from respiration or, in photosynthesis, from sunlight is used to convert ADP into ATP. It is then reconverted to ADP in various parts of the cell, liberating energy which is used to drive cell reactions. ATP is a nucleotide which consists of adenine and ribose with three phosphate groups attached. Hydrolysis of the terminal phosphate bond evolves energy (30.6k $Jmole^{-1}$) and is coupled to an energy-requiring process. Further hydrolysis of ADP to AMP sometimes takes place, releasing more energy. The pool of ATP is small, but the faster it is used, the faster it gets replenished.

Atoll. Refers to the horse-shoe-shaped or ring-shaped island of a belt of coral reef which is surrounding a central lagoon.

Atrial Cavity. Used for describing a cavity in protchordates and some caudate larvae (e.g. tadpole) through which water passes after moving out of gill slits.

Atrichous Isorhizas. Also known as steroline glutinants; oval in shape having small thread but bearing no spines. Takes part in locomotion of certain animals like *Hydra* by adhering with the substratum through hypostome.

Atriopere. Used for the external opening from atrial cavity in *Amphioxus* and related animal forms.

Atrioventricular Node (AV-Node). Refers to one of the conducting systems of heart situated in the right posterior portion of inter auricular septum. There are three bundles of fibres which have Purkinjie fibres and fond to conduct impulses from SA-node to this node (anterior) internodal tract of Bachman, middle internodal of Wenokebach and posterior internodal tract of Thorel). AV-node is continuous with bundle of His.

Atrium (1) (auricle). Used for a cavity or chamber in the body, especially in the vertebrate heart. The mammals, are having two atria which are forming the upper chambers of the heart; the left atrium receives oxygenated blood from the lungs while the right atrium receives deoxygenated blood from the body.

(2) Refers to a cavity which is surrounding the gills of primitive chordates; for example, in *Apmhioxus* the atrium connects with the exterior via a small pore (an *atriopore*).

(3) Refers to any cavity which is opening to the exterior; for example, the genital cavity of *Helix*.

Atrophy. Used for describing the shrinking in size of a tissue or an organ.

Atropous. *See *orthortropus*.

Attenuation. Refers to the loss of virulence of a pathogenic microorganism after several generations of culture in *vitro*. Attenuated microorganisms find use in the preparation of vaccines.

Auditory Capsule (otic capsule). Said of the cartilaginous or bony part of the vertebrate skull which is enclosing the inner ear. It has a single periotic bone in adult mammals.

Auditory Nerve. (Vestibulocochlear nerve, acoustic nerve, cranial nerve VIII). The nerve supplying the inner ear and arises from the dorsal region of the medulla oblongata in the vertebrate brain.

Auditory Ossicles. The small bones which connect the tympanic membrane with the *fenestra ovalis* or inner ear drum in the terrestrial vertebrates. Their function is to convey sound vibrations across the cavity of the middle ear and to magnify the force of these vibrations by a lever action.

Auerbach's Pleuus. Used for describing a network of small nerve *fibres and* ganglia between the two muscular coats of the intestine; able to cause local reflex arcs by which pressure of food in the intestine can bring about peristaltic movement.

Auricle (Auricula). (1) Refers to any ear-like structure (Latin dim. of *Auris*: an ear).

(2) Used for describing the first compartment (or pair of compartments) of the heart, into which blood flows from the veins.

(3) Refers to five calcareous arches which are situated round the circumference of the jawskeleton called *Aristotle's lantern* in sea-urchins (Echinoidea). Muscles from these arches help to pump water to the gills.

Auricular Feathers. Refers to a tuft of large feathers which are surrounding the ear of an owl or similar bird and giving the appearance of a pinna.

Auricularia Larva. Used for describing the typical larval form of a holothuroid. It has a band of cilia arranged in a complex pattern down each side and folded back at each end where the body has been extended to form a pre-oral and a post-anal lobe.

Autograft. A type of graft which involves transplantation of tissue or an organ from one part of an individual to another part of the same individual.

Autoimmunity. Refers to a diseased state in which there occurs the formation of antibodies that react against a normal component of the animal's own tissues. Autoimmunity is a contributory factor of a number of diseases (*autoimmune diseases*) like rheumatoid arthritis and some forms of gout.

Autolysis. Refers to the self-destruction of cells by digestive enzyme activity. It is the final stage of cell senescence thereby causing complete digestion of all cell components. See *lysosome.*

Autonomic Movements (Autogenic movements, spontaneous movements). Refers to the movements of plants in response to internal rather than external stimuli, e.g., cytoplasmic streaming, chromosome movement during nuclear division, and growth itself.

Autonomic Nervous System. A division of the vertebrate nervous system that is supplying motor nerves to the smooth muscles of the gut and internal organs and to heart muscle. It is made up of *sympathetic nervous system,* that (when stimulated) increases heart rate, breathing rate, and blood pressure and slows down digestive processes, and *parasympathetic nervous system,* that decreases heart rate and promotes digestion.

Autophagy. Refers to the process whereby redundant, faulty, or ageing cell organelles get destroyed. The organelle or cell portion is surrounded by a membrane which is derived either from the endoplasmic reticulum or a vacuole. Lysosomes then undergoes fusion with the compartment thus formed, releasing their digestive enzymes and thereby destroying its contents.

Autorhytemicity. Refers to the generation of rhythmic activity without involving any extrinsic control e.g., activity in cilia and flagella and pace-makers (in the myogenic hearts). These structures beat even in the absence of nervous control.

Autosome. Used for describing any chromosome except sex chromosomes e.g., man has 22 pairs of autosomes.

Autostylic Jaw. Refers to a type of jaw suspension when lower jaw gets articulated with the quadrate bone e.g., jaw suspension in frog, lizards and birds.

Autotomy. Self-mutilation i.e., the autonomic voluntary breaking off a part of body of an animal, as we see in the tail of a lizard.

Autotrophic Nutrition. Refers to the process of nutrition in which an organism is manufacturing is own food e.g., nutrition in Euglena where animal is able to synthesise carbohydrates with the help of its chloroplast from CO_2 and water in presence of sunlight (as seen in the plants).

Aves. Refers to a group of animals to which birds belong. They are having feathered bipeds. They are warm blooded animals capable of aerial life; have high temperature metabolism and normal body temperature ranges from 110-120°F (often 140°F). Their heart is four chambered. There are oviparous and amniotes.

Axenic. Refers to the type of culture medium in which available food has been specially identified.

Axial Skeleton. Said of the skeleton which is supporting body axis; skull and vertebral column constitute this skeleton.

Axis. Uses for describing the second cervical vertebra. Also refers to a line along anteroposterior axis of body around which parts have been symmetrically arranged.

Axis Cylinder. Used for rod-like central portion of a medullated nerve fibre.

Axolemma. A thin membrane which is binding axon. It is revealed only under electron microscope to have a few large fibrils.

Axolotl Larva. Used for describing the pedogenic larva of any of several species of salamander of genus *Ambystoma.* They are common in Mexico and Southwest part of United States. They bear external gills, tail, limbs etc., and are capable of reproducing sexually without involving metamorphosis.

Axon. Used for the fibre of a nerve cell/neuron which is conducting nerve impulses away from cell body or cyton of which it is a part; elongated cylindrical process of neuron along which there occurs conduction of action potentials myelinated or non-myelinated both (i.e., with lipoprotein sheath or without it).

Axoneme. Refers to the complex of microtubules and associated structures which are found in flagellar or ciliary shaft. It has nine doublets and two central microtubules and is enclosed in an evaginated extension of cell membrane.

Axoplasm. Used for the protoplasm of axon; in some cases it is gel and sometimes a viscous liquid and in it there are extremely thin ultramicroscopic neurofibrils.

Azygos. Used for describing the unpaired anatomical structure e.g., azygos lone of lung, azygos vein of mammals.

B

Back Bone. *See vertebral column.*

Back Cross. The mating of a hybrid or heterozygous individual with either of its homozygous parents. The method is used as a test for the dominance of one quality over another and as a means of determining whether a recessive gene is present or not.

Balanced Carrier. An individual whose karyotype is abnormal but who nevertheless is having a normal chromosomal complement. Usually the result of translocation.

Back Cross. The mating of a hybrid or heterozygous individual with either of its homozygous parents. The method is used as a test for the dominance of one quality over another and as a means of determining whether a recessive gene is present or not.

Balanced Carrier. An individual whose karyotype is abnormal but who nevertheless is having a normal chromosomal complement. Usually the result of translocation.

Barbs (of feathers). Lateral lamellar forming the vane of a feather. (Fig. 1)

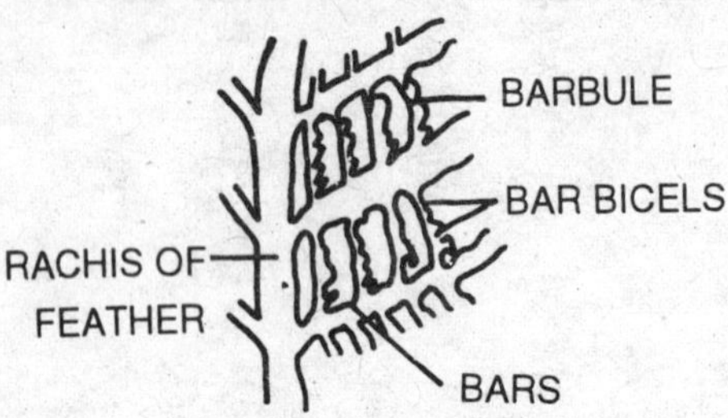

Fig. 1.

Barbules (of Feathers). Pertaining to small processes on each side of the barbs of a feather, interlocking by means of hooks called *hamuli* or *barbicels.*

Baroceptors. Receptors sensitive to changes in the fluid pressure which is exerted on them. In man, these accur in the *carotid sinus* where a slight increase of blood pressure will result in a nervous stimulation thereby causing a reflex slowing of the heart.

Barr Body. In human female's egg cells, it refers to a dark staining portion which is visible under light microscope. It is actually one of the two X-chromosomes and is inactivated.

Barrier. Pertaining to any type of obstruction-physical, chemical or biological that does not allow the migration of animals or gradual extension of their territories.

Barrier Reef. Reef which is separated from shore by a wide, deep channel.

Basal Body (blepharoplast, kinetosome). A body which is found at the base of all cilia and flagella. It is identical in structure to the centriole.It is essential for the formation of cilia and flagella.

Basal Metabolic Rate (BMR). The minimum rate of energy expenditure by an animal which is required for maintaining the vital processes, e.g. circulation, respiration, etc. It is represented as the output of joules per square meter of body surface area per hour and is measured either directly from heat production or indirectly from oxygen consumption. The thyroid hormones act as the prime regulators of the BMR.

Basapophyses. The ventrolateral processes of outstrum, remnants of a haemal arch.

Basement Membrane. A structureless or fibrillar membrane to which epithelial cells get attached and on which they rest.

Basilar Artery. Blood flow to brain in human being takes place via 2 vertebral and 2 internal carotids. The vertebral arteries get united to form basilar artery and also part of circle of Willis.

Basilar Membrane. A membrane which is forming the base of the *organ of Corti* and supporting the sensory cells in the cochlea of the mammalian ear. It separates the *scala tympani* from the *scale media.*

Basophil. A white blood cell (leucocyte) which is having granules that stain with basic dyes. It has a lobed nucleus. Basophils constitute only 0.5% of all leucocytes. They move about in an amoeboid fashion and may ingest bacteria.

Bathyal. Ocean zone which comprises of upper 200 meters of water.

Bed-bug. Common name for insect *Cimex.* Ectoparasite with piercing mouth parts. No wings, found in cots, wooden furniture etc. Carrier for several diseases.

Beetles. A type of insect whose wings are covered by two hard protective structures called elytra.

Behaviour, Animal. Used for describing any observable activity of a whole animal. Behaviour incorporates all the processes by which an animal senses its external surroundings and the internal state of its body and responds to any changes it perceives.

Bell-Magendie Law. This law states that the dorsal root of spinal cord has only sensory axons while the ventral root has only motor axons.

Beri-Beri. A form of polyneuritis (degeneration of the peripheral nerves causing paralysis) which is caused in mammals, birds and possible other animals by a deficiency of vitamin B_1 (*thiamin*).

Bertholin's Glands. See *bulbourethral glands or Cowper's glards.*

Beta-Cells (β-cells). One of the cells found in islets of Langerhans constituting about 75% of its part; having granules which stain purple (bluish-purple) with Mallory's stain; secreting insulin which is able to control excess of blood sugar by converting it into glycogen. (Also see *glycogenesis*).

Beta-Keratin. Refers to the insoluable sulphur-rich scleroprotien; constituent of epidermis, horns, nails, feathers and tooth enamel. Beta Pertains to protein's secondary structure which is in pleated sheets (e.g., silk produced by caterpillers).

Beta-Oxidation. Pertains to oxidation at beta-position, a pathway in which there occurs removal of two carbon atoms at a time from carboxyl group (end) of molecule. Knoop postulated such oxidation for fatty acids. β-oxidation of fatty acids gives rise to the formation of two molecules of acetyl Coenzyme-A.

Bicuspid Valve. Said of the mitral valve or left auriculoventricular valve in the mammalian heart; so-called due to the two cusps or flaps by which it can close.

Bidder's Canal. In male amphibia e.g., frog, some urinary tubles connect with testes and transport sperms into a duct which is called Bidder's canal. It runs along medial edge of and outside the kidney and communicates with ureter.

Bidder's Organ. A vestigial ovary, sometimes having a mass of immature ova, in front of the testis or ovary in certain toads of both sexes.

Bilateral Symmetry. Pertaining to the arrangement of the parts of an organism in such a way so that the right and left halves become mirror images of each other.

Bile. A fluid which is secreted by liver in the vertebrates; having about 97% water, 0.7% bile salts, 0.2% bile pigments, 0.6% cholesterol, 0.7% inorganic salts and 0.1% fats alongwith little alkaline phosphatases; making food alkaline, preventing putrifaction and reducing surface tension over fat moelcules for making then emulsified.

Bile Duct. The duct through which bile is flowing from the gall bladder to the duodenum in a vertebrate.

Bile Pigments. Refer to *bilirubin* ($C_{33}H_{36}N_4O_6$), an orange-coloured substance and *biliverdin*, a green substance, an oxidation product of bilirubin. Both are obtained by the breakdown of haemoglobin in the liver (see *Bile*).

Bile Salts. Components of bile which aid digestion in the duodenal region of the gut. Sodium taurocholate and sodium glycocholate emulsify fats by decreasing their surface tension, making them to break up into numerous tiny droplets and facilitating the digestive action of lipase.

Bilirubin. A brown or orange bile pigment which is formed from the breaking down or red blood corpuscles.

Biliverdin. A green bile pigment which is formed from the break down of red blood corpuscles.

Binocular Vision. Refers to the type of vision in which the eyes point forwards so that it becomes possible to focus the image of a single object onto the fovea of both eyes at once. This permits perception of depth and distance. Binocular vision occurs in primates and other vertebrates, especially active predators, such as owls.

Binomial Nomenclature. A system of classification with was introduced by Linnaeus, the Swedish botanist. In this system each species is assigned two names..The first is the generic name, written with a capital letter, which designates the genus to which the species belongs. The second is the specific name, which indicates the species. The generic and specific names are in Latin. For extempro, Man belongs to the species *Homo sapiens. Homo* is the generic name while *sapiens* is the specific name.

Bioassay. An experimental technique which is used for measuring quantitatively the strength of a biologically active chemical by its effect on a living organism.

Biochemical Oxygen Demand (BOD). Refers to the standard measure for determining the level of organic pollution in a sample of water. It may be defined as the amount of oxygen which is used by micro-organisms feeding on the organic material over a given period of time. Sewage effluent has to be diluted to comply with the statutory BOD before it could be disposed of into clean rivers.

Biochemistry. Deals with the study of chemical reactions taking place in living organisms.

Bioengineering. (1) Refers to the design, manufacture, and use of replacements or aids for body parts or organs that have been removed by accident or so or are defective, e.g., artificial limbs, hearing aids, etc.

(2) The design, manufacture, and use of equipment for industrial biological processes, such as fermentation.

Biogenesis. The theory postulating that living things originate only from other living things as opposed to non-living matter. The theory became accepted due to the work of Redi and Pasteur, who proved that dirt, for example, does not itself produce bacteria or maggots, but that bacteria and maggots only come from spores or eggs already present in the dirt.

Biogenetic Law. This law postulates that an animal is able to repeat corresponding stages of its ancestors during its embryonic development or 'Ontogeny Repeats Phylogeny'. This theory was given by

Von Hackel and also called recapitulation theory.

Biological Clock. The internal mechanism of an organism which is able to regulate circadian rhythms and various other periodic cycles.

Biological Control. Refers to the use of natural predators, instead of chemicals, for controlling the pests. An interesting example was the introduction of the grey moth, *Cactoblastis cactorum*, into Australia for controlling the prickly pear, *Opuntia intermis*, which was overrunning vast tracks of land.

Biological Membrane. The plasma membrane which is usually 7.5 nm(75 A) thick and made up of protein and lipids which are usually phospholipids e.g., phosphatidylcholine and phosphatidyl ethanolamine.

Biological Oxidation. Refers to such processes which deal with the production of energy in cell. Catalysed by enzyme and a particular protein enzyme being responsible in most cases.

Biology. The branch of science dealing with the study of living things; divided in to zoology and botany which are concerned with study of animals and plants respectively.

Bioluminescence. Refers to the production of light by living organisms. Bioluminescence is common in many marine organisms, especially deep-sea organisms. It is also common in some insects, e.g. fireflies, and certain bacteria. The light is produced due to a chemical reaction whereby the compound luciferin gets oxidized. An enzyme, luciferase, catalyses the reaction in which ATP supplies the energy.

Biomass. Pertaining to the weight or volume of living organisms of one animal or plant species per unit area (*species biomass*), or of all the species in the community (*community biomass*).

Biome. Refers to the major regional community of plants and animals having similar life forms and environmental conditions. It forms the largest geographical biotic unit, and has been named after the dominant type of vegetation, such as tropical rain forest, grassland, or coral reef.

Biopoiesis. Refers to the origin of organisms from replicating molecules. Biopoiesis is a cornerstone of abiogenesis. Deoxyribonucleic acid

(DNA) is the best example of a self-replicating molecule.

Bisphere. Includes the part of the earth and its atmosphere that is inhabited by living organisms. The Earth's surface and the top layer of the hydrosphere (water layer) possess the greatest density of living organisms.

Biosynthesis. Refers to chemical reactions in which a living cell build up its necessary molecules from other molecules present.

Biota. A term which is used to describe collectively all the organisms present in a particular blome.

Biotin. A member of vitamin B group that has been found to catalyse CO_2 fixation in fatty acids synthesis etc. Its deficiency causes dermatitis and gastrointestinal diorders. It mainly occurs in egg, yolk, liver, tomatoes etc.

Bipinnaria. A form of dipleurula larva that is characteristic of starfish. It bears lobes carrying ciliated bands which are used for feeding and locomotion.

Bipolar Cells. A neuron having two axons which are emerging from opposite sides of the soma. One such class exists in retina of the eye where they transmit signals from visual receptor cell to ganglionic cells of optic nerve.

Biradial Symmetry. Pertains to the condition in which an animal is having radially arranged parts that lie half on one side and half on other side of a median longitudinal plane. It occurs in ctenophores.

Biramous Limbs (of Crustacea). The typical forked appendages of a crustacean, sometimes modified for various functions but essentially consisting of two branches arising from a basal portion.

Bisexual. Hermaphrodite: having both male and female organs on the same individual.

Bivalent. A pair of homologous chromosomes which are in close association with each other during meiosis.

Bladder Worm. The bladder like larva of a tapeworm which, because it is present in a different host from the adult worm, was orginally considered to be as a distinct species.

Blastocoel (Embryology). A fluid filled cavity within a blastula.

Blastocyst (Embryology). A blastula or hollow ball of cells which are formed by repeated cleavage of the fertilized mammalian ovum.

Blastoderm (Blastodisc). A superficial layer of cells which are formed by partial cleavage of a large egg such as that of a bird. The presence of a large amount of yolk does not allow complete cleavage and the disc is produced by the rapid and repeated division of cells on the side away from the yolk.

Blastokinesis. Refers to the movements of an embryo in an egg during development.

Blastomeres. The cells of a blastula (see *Cleavage*).

Blastopore (Embryology). Refers to an opening by which the cavity of a gastrula (the *archenteron* or primitive gut) is communicating with the exterior.

Blastosphere. An alternative name for the blastula (*q.v.*).

Blastostyle. A polyp which is modified for reproducion in certain colonial coelenterats such as *Obelia*. It is generally a hollow elongated rod from which separate individual 'jelly-fish' or *medusae* bud off, detach themselves and swim away.

Blastozooid. A zooid which arises by budding from the parent, as in the coelenterates, tunicates etc.

Blastula (Embryology). A hollow ball of cells formed by repeated cleavage of a fertilized egg-cell.

Bleeding Time. Used for describing the time when bleeding from some injured part ceases. Normally it is 2-6 minutes in most of the cases except under purpura when it is exceeded from 10-20 minutes.

Blepharoplast. The basal granule that is associated with flagellum. (Also called basal granule).

Blind Spot. Refers to a point on the retina of a vertebrate eye which is not sensitive to light. It has no rods or cones, but only nerve fibres passing to the optic nerve, which is joining the eyeball at this point.

Blood. A fluid which is circulating in vascular system of man and other animals; liquid connective tissue having 60% fluid portion called plasma and 40% cellular portion called corpuscles (RBCs, WBCs and platelets). Its main function is to transport substances from one part to the other. (Fig.2)

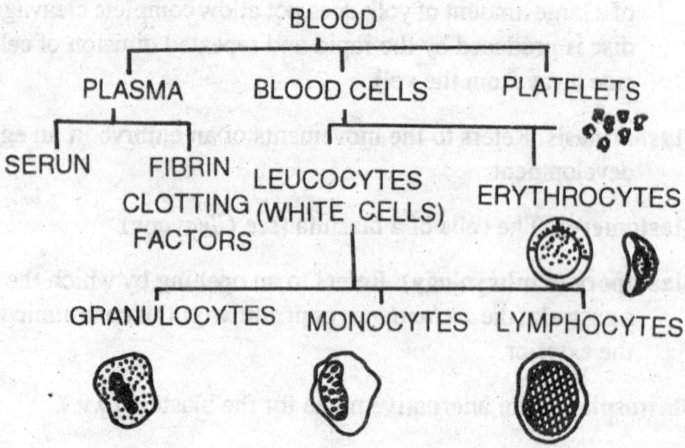

Fig. 2 : Composition of mammalian blood

Blood-brain Barrier. Refers to a biological barrier which exists between brain ECF and blood. This barrier permits water, carbon dioxide and oxygen to cross cerebral capillaries with ease and other substances slowly. This occurs due to H+ gradient which exists between brain ECF and blood, (pH of ECF is 7.33 and that of blood is 7.40). Develops since early stage. Able to maintain constancy of environment of neurons in CNS. It breaks down in areas of brain that are infected, irradiated or the sites of tumours. Thus, it is having application in diagnosing such diseases.

Blood Cell (blood corpuscle). Any of the cells within the fluid plasma of blood. 45% of blood volume is made up of red cells (erythrocytes) and 1% of white cells (leucocytes).

Blood Clotting. Refers to a phenomenon by which blood is immediately converted to a clot-like structure where there occurs some injury and bleeding. Faciliated by different factors called clotting factors. The brief events of this process are summarised as follows:

Intrinsic Extrinsic
Inactive F-XII Tissue Thromboplastin
↓ Collagen (on injury)
Active F-XII |
↓ |
Inactive F-XI→ Active F-XI |
↓ |
Inactive F-IX→ Active F-IX Active F-VII←Inactive F-VI
F-VIII ↓ ↓
Inactive F-X→ Active F-X←Inactive F-X Fibrinogen
Platelets F-V ↓ Ca^{2+} ↓
Fibrin
Prothrombin——→Thrombin ——→ ↓ F-XI
Clott

Blood Clotting

Blood Groups. Types into which the blood has been classified. Since 1900 it has been known that human blood can be divided into four groups. A (42% of the population), B (9%), AB (3%), and O (46%), based on the presence or absence of certain (antigens), called A and B, on the surface of the red blood cells. In group AB, for example, both antigens are present, while group O contains neither antigen.

Blood Plasma. The straw-coloured liquid left when all the cells have been removed from blood. It consists of 91% water and 7% proteins, which are albumins, globulins (mainly antibodies), and prothrombin and fibrinogen (concerned with clotting). Plasma also consists of the ions of dissolved salts, especially sodium, potassium, chloride, bicarbonate, sulphate and phosphate. Plasma is slightly alkaline pH 7.3) and the proteins and bicarbonate act as buffers to keep this constant. Its main function is to transport dissolved food (as glucose, amino acids, fat, and fatty acids), excretory products, (urea and uric acid), dissolved gases (about 40 mm^3 oxyen, 19 mm^3 carbon dioxide, and 1mm^3 nitrogen in 100mm^3 plasma), hormones and vitamins.

Blood Platelet. See *platelet*.

Blood Pressure. Refers to the pressure of the blood which is acting against the walls of the blood vessels, in particular the main arteries, that results from the pumping action of the heart and the elasticity of the arterial walls. In a resting human, it normally oscillates between 120 mmHg at systole and 80 mmHg at diastole.

Blood Serum. The pale fluid left after blood has clotted. It consists of plasma which does not have any of the substances involved in clotting plasma.

Blood Vascular System. In mammals, it is a continuous system of vessels having blood, which transports food materials, excretory products, hormones, respiratory gases, etc., from one part of the body to another. Certain invertebrates (molluscs and arthropods) possess and open system in which blood flows in blood spaces (e.g. the haemoceol of Crustacea). Vertebrates and most invertebrates possess a closed system with the blood contained in blood vessels and generally circulated by muscular contractions of the vessels or a heart.

Blood Vessel. A tubular structure present in vertebrates and some invertebrates that is able to transport blood throughout the body. Blood vessels vary in diameter and hence are able to regulate the blood flow to different parts of the body; for example to the surface layers to regulate heat loss.

Body Cavity. In most metazoan animals, it refers to the cavity which is bounded by the body wall. The body cavity has the heart, viscera, and many other organs.

Bohr Effect. Used for describing the phenomenon whereby the affinity of the respiratory pigment of the blood (haemoglobin in vertebrates) for oxyen gets decreased as the level of carbon dioxide is increased. This facilitates gaseous exchange, because more oxyen will be released in the tissues where the amount of carbon dioxide is increasing due to metabolic activity. At the same time, more oxyen will be taken up at the lungs or gills where the amount of carbon dioxide is low.

Bolus. Lump of masticated food with saliva.

Bombykol. Refers to the sex attractant pheromone of female silk-worm moth *Bombyx*.

Bone. Used for describing the hardest tissue in the body; connective tissue composed of collagen, calcium, phosphorous salt and ossein. Covered by periosteum. (Fig. 3)

Fig. 3 : Structure of compact bone

Bone Marrow. Vascular connective tissue present at the ends of long bones. Active cellular marrow is known as red bone marrow whereas that infiltrated with fat is known as yellow bone marrow. Red bone marrow comprises of myeloblasts, haemocytoblasts and megakaryoblasts which form WBCs, RBCs and platelets respectively.

Book Lungs. Paired respiratory sturctures that are present in various arachnids e.g., scorpion.

Boreal Forest. Ever green forest also named as coniferous forest.

Bottle-neck Effect. Type of genetic drift in which due to the rapid decrease in population size the genetic pool has been drastically diminished thus gene frequencies of remaining genes are greatly enhanced with respect to original pretrauma population values.

Bowman's Capsule. Refers to the cup-shaped end of a tubule of uriniferous tubule of the vertebrate kidney. It is surrounding a knot of blood capillaries (glomerulus) and together they form a Malpighian body.

Brachium. Used for describing the upper part of fore limb of a vertebrate. It is anteriorly articulated with shoulder and posteriorly with antibrachium. Skeleton support is provided by humerus bone.

Bradycardia. A condition in which there occurs heart rate reduction from normal level.

Bradykinin. A hormone formed from precursor which is normally circulating in blood. It is an important vasodialator.

Brain. Refers to one of the divisions of central nervous system that is formed by anterior part of neural tube; comprises fore, mid and hind brains. Used as a seat and controlling apparatus and co-ordinator of nervous activities of an individual. (Fig. 4)

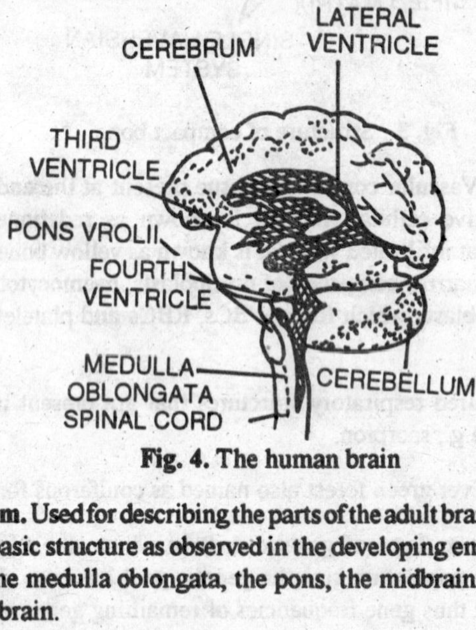

Fig. 4. The human brain

Brain Stem. Used for describing the parts of the adult brain that constitu its basic structure as observed in the developing embryo. It consis of the medulla oblongata, the pons, the midbrain, and part of tl forebrain.

Branchia. The gill in an animal.

Branchial Arches. Refers to the five pairs of embryonic visceral arcl in the vertebrates. They support the gills and pharynx permitti muscular movements of pharynegeal wall for a respiratory curre A typical branchial arch comprises 9 cartilageneous pieces, a lov median basibranchial to which a hypobranchial is attached on e:

side, a ceratobranchial and a pharyngobranchial cartilages. The ceratobranchials support gill rays.

Breast. Refers to the human lactating apparatus. It arises as an invagination from surface epithelium which dips down into underlying connective tissue as a solid column; these gradually hollowed out to form ducts of mammary glands. It bears a nipple at which the various ducts open and surrounding it is areola. It is rudimentary at birth, little further development take place until puberty and fully developed after delivery. It further undergoes proliferative change with recurrence to each menstrual cycle.

Breathing. Means a physical process that is affected either by osmosis (aquatic animals) or diffusion (terrestrial animals) through which an exchange of oxyen and carbon dioxide takes place between body and environment of an individual. Also termed as external respiration.

Brittle-Stars. Marine echinoderms having a small central disc and five snake-like arms arising from it. As these arms are broken when touched/disturbed, animal is termed as brittle-star; common name for *Ophiothrix*.

Bronchiole. Refers to one of the smaller air passage in the lungs of mammals. A branching system of bronchioles carries air from the bronchi to the different regions of the lung. The finest bronchioles end in alveoil. The cells lining the walls of bronchioles secrete mucus, which traps dirt and bacteria, and cells with cilia, beat to carry foreign matters out of the lungs.

Bronchus. An air passage which is leading from the trachea and entering the lung in a tetrapod. The main bronchi branch to form a number of smaller bronchi, which lead to the bronchioles.

Brood Spot. Refers to a prolactin induced bald area present on the ventral surface of some brooding birds which is richly supplied with blood for incubation of eggs.

Brown Fat. A special type of fat present in infants but not in adults of human beings. Its metabolic rate is vary high and produces heat that aids thermoregulation. It is situated between and around the scapulas. Its function can be compared to that of an electric blanket.

Brownian Movement. Used for describing the movement of cytoplasm and/or karyoplasm which takes place due to intracellular molecular collision specially observed in *Amoeba*.

Bruits. Refer to the abnormal sounds which are heard in various parts of vascular system; major cause is attributed to the disease of valves of heart.

Brush-Border. A free epithelium or epithelial cell surface that is bearing numerous, small microvilli e.g., epithelium of proximal urinary tubule.

Bruxism. Teeth grinding which takes place during sleep in some individuals; also associated with dreaming.

Buccal Cavity. Refers to the mouth cavity, by which food is taken into the alimentary canal. In mammals it is surrounded by the lips and cheeks. Ducts from salivary glands open into the mouth so that food gets moistened and lubricated and, in some cases, digested. Other vertebrates are also having buccal cavities but without cheeks and cannot therefore chew their food, but must swallow it whole. Invertebrates are also having a variety of buccal cavities.

Bud. An outgrowth of an animal that is capable of vegetative reproduction. In lower animals the production of buds that grow into new individuals and then break away from the parent is a common form of asexual reproduction.

Budding. Refers to the type of asexual reproduction in which a new individual is produced as an outgrowth (bud) of the parent organism. It takes place in certain animal groups, such as coelenterates, sponges and urochordates, where it is also called *gemmation*. It also takes place in the unicelluar fungi, especially the yeasts.

Bulbourethral Glands. Refers to the paired glands which are associated with urinogenital organs in mammals and are present surrounding urethra. They secrete a mucous-like substance that comes out before ejaculation. It lubricates vagina, neutralises acidity a of vagina, aids fertilisation by dissolving certain cells around the ovum. Also known as Cowper's glands.

Bulbus Arterious. Enlargement (bulb like) at the base of ventral aorta in bony fishes.

Bulla. Hollow bony growth such as bursa of a joint.

Bundle of His. Refers to a specialised type of conducting tissue present within the heart of mammals, which is continuous anteriorly with AV-node and posteriorly gives off a left bundle branch. The left bundle branch and fassicles are running subendocardially down either side of septum and coming in contact with Purkinhjie's system.

Bundle of Kent. Individuals having Wolff-Parkinson-white syndrome probably possess and additional aberrant muscular or nodal tissue connection between ataria and ventricle. It conducts more rapidly than slow conducting AV-node and is responsible for accelerated conduction.

Bungarotoxin. A group of neurotoxins which are isolated form venom of snake kraint (*Bungarus*). It binds selectively and irreversibly to acetylcholine receptors.

Bunner's Glands. Refers to small coiled acinotublar glands in the duodenum that secrete a thick alkaline mucous which probably projects the duodenal wall form gastric acid. They open in the krypts to Loiberkuhn.

Bunodont Teeth. When cusps in molar tooth are distinct from one another, the tooth is turmed as bundont e.g. man, monkey etc.

Bursa Entiana. A small sac-like structure which is present between pyloric stomach and intestine in cartilagenous fishes like, *Scoliodon*. Its definite function is not known.

Bursa Fabricii. Refers to a thick glandullar pouch of lymphatic tissue lying just above cloaca in case of many birds, e.g., pigeon; serves to fight infections of cloaca.

Bursicon. Neurosecretory cells in brain and nerve cord of insects secrets a hormone that affects the cuticle development, including process of tanning which gets completed several hours after each moult; small molecular protein having molecular weight of about 40,000.

Byssus. Describing the silk-like threads which are secreted by a group of molluscs e.g., mussels for attachment to rock surfaces.

C

Calcitonin. A hormone which is secreted by 'C' cells of thyroid gland. Its molecular weight is 3600. The main function of this hormone is to lower the circulating levels of calcium by inhibiting bone reabsorption.

Caecum. Refers to a cavity which goes at one end e.g., pouch which is the beginning of large intenstine.

Caenozoic (Cainozoic, Cenozoic). The present geological, era, which began some 65 million years ago, and divided into two periods, the Tertiary and the Quaternary. This era is characterized by the rise of modern organisms, especially mammals and flowering plants.

Calamus. Refers to the hollow semi-transparent quill or stalk of a feather.

Calamus Scriptorius. Mean a pointed depression in the floor of the fourth ventricle of the mammalian brain, which leads into the central canal of the spinal cord.

Calanoida. Minute free-living planktonic copepods existing in enormous numbers in the sea an forming an important part of the food of many fish.

Calcarea. A class of sponges whose skeletons have solely of calcareous and none-siliceous spicules.

Calcaronea. Refers to a group of calcareous sponges that comprise of most of those which are classed by *Heterocoela*, i.e., those whose central cavity is lined by flattened cells and which are having collared cells in special flagellated canals.

Calciferol, $C_{28}H_{43}OH$. Refers to a substance of the steroid group which is closely related to Vitamin D. It can be produced by irradiation of *ergosterol* with ultra-violet light. Ergosterol is present in many animals and plants. It is concerned with the regulation of calcium and phospate absorption and the calcification of bones. Lack of

Vitamin D produces *rickets* or softening of the bones.

Calciferous Glands. Glands secreting calcium carbonate or other calcium compounds, e.g., the oesophageal glands of an earthworm.

Calcoblasts. Refers to the specialised mesogloeal cells that are present in the body of sponges. They secrete calcareous spicules and are amoeboid in shape having calcareous inclusions in cytoplasm.

Canaliculi. Used for describing fine channels which are communicating with lumen of a canal or gland.

Canal of Schlemm. A passage which is connecting the aqueous humour with the scleral veins of the vertebrate eye and running through the junction between the cornea and the sclerotic.

Cancer. Used for describing a group of diseases which have common characteristic of uncontrolled cellular growth.

Capillary. Refers to the branches of metarteriole of arteriole. The opening of this has been surrounded by minute smooth muscles which are called precapillary sphincters. It connects metarteriole with a vennule.

Capitular Facet. A facet or demi-facet on the centurm of a throacic vertebra to which the capitulum of a rib has articulated.

Capitulum (of Rib). One of two processes (the other being the *tuberculum*) by which the rib of a mammal or bird is articulating with the thoracic vertebrae. The capitulum is articulating with the centrum of the vertebra and the tuberculum with the transverse process.

Carapace (1) In some arthropods (e.g. crustaceans, spiders, and king crabs), it refers to a shieldlike protective covering of the dorsal and lateral surfaces of the cephalothorax.

(2) In reptiles of the order Chelonia (turtles, tortoises, etc.), it refers to the domed dorsal part of the shell, consisting of platelike bones covered on the outside by horny plates (*scuta*).

Carbohydrase. An anzyme which promotes the breakdown or synthesis of a carbohydrate.

Carbohydrates. A term used for describing a class of compounds occurring widely in nature and having the general formula type

$C_X(H_2O)_Y$. Carbohydrates are both stores of energy and structural elements in living systems; plants having typically 15% carbohydrate and animals about 1% carbohydrate. The body is able to synthesis polysaccharides from simple units (anabolism) or break the larger units down to more simple units for releasing energy (catabolism).

Carbonic Anhydrase. An enzyme splitting carbonic acid into water and carbon dioxide. It is found in red blood corpuscles and also in some cells of the pancreas, the brain and the mucous membrane of the stomach.

Carboxyamino Haemoglobin. Haemoglobin on deoxygenation forms carbamine haemoglobine by reacting with carbondioxide and thus the transport of carbondioxide gets facilitated in the venous blood. About 20% carbondioxide is carried to lungs by this process. At respiratory surface the high partial pressure of oxygen breaks this association and carbon dioxide is liberated.

Carboxy-Haemoglobin. This compound is formed by reacting carbon monoxide with haemoglobin. The affinity of haemoglobin for oxygen is much smaller than carbon monoxide and this substance tends to displace oxygen on haemoglobin reducing oxygen carrying capacity of blood thereby giving rise to the tissue anoxia.

Carboxy-peptidase. A type of proteolytic enzyme which occurs in pancreatic juice. It attacks peptide chain by stepwise removal of amino acid residues from carboxyl group at the end of the chain. Two such enzymes are known, carboxypeptidase which is able to cleave carboxyterminal amino acids with aromatic or branched aliphatic side chains and carboxypeptidase which is able to cleave carboxyterminal amino acids with basic chains.

Carcinogen. Any agent bringing about a transformation in cell to oancrerous state.

Cardiac Arrest. As a result of abnormally in heart beating ventricular extrasystoles take place, thereby leading to ventricular fibrillation due to ventricular tachycardia. The fibrillating ventricles fail to pump blood effectively and circulation of blood stops thereby resulting cardiac arrest. In absence of emergency treatment, death may take place in few minutes.

Cardiac Cycle. Refers to the contraction and relaxation of heart which are responsible for receiving of blood and its supply out of heart. It involves atrial systole isovolumetric ventricular contraction, ventricular ejection isovolumetric ventricular relaxation and ventricular filling. It gets completed in 0.8 second (artial systole 0.1 sec., ventricular systole 0.1-0.3 sec. and diastole 0.4-0.6 sec.).

Cardiac Gland. A large digestive gland which is present near the cardiac end of the stomach in certain mammals.

Cardiac Muscle. Muscular tissue of a type which is present only in the wall of the heart. Its fibres somewhat resemble those of striated muscle but, unlike them.

Cardiac Output. Refers to the total volume of blood which is pumped out by heart per unit time. Cardiac output equals heart rate time stroke volume.

Cardinal Vens. Used for describing the large paired longitudinal veins or sinuses in fish and in the embryos of higher vertebrates. An anterior and a posterior pair are collecting blood from most part of the body returning it by way of the *Cuvierian ducts* to the heart.

Cardio-accelerator Centre. Refers to a region of the medulla oblongata from which impulses can accelerate the heart by involving the sympathetic nervous system.

Cardioblasts. Specialized cells that give rise to the wall of the heart in arthropods.

Cardio-inhibitory Centre. Refers to a region of the medulla oblongata from which impulses can slow down the rate of heart-beat by way of the vagus nerve.

Carnassial Teeth. Used to describe specialised teeth found in dogs and other carnivorous mammals. They comprise the last premolar in the upper jaw and the first molar in the lower jaw. These are very large and are used to cut up meat and shear it from bones.

Carnivora. The order that have the flesh-eating mammals, including *Canis* (wolf, dog), *Felis* (cat), *Meles* (badger), and *Lutra* (otter). The teeth of carnivores are specialized for biting and tearing flesh.

Carnivore. A flesh-eating animals which is especially a mammal of the order Carnivora (e.g., cats, wolves, seals, etc.) Carnivores generally possess powerful jaws, teeth modified for tearing flesh and cracking bones, and well-developed claws.

Carotid Artery. Used for describing one of the pair of blood vessels that supplies oxygenated blood to the head and neck. They are derived from the third aortic arch.

Carotid Body. A vascular structure which is present at the base of the external carotid artery. It has chemoreceptors, which monitor carbon dioxide and oxygen concentrations and pH of the blood.

Carotid Sinus. Refers to a swollen portion of the internal carotid artery, near its origin in the neck, having sensory receptors that monitor changes in blood pressure.

Carpal Bones. Bones in the distal region of the forelimb of tetrapods; in man they constitute the wrist (carpus). In the typical pentadactyl limb there are 12 carpal bones, which are arranged in three rows. However, there are many modifications and reductions to this basic pattern; in man there are only eight which articulate with each other, and with the metacarpal bones distally. Three carpal bones are forming the wrist joint the radius of the forearm.

Carps. A group of edible herbivorous fishes which lack teeth in mouth. Their body is covered with cycloid scales and they have an air bladder. Examples are *Labeo rohita, Cirrhina mrigala, Labeo calbasu* and *Catla* which are known as major carps of India.

Carpus. The collection of carpal bones forming the wrist in man.

Carrier. In genetics, it refers to an organism that carries a recessive, often harmful, gene which is masked in the phenotype by a normal dominant gene.

Cartilage. Gristle : a firm but flexible skeletal material that is making up the entire skeleton of the cartilaginous fishes (sharks, etc.).

Cartilage Bone (replacing bone). The type of bone which is formed from cartilage in the embryo. The cartilage is invaded by bone-forming cells (osteoblasts), which are able to convert it into bone in the process of ossification.

Cartilage of Santorini. Small, nodule-like structure which are present at the tips of arytenoid cartilages of larynx in mammals.

Caste. Any one of the types that belong to any single species of social insects (like honey bees, termites etc.).

Castration. Refers to the removal of gonads specially testes from an individual.

Catabolism. Refers to the breakdown or destructive changes taking place during metabolism; metabolic processes in which breakdown takes place leading to the formation of energy e.g., respiration.

Catadromous. Used for describing fishes that live in fresh water and go to the sea for spawning e.g., eels. *Opposite* to andromous.

Catecholamines. A group of chemicals (amine derivatives of catechol) that are acting as neurohormones or neurotransmitters. Examples are noradrenaline, adrenaline and dopamine.

Cat-fishes. A general name for describing bony fishes that have barbels and look like whiskers of cat (family siluridae). Body nacked and fins usually having spines. Also having an air bladder. Mostly predators.

Cauda Epididymis. Refers to an enlargement at the distal end of the epididymis or tube which is leading from the testis to the vas deferens in mammals (see also *Caput epididymis*).

Cauda Equina. Used for describing the terminal spinal nerves of a mammal which together with the *filum terminale* form a bunch. The name has been derived from its resemblance to a horse's tail.

Caudae Lobe. Refers to a small lobe in the liver of mammals. (Also see *liver*).

Caudal Spine. A spine which is forming the 'tail' or telson on some arthropoda, e.g., scorpions.

Caudal Styles. Stylets or filaments, which are usually paired, on the hind-segments of many arthropods.

Caudal Vertebrae. Bones of the vertebral column which are protecting the spinal cord in the tail region.

Cell. Refers to the basic unit of structure of all living organisms, except viruses. Cells were discovered by Robert Hooke in 1665, but Scheiden and Schwann in 1839 were the first to put forward a clear *cell theory* which stated that all living things were cellular. Prokaryotic cells (typical diameter 1 μm) are significantly smaller than eukaryotic cells (typical diameter 20 μm). The largest cells are egg cells (e.g. ostrich, 5cm diameter); the smallest are mycoplasmas (about 0.1μ m diameter). All cells have genetic material in the form of DNA, which is able to control the cell's activities; in eukaryotes it is enclosed in the nucleus. (Fig. 1)

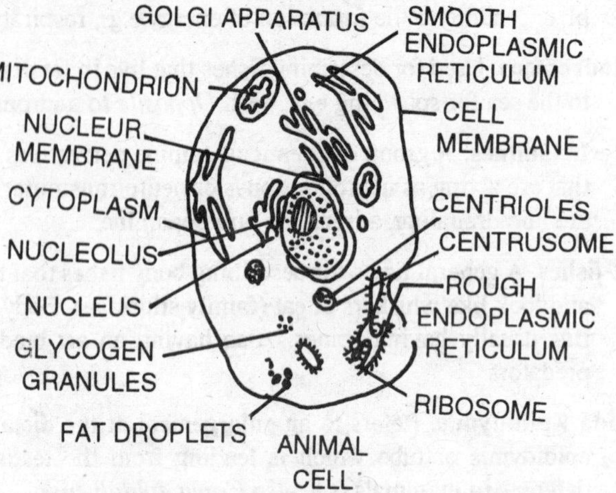

Fig. 1 : The various differences between Prokaryotic and Eukaryotic Cells

	Prokaryote	**Eukaryote**
Occurrence	blue-green algae and bacteria	all higher animals and plants
Average diameter	1μm	20 μm
Nuleolus material	not separated from cytoplasm by memberane	bounded by nuclear membrane
DNA	circular and forming only one linkage group	linear and divided into a number of chromosomes
Nucleolus		+
Cell division	amitotic	usually by mitosis or meiosis
Sexuality	—	+ and — strains, male and female forms
Cytoplasmic streaming	—	+

Vacuoles	—	+
Plastids	—	+
Ribosomes	smaller (70S)	larger (BOS)
Endoplasmic reticulum	—	+
Golgi apparatus	—	+

+ presence
— absence

Cell Body. Refers to the part of a nerve cell (neurone) that has the nucleus. It has a swollen appearance and possesses Nissl granules. The cell body is a centre of synthesis, supplying materials to the rest of the neurone.

Cell Cycle. Involves ordered sequence of phases through which a cell has to pass between successive mitotic events, i.e., during interphase. It is divided into three phases G_1 S, and G_2. Synthesis of messenger RNA, transfer RNA, and ribosomes takes place in G_1. and replication of DNA takes place during the S phase. The materials required for spindle formation are probably formed in G_2. The time taken to complete the cell cycle is different in different tissues. For example, epithelial cells of the intenstine wall may divide every 8-10 hours.

Cell Division. The process by which a cell gets divided into daughter cells. In unicellular organisms it is a method of reproduction. Multicellular organisms grow by cell division and expansion, and division may be very rapid in young tissues.

Cell Fractionation. Refers to the separation of the different constituents of the cell into homogenous fractions. This is carried out by breaking up the cells in a mincer or grinder and then centrifuging the resultant liquid. The various components settle out at different rates in a centrifuge and are thus separated by appropriately adjusting the speed and or time of centrifugation.

Cell Lineage. The theory which states that cells arise only from pre-existing cells. The cell lineage of a structure traces the successive stages that the cells have to pass through from the time of their formation in the zygote to their appearance in the mature functional structure.

Cell Theory. The theory stating that all organisms are composed of cells and cell products and that growth and development take place by the division and differentiation of cells. This idea was obtained from numerous investigations that started at the beginning of the 19th century, and it was finally given in this form by Schleiden and Schwann in 1839.

Cement (Cementum). A bonelike substance covering the root of a mammalian tooth. It helps to fix it in the socket (alveolus) of the jaw bone. It is very compact and hard and has a higher mineral content than bone. Cement normally does not have a system of Haversian canals.

Central Nervous System (CNS). The part of the nervous system that is able to receive sensory information from all parts of the body and, through the many interconnections that are possible, brings about the appropriate message to be sent out to muscles and other organs. In vertebrates CNS is having the brain and spinal cord.

Centriole (Centrosome). An organelle taking part in spindle formation during nuclear division. Each cell normally has a pair of centrioles which are laying with their axes at right angles to each other. Most animals and some lower plant cells have centrioles, but they are not present in higher plant cells.

Centrolecithal. Egg with yolk confined to centre leading to superficial cleavage. It takes place in insects.

Centromere. A point to which the spindle fibres attached; the constricted region of chromosome which is not coloured by usual chromosomal stains.

Centrum. Refers to the ventral portion of vertebra, which is formed by ossification of embryonic notochord form which neural spines, transverse processes and zygapophyses are coming out.

Cephalic. Referring to or situated near head.

Cephalisation. Used for describing the tendency of centralisation of important parts such as sense organs and mouth in the head region.

Cercopithicoid. Old world monkeys having a tail which was not employed as limb.

Cerebellum. A part of the brain which is having a pair of greyish deeply-folded hemispheres lying dorsal to the medulla oblongata and partially hidden by the overlying cerebral hemispheres. It has been lound to monitor the position of limbs and tension of their muscles and makes any necessary adjustments to the messages which are sent out to voluntary muscles by the cerebral cortex. It thus finds importance in maintaining balance, in locomotion, etc.

Cerebral Cortex (Pallium). Used for describing the surface layer of the cerebrum of the brain. It has billions of nerve cell bodies, collectively known as grey matter. It is responsible for the senses of vision, hearing, smell and touch, for stimulating the contraction of voluntary muscles, and for higher brain activities, such as language and memory. Many of these activities take place in specific regions of the cortex.

Cerebral Hemisphere. A pair of structure that are originating from the forebrain. They have the centres which are concerned with the major senses, voluntary muscle activities, and higher brain functions, like languages and memory.

Cerebrospinal Fluid. A lymph-like liquid which does not have blood cells and large molecules. It is present in the internal cavities and between the surrounding membranes of the central nervous system. It is filtered from blood by the choroid plexuses in the brain and eventually returns via lymph vessels or in venous blood. It cushions and protects nerve tissues.

Cerebrum. Refers to the most prominent region of the brain in man. It consists of two cerebral hemispheres which are joined by a band of nerve fibres (the *corpus callosum*). It forms the site of such functions as vision, hearing, touch, smell voluntary muscle activity, speech and memory.

Ceruloplasmin. Refers to copper containing plasma protein. Its deficiency causes Wilson's disease, a familiar disorder of copper metabolism.

Cervical. The neck region of a vertebrate.

Cervical Vertebrae. The small vertebrae of the neck. They form a very flexible portion of the vertebral column. The mammals have usually seven cervical vertebrae the first and second are modified to form the atlas and axis. They are all characterized by relatively large neural canals.

Cervix. Neck-like structure present between uterus and vagina in a typical mammals; opening of uterus in a mammal.

Cestoda. A class of endoparasitic Platyhelminthes, the tapeworms. They have a complex life cycle which is involving one or more intermediate hosts; for example, Taenia uses the pig as an intermediate host and man as the final host.

Cetacea. The order including the only completely marine mammals—the whales, dolphins, and porpoises. They loses a hairless streamlined body, no hind limbs, forelimbs modified as flippers, and a tail with horizontal flukes employed for propulsion.

Chaeta (Seta). A stiff bristle-like structure present in some animals. Chaetae are present on the outside of the body of an earthworm, in which they are employed for gripping the soil during locomotion.

Chain Terminating Codon. Refers to a sequence of three nucleotides which indicate the end of the information for assembly of amino acids in a polypeptide chain. It has three chain terminating codon each of which can bring translation to a halt. Also termed as non-sense codons.

Chalone. A substance present in mammalian tissue that can inhibit mitosis in the presence of, for example, adrenaline. It has been suggested that these substances play a role in the ageing process and in cancer.

Character. Refers to the distinguishing structure of function, any trait of an organism.

Chela (1) The large pinching claw which is present on the ninth pair of appendages in such crustaceans as carbs and lobsters. It is used for attack and defence.

(2) The large claw which is present on the second pair of appendages in scorpions. It is used for catching prey and tearing it apart.

Chelicera. A paired appendage which is present on the third segment of the head of arachnids, in the form of a sharp claw or pincer. Ducts lead from poison glands to these appendages, which are able to inject paralysing poisons into the bodies of prey.

Chelosis. Inflamation and soreness of the corners of mouth. It is associated with deficiency of vitamin B_2.

Chemical Fossils. Particularly resistant organic chemicals present in geological strata that reveal the existence of life in the period when the rocks were formed. Chemical fossils (e.g. alkanes and porphyrins) cite the only evidence for life in rocks of pre-Cambrian age.

Chemoautotrophism (Chemosythesis). See *autotrophism, chemotrophism.*

Chemoheterotrophic. Refers to type heterotrophic nutrition in which energy requirements for the synthesis of organic molecules from organic precursors are given by the chemical breakdown of organic moleeules.

Chemoreceptor. A receptor responding to chemical compounds, e.g., the taste buds.

Chemotropism (Chemotropic movement). A tropism in which the stimulus is chemical.

Chiasma. Used for describing connection between homologous chromosomes which is seen during the prophase stage of meiosis. Chiasmata involve a mutual exchange of material between homologous, non-sister chromatids (crossing over) and offer one mechanism by which recombination takes place through the splitting of linkage groups.

Chilopoda. The class of arthropods containing the centipedes (e.g. *Lithobius*). They are characterized by a flat body that is divided into numerous segments, each bearing one pair of walking legs. They are terrestrial, and breathe air through tracheae. Excretion takes place by a Malphighian tubules.

Chiropteara. The order containing the bats the only flying mammals. Bats have a thin elastic hairless flight membrane (patagium) which extends from the elongated forearm and four of the elongated fingers to the hind limbs and, usually, the tail.

Chitin. A nitrogen-containing heterosaccharide present in some animals and the cell walls of most fungi. The outer covering of arthropods, the cuticle, gets impregnated in its outer layers with chitin, which makes the exoskeleton mane rigid.

Chloride Cells. Epithelial gill cells in fishes that take part in active transports of salts and help in osmoregulation.

Chloride Shift. The movement of chloride ions which takes place across the RBCs membrane to compensate the movement of bicarbonate ions.

Chilocruorin. A green pigment which is present in some marine polychaetes: its action is similar to that of haemoglobin.

Chlorogogues. Refers to the excretory cells of certain invertebrates including annelids that are able to carry wastes from body fluid to epithelium also called chlorogogen cells.

Chlorophyll. Green pigment which is present in plants and possible some animals. It is essential for photosynthesis.

Chloroplast. A chlorophyll having chromatophore.

Choanae. These are the structures through which nasal chambers in frog communicate with mouth cavity. (Also called internal nostrils).

Choanocytes. Flagelleted collar cells which are present in sponges. They are responsible for setting water current through their body.

Cholecystokinin. A hormone which is liberated by upper intestinal mucosa. It induces gall bladder contribution and the release of pancreatic juice. Also known as pancreozymin (CCK-PZ).

Cholesterol. A natural sterol which is precusor to the steroid hormone (aldosterone, estrogen progesterone and testosterone etc.), vitamins D_1, D_2, D_3 and cholic acid.

Cholinergic Discharge. Means the discharge/release of acetylcholine or substances having similar actions.

Cholinergic Neuron. A nerve fibre releasing acetylcholine from axon terminal when transmission of impulses takes pace across the synapses during transmission of nerve impulse.

Chondrichthyes. Class of fishes e.g., sharks and rays having endoskeleton of cartilages; exclusively marine.

Chondrin. Material present in the matrix of simple cartilage.

Chondroblasts (Chondrocytes). The cells embedded in the matrix of cartilage.

Chondrocranium. Refers to the first part of the skull to form invertebrate embryos. It is having cartilaginous structures—plates and capsules—that protect and support the-brain, olfactory organs, eyes and the inner ear. It usually gets ossified in adults, although it remains cartilaginous in Chondrichthyes (cartilaginous fish).

Chordae Tendinae. Refers to string-like cords which are attached to myocardium wall and atrioventricular valves, which on concentration of ventricles keep those valves shut thereby making blood from flowing back into the auricles.

Chordata. A major phplum of bilaterally symmetrical metamerically segmented coelomate animals that are characterized by the possession at some or all stages in the life history of dorsal supporting rod, the notochord.

Chorion (1) Refers to one of the three embryonic membranes of reptiles, birds and mammals. It is arising from the outer layer of the amniotic hood and enclosing the amnion (with embryo or foetus inside), yolk sac and allantois. The trophoblast of mammals forms part of the chorion.

(2) Refers to the tough outer membrane of some eggs, notably those of insects. These is usually a pore, the micropyle, to admit spermatozoa.

Chorinic Villi. Villi-like structure which are produced by extra-embryonic membrane chorion in case of reptiles, birds and mammals.

Choroid. The middle layer of the vertebrate eye which is present between the sclerotic and the retina. It is rich in blood vessels and has pigment that absorbs light and stops it going through the eye past the retina. Anteriorly, it projects into ciliary body and the iris.

Choroid Rete. Used for describing a counter current arrangement of aterioles and venules, behind retina in eyes of bony-fish (generally a teleost).

Choroid Plexus. Refers to one of two thin-walled greatly folded highly vascular regions in the inner wall of the brain through which there occurs exchange of materials between blood and cerebrospinal fluid. It is situated in the roof of the fourth ventricle and at the junction of the first and second ventricles with the third ventricle, the choroid plexuses (which form the *blood-brain barrier*) bar the entry of blood cells and large molecules (including potentially harmful foreign materials and certain drugs) into the cerebrospinal fluid.

Christmas Disease. Refers by haemophilia-B.

Christmas Factor. Also known as plasma thromoplastic component (PTC) or antihaemophilic factor-B. This factor in presence of factor-VIII activates factor-X which is catalysing conversion of prothombin that further aids in the formation of blood clot.

Chromaffin Gland. Epinephrine secreting cells of adrenal medulla which are named after their high affinity for chromium stains.

Chromatid. Refers to either of the two identical halves in which a chromosome starts splitting longitudinally in the cells division; actually each chromosome duplicates another copy next to itself.

Chromatin. Used for describing the stainable protoplasmic substance which is present in the cell. It gives rise to conspicuous chromosomes during mitosis/meiosis.

Chromomere. Small unequally sized regions of denser material which are arranged at irregular intervals along the chromosomes. They can be observed during prophase along the extended chromosomes. The pattern of these densely coiled regions may vary between individuals of a population and reflects differentiation along the chromosome.

Chromonemata. A number of thread-like structure (at least four) in chromosome which can be seen during prophase stage of meiosis. Difficult to see except in forms like *protozoa*.

Chromphore. Used for describing the specialised pigment bearing body or cell such as chromatophore which is responsible for colour making many animals.

Chromoprotein. A protein which is conjugated with pigment e.g. haemoglobin, myoglobin etc.

Chromosome. Refers to one of a group of thread like structures of different lengths and shapes in nuclei of eukaryotic cells. During nuclear division they are tightly coiled and can be easily seen through the light microscope. After division, they uncoil and are difficult to see. They control cell activity because they carry the genes.

Chromosome Map. A diagram which shows the order of genes along a chromosome. Such maps could be constructed from information which is gained by both linkage studies (to give a *linkage map*) and observations made on the polytene (giant) salivary-gland chromosomes of certain insects, e.g. *Drosophila*.

Chromosome Mutation. Refers to a change in the number or arrangement of genes in a chromosome. If chromosome segments break away during nuclear division, they may rejoin the chromosome in a wrong way, giving an *inversion*. Alternatively, they may rejoin a different part of the same chromosome, or another chromosome, yielding a *translocation*. It the segment is lost, it is called a *deficiency* or *deletion*; it is often fatal.

Chrysalis. Refers to the pupa of insects of the order Lepidoptera (butterflies and moths). See *pupa*.

Chyle. Used for describing the fluid food within the intestine (the food during its passage to intestine gets a large supply of water which is brought through various juices rendering it more fluidy; it also occurs due to digestion of various food stuffs).

Chyme. Food is the stomach is reduced to thick creamy paste called chyme. It is formed due to addition of water, churning of food by stomach walls and by the action of gastric juices and spurted into duodenum in form of small jets.

Chymotrypsin. An enzyme which is used to carry out the partial hydrolysis of peptide chains. It occurs in pancreatic tissue and in pancreatic juices as an inactive form, chymetrysinogen. Chymotrysin catalyses hydrolysis of only the peptide bonds which are formed by the carbonyl groups from phenylalanine, tryosine, and tryptophan residues.

Cilia. Microscopic hair-like protroplasmic processes that project from free surface of certain cells and capable of vibration.

Ciliary Body. A ring of muscular tissue present at the junction of the choroid with the iris in the vertebrate eye. It surrounds the lens, to which it gets attached by the suspensory ligament. It is helpful in altering the curvature of the lens, to accommodate for near or distant vision. It has been ciliary glands, which secrete the aqueous humour.

Ciliary Feeding. Means a method of feeding in some invertebrates, in which cilia, after creating a current of water towards the body, filter out food particles and transport them into the mouth. Most bivalve molluscs and the protochordates are ciliary feeders.

Ciliary Muscles. A modification of choroid coat to which zonules of Zinn attached; secretes aqueous and vitreous humors in the eye.

Cilium. Small whip-like structure having 9+2 arrangement of inner filaments; locomotory organelle in *Paramecium*. Also known as undulipodium.

Ciomvallate Papillae. The paillae lying at the base of tongue of man. They are having taste buds which are large but few in number and each is surrounded by a groove (Also see *tongue*).

Circadian Rhythm. A cycle with period of 24 hours.

Circle of Willis. In man the internal carotid and vertebral arteries (forming basilar artery) get united at the base of brain at the circle of willis from this anastomsis six cerebral branches go to the cortex, subcortex and brain stem.

Circulatory System. Constitutes continuous series of vessels or spaces in nearly all animals that are table to transport materials around the body. The system is best developed in mammals. They enable all parts of the body to get constant supply of oxygen, food, etc., and to have waste products removed promptly.

Cirrhosis. Refers to the diffuse hepatic fibrosis (enlargement) which is generally caused by fatty liver or either liver infections. The gross chronic condition of fatty liver results in death of many liver cells and their replacement by scar tissues. The scarring by interferring with the hepatic blood supply gets aggrevated the condition thereby, resulting more recrosis and more scarring. The disease occurs either due to severe malnutrition or due to alchoholism.

Cirri. Small slender projections or appendages that appear almost like tentacles except for their position.

Cisternamagna. Cavity dorsal to coelom of vertebrates in which there are kidneys.

Cis-Trans Effect. The phenomenon occurring due to recombination within a gene (cistron), in which a mutation is only expressed in the phenotype provided the mutant pseudoalleles are present on different homologous chromosomes. See *cis-trans test*.

Cis-Trans Test. A test determining whether two mutations that have the same effect take place in the same gene or in different gene. The mutations may be in the either the cis position (i.e., on the same chromosome) or the transposition (one one each homolouge). If the mutations are in different genes then a normal phenotype is formed whether the mutations are in the cis or trans position, because they get masked by corresponding dominant genes on the other homologue. However, when the mutations are in the same gene then a normal phenotype will be formed only if the mutations are in the cis position. In the transposition the mutant phenotype is expressed, because both alleles of the locus are mutants.

Cistron. A unit of function, i.e., a segment of DNA which determines a single polypetide chain of a protein molecule. Its extent could be defined by the cis-trans test. See *gene*.

Citric Acid Cycle. Refers to the metabolic cycle which brings about the complete oxidation of acetyl portion of acetyl Coenzyme-A and regenerates 36 molecules of ATP. Also termed as Kreb's cycle/TCA cycle.

Clams. A common name used for describing the bivalved molluscs e.g. *Unio, Lomellidens* etc., that usually lie partly burried in muddy/sands bottoms of lakes or streams. They burrow and move from place to place by using the foot which can be extended from anterior end of the shell. Water along with food and oxygen enters the body through a slit-like opening at the posterior end of shell (called incurrent siphon) and water goes out with wastes and faecal matter etc., through same end of shell (excurrent siphon).

Claspers. Refers to a pair of hard elongated structures which are attached to pelvic fins of some cartilagenous fishes like *Scoliodon*. They possess a groove internally and are connected with cloaca and used for transferring sperms into the cloaca of female.

Class. A term used for describing a collection of similar orders. Algal classes end in *phyceae* (e.g. Bacillariophyceae) and fungal classes end in *mycetes* (e.g. Ascomycetes). Animal and higher plant classes are having a variety of endings, e.g Pisces, Amphibia, Musci, Gymnospermae. Classes are divided into subclasses. Similar plant classes together are said to constitute a division, while similar animal classes are laid to constitute a phylum.

Classical Conditioning. A learning paradigm under which a conditioned stimulus gets paired with unconditioned stimulus to elicit the same response.

Classification (1) Used for describing the grouping and arrangement of organisms into a hierarchical order.

(2) Refers to the arrangement of organisms which is resulting from classification procedures. An important aspect of classifications is their predictive value.

Clavicle. One of a pair of bones which are lying anteriorly, on each side of the base the neck, in some vertebrates; they form the collar bones in man.

Cleavage. Used for describing series of early divisions of an egg into many cells, eventually forming a morulla. (Also see *holoblastic and meroblastic cleavages*).

Cleavage Furrow. Line of cytoplasmic division during division of an egg into many cells.

Climatic Migration. Used for describing a locomotory activity which brings the organism to a preferred climatic zone.

Climax Community. An ecological community that has reached a state of dynamic equilibrium.

Cline. A graded series of characters which are exhibited by a species or other related group of organisms, usually along a line environment or geographical transition. The populations at each of the cline may be substantially different from each other.

Clitellum (Saddle). Refers to a glandular and vascular swelling of the epidermis of some Annelida (earthworms and leeches), which is involved in reproduction. In the earthworm, it is present over segment 32 to 37 and forms the site of attachment for segment 9 to 15 of a mate at copulation. It secretes a mucus tube which is able to bind the two worms together. After copulation it secretes a cocoon, which is later deposited having fertilized eggs.

Clitoris. Used for describing a small erectile rod of tissue in female mammals that is situated anterior to the vagina and urethra. It is homologous with the male penis.

Cloaca. In most vertebrates it refers to the single chamber into which the alimentary canal, kidney ducts, and genital ducts discharge their contents. The placental mammals have two (or females have three) separate openings for faeces, urine and gametes.

Clone. A group of organism or cells derived from a single parental organism or cell by involving asexual reproduction or parthenogenesis and are thus genetically identical.

Clotting Factors. Refers to a group of twelve substances that get activated when blood leaves the circulatory system, usually by injury and brings about blood clotting. They include various proteins, vitamin K, calciumions and platelets.

Clypeus. A hard plate on the middle of the front of head in insects before the flons.

Cnidaria. A group of animals all having cnidoblasts e.g., *Hydra.*

Cnidoblast. Type of cell the cnidarians have; forms nematocysts.

CNS (Central Nervous System). Used for one of the principal divisions of nervous system in an animal. In vertebrates it has brain and spinal cord.

Coagulation. See *clothing of blood.*

Coccyx. Caudal vertebrae of man (3-4) fused into a single mass; a vestigeal organ in man.

Cochlea. Structure within internal ear of mammals where sound waves are changed into the nerve impulses effecting hearing.

Cocoon. Used for describing protective covering for eggs and/or larvae produced by many invertebrates, e.g. spiders and earthworms. The larvae of many insects also spin cocoons for the pupal stage.

Co-dominance. Refers to the situation in which two different alleles are equally dominant. If they occur together the resulting phenotype will be intermediate between the two respective homozygotes.

Codon. Used for describing a group of three nucleotide bases (i.e., a nucleotide triplet) that codes for a specific amino acid. As four different bases are present in nucleic acids, there are 64 ($4 \times 4 \times 4$) possible triplet combinations. The arrangement of codons along the DNA molecule constitutes the *genetic code.*

Coelacantini. An order of bony fish having only one living genus, *Latimeria.* Coelacanths were considered to have become extinct in the Cretaceous until the first live coelacanth of modern times was caught off south Africa in 1938.

Coelenterata (Cnidaria). Refers to a large and successful phylum of aquatic, mostly marine, invertebrates and the most primitive of the truly multicellular (metazoan) animals. Coelenterates are typically radially symmetrical and are diplobalstic.

Coelenteron. Used for describing body cavity of coelenterates. It is having a single opening (the mouth) through which food is ingested and waste products are egested.

Coelom. A fluid-filled cavity which arises in the mesoderm of the more advanced animals. It functions as a hydrostatic skeleton in some worm (e.g. earthworm) providing an incompressible barrier for the muscles to act against.

Coenosarc. Refers to the inner cellular part of a hydranth as distinguished from the outer surrounding perisarc.

Coenozoic. Geological era which was following mesozoic, from approximately 75 million years ago upto and including the present time.

Coenzyme. An organic molecule combining with an apoenzyme to form the functional holoenzyme.

Coezyme A (CoA). A coenzyme which is important in the synthesis and reactions of fatty acids. In the Krebs cycle it reacts with pyruvic acid, resulting the loss of carbon dioxide. It is a complex nucleotide having an active SH group. The compound gets readily acetylated to $CoAS\text{-}COCH_3$.

Coenzyme Q (Ubiquinone). An electron-carrying coenzyme which is active in the electron-transport chain.

Cofactor. A non-protein substance which makes an enzyme to carry out its activity. Cofactors may be cations or organic molecules, called coenzymes. Unlike enzymes they are, in general, stable to heat. When a catalytically active enzyme forms a complex with a cofactor, a *holoenzyme* is formed. An enzyme without its cofactor is called an *apoenzyme.*

Colchicine. A drug obtained from the autumn crocus *Colchicum autumnale* which is able to prevent spindle formation in mitosis. It halts cell division at metaphase, the stage at which the chromosomes get duplicated to give four homologues for each chromosome.

Coefficient of Consanguinity. A value expressing the chance that an individual has received from both parents a specific gene which is transmitted to them by an ancestor which they have in common.

Coefficient of Relationship. The value expressing the chance of carrying a specific gene, by both of two related persons, which traces back to an ancestor that they are having common. The value also reveals the proportion of genes in any of the two related persons which on average becomes identical. The value increases with closeness of relationship. It is exact ½ only for a parent and child relationship as child must carry one half of his/her genetic component in common with a parent.

Cold Blooded. (Of animals) having the degree of body heat not fixed but changing with that of the air or water around, as fish and snakes.

Coleoptera. The largest order of insects and possibly the order in the animal kindom, having the beetls and weevils. Their forewings form hard leathery *elytra*, which protect the membranous hind wings and soft abdomen when resting.

Colinearity. The corresponding relationship existing between order of amino acids along a polypeptide chain and order of codon in DNA of the gene which is specifying that polypeptide.

Collagen. A substance which is making up parts of tendons; having fibres of white variety.

Collecting Duct. Refers to the portion of renal tubule in which final concentration of urine takes place; distal part of convoluted part of renal tubule opening into ureter.

Colon. The anterior part of large intestine of a vertebrate e.g., mammal.

Colony. Group of individuals unicellular/multicellular, of same species that are developed from a common parent and are organically attached together; also used for describing insect societies.

Colostrum. Small quantity of milk from 100th day before delivery occurs in mammals. (Also see *mammary glands*).

Columella Auris. A rod-like structure that extends across the cavity of middle-ear from ear drum to inner ear; transmits the vibrations to inner ear.

Commensalism. Used for describing association of two individuals in which at least one gets benefited and other neither benefited nor harmed e.g., sparrow in the house.

Commissure. Nerve connection which gives two similar ganglia.

Community. A more or less complex group of plants/animals that are occupying a particular area.

Complement. A group of nine proteins which are normally present in vertebrate blood that react in an ordered sequence when an antigen antibody complex is formed. The reaction brings about lysis of the foreign cells or bacteria.

Complemental Males. Small and usually degenerate males of certain animals (e.g., some baranaces and angler fish) that are living on or in the body of the female. They depend on the female for nutrition and are sometimes reduced to little more than an attached testis. Their attachment ensures that fertilization (cross-fertilization in herma-phrodite barnacles) takes place.

Complementary Genes. Genes that could only be expressed in the presence of other genes; for example if one gene is able to control the formation of pigment precursor and another gene is able to control the transformation of that precursor into the pigment, then both genes must be present for the colour to get developed in the phenotype. Such interactions between genes result in apparent deviations from the 9:3:3:1 dihybrid ratio in the F_2.

Complementation. Refers to the production of a normal phenotype from a genotype that is apparently having two mutants alleles at a given locus.

Complement Fixation. The combination of complement with antibody antigen complexes. It is a property which is used to test for the presence of a specific antigen or antibody.

Compound Eye. The type of eye present in crustaceans and insects. It is having several thousands of units (*ommatadia*). The spots of light focused by these give a mosaic immage, which does not have visual acuity. However, the compound eye is very efficient at detecting the slightest movement over a wide area.

Conditioned Reflex. A reflex action by an animal which becomes modified by experience (or conditioning) so that instead of taking place in response to the original stimulus it follows a different 'learned' stimulus. However, it is not permanent and needs periodic reinforcement with the original stimulus.

Condyle. A round ball-like structure which is formed at the extermity of some bones e.g., femur, humerus; also occipetal condyle when it is associated with skull.

Cone Cells. Refers to vertebrate visual receptor cells having tappered outer segment in which lamellar membranes are continuous with surface membrane.

Congenital. Present at birth which may be due to genetic or solely to environmental causes.

Conjugation. Refers to the temporary union of two sexes of a same animal for exchange of genetic material or sex cells.

Conjunctiva. Refers to the continuation of mucous membrane over the cornea of eye which connects the cornea with inner covering of the eye lid and keeps eye separated from external world.

Conn's Syndrome. A disease in which person is suffering with severe K^+ depletion.

Connective. The tissue, which contains the vascular strand, that joins the two lobes of the another.

Consanguinity. Term used for describing mating of genetically related individual; also pertaining to the commonality of atleast a portion of the chromosomal complement of individuals (blood relatives).

Constipation. Means the slowing of rate of movement of faecal matter in the intestine/colon thereby resulting accumulation of hard and dry stool; occurs due to untimely and improper habits of defaecation.

Consumers. In an ecosystems those which derive energy from producers (primary consumers) or from other consumers (secondary consumers).

Contact Inhibition. The property of normal cells placing a restriction on cell proliferation in response to contact with neighbouring cells.

Continental Island. Island which was once attached to nearby continent. Generally fauna are characteristic of that continent.

Continuous Variation. Refers to the phenotypic variation seen in many characteristics which does not fit into distinct sharply defined catagories but which grades from one extreme to the other.

Contonr Feather. Feather having stiff shaft and vane. Appears on exterior of body and gives a definite shape to the body of bird.

Contractile Vacuole. Osmoreglatory apparatus which secretes water from cytoplasm.

Conus Arteriosus. Refers to the expanded cone-like structure over the ventricle in some vertebrates e.g., fishes that empties blood into aorta. It is having valves that check back flow of blood.

Coprophagy. Eating faeces by an animal e.g., rabbit.

Copulation. A temporary union taking place between male and female for getting sperms leading to fertilisation.

Coral Reefs. Refers to one of th formations of coral polyps that are confined to water of at least 20°C. They are built by epidermal cells of countless numbers of small polyps each one secreting its cup-shaped skeleton. The polyps die and new generations secrete cups upon old ones, only surface of coral mass is alive. Three types of coral reefs are known:

Fringing reef. Ridge of coral which is built up from sea bottom. It is located so near to land that no navigable channel is present between it and the shoe. Frequently breaks take place and irregularpools and channels are created in the reef which often get inhibited by many kinds of animals (often brilliantly coloured).

Barrier reef. It is separated from shore by a wide deeply formed channel (usually affording passage for relatively larger boats) such as great barrier reef of Australia (2161 km. in length and channel is 20-50 meters deep). It is entirely surrounding island.

Atoll reef. It is or more islands having a belt of coral reef which is surrounding a central lagoon. These are situated in mid Pacific with lagoon of few hundred meters to several kilometers in diameter. Bikni atoll is having a lagoon area of 7.25 sq. kilometers and land area of 7.43 sq. kilometer.

Corium. The inner dermal portion of skin.

Cornea. The outer very transparent layer of eye which is secreted by corneagen cells.

Corona. A ring/circle of tentacles or cilia.

Coronary Thrombosis. Refers to the occlusion in the branch of coronary artery of heart by a thrombus due to plaque formation as a result of arteriosclerosis resulting no-supply of blood to myocardium and may cause myocardinal infraction. If major vessel is blocked, there may occur instant death.

Corpora Cardiaca. Term used for describing the neurohaemal organ in insects which is situated between brain and corpus allatum, Neurosecretory axons are extending from brain to this part. It is known to secret its own hormones but its main function is mainly to store and release neurohormones from brain. Its hormones acclerate heart beating, peristalsis in the fore gut, also regulate protein metabolism in fat body and stimulate secreting of moulting hormones.

Corpora Cavernosa. Refers to the two lateral columns of erectile tissue in penis of mammals separated by a septum in the middle attached to ischium or pubis. There is a small penis bone or ospenis in the septum of these columns.

Cor Pulmonale. When a person suffers from degenerative pulmonary disease, the alveoli of lung are replaced by large air sacs as a result inspiration and expiration become difficult and breathing gets greatly increased. Pulmonary hypertension is developed and right side of heart enlarges and then fails. The enlarged large side is known as corpulmonale.

Corpus Albicans. When pregnancy fails, the corpus luteum regresses and gets replaced by a scar tissue called corpus albicans.

Corpus Allatum. Endocrine gland present behind brain in insect's head. It secretes hormone which is associated with larval moults.

Corpus Callosum. Tracts of transverse nerve fibres in mammals which are uniting the two cerebral hemispheres.

Corpuscles. Cellular structures of blood which constitute 40% of it. Its three types are known—RBCs, WBCs and platelets (RBCs 400000-

500000/ml, WBCs 7000/cu mm and platelets 250000/cu mm).

Corpuscular Radiation. Refers to a stream of subatomic particles which can travel through space and lose their energy upon contact with matter.

Corpus Haemorrhagicum. After release of ovum from Graffian follicle the entire follicle with little blood inside it, may develop to form corpus luteum, provided pregnancy has taken place but if pregnancy fails it is developed into scartissue called corpus haemorrhagicum.

Corpus Luteum. The Yellow Body; an endocrine organ which is formed by enlargement of a Graafian follicle after the release of an ovum in mammals. In the normal course of events these bodies develop for a short time and then die down. If pregenancy has taken place, however, they continue to grown and to secrete the hormone *progesterone* which makes enlargement of the uterus and other changes.

Corpus Spongiosum. Spongy erectile tissue which is running along the middle of the mammalian penis. The urethra runs through the centre of it and on each side there are sheets of muscle forming the boundaries of blood sinuses. It the latter are filled with blood, they become turgid and bring about erection of the penis prior to copulation.

Corpus Stratum. Refers to the thickened ventrolateral sides of floor of cerebral hemispheres.

Cortex. Used for describing the outer portion of a structure of an organ or gland i.e., adrenal cortex.

Corticosteroids. A family of steroids which are derived from cholesterol. They are secreted by cortex of adrenal gland and included glucocorticoids like corticosterone, II-deoxy-corticosterone and mineralocorticoid aldosterone. They control carbohydrate metabolism and mineral balance which is also associated with conservation reactions during stress.

Cortisol. Refers to one of the glucocorticoids from adrenal cortex. It is metabolised in the liver where most of it gets reduced to dihydrocortisol which on conjugation with glucuronic acid forms tetrahydrocortisol glucuronide and some of it gets converted to cortisone.

Cortisone. A drug which is able to stablise membranes. It is formed in the body by conversion of cortisol within the liver.

Coughing. Refers to a body disturbance which is caused by irritation of trachea or large bronchi. The epiglottis in this process closes glottis tightly and air is forced through tubes which causes clearing off the blockade within the tubes. It is a reflex response to irritation of receptors in mucosa of large respiratory passages.

Cowper Glands. Two glands, one on the right side and one on the left, of the male ure thra, into which they send their produce.

Cranial Nerves. Nerves comprising of sensory, motor or both fibres coming out of brain. Their number being 10 pairs in fishes and amphibians and 12 pairs in mammals. Their names are—olfactory, optic, occulomotor, pathetic, trigeminal, abducens, facialis, auditory, glossopharyngeal, vagus. In addition spinal accessory and hypoglossal occur in mammals only.

Craniostylic Jaw Suspension. When jaw (lower jaw) gets articulated at squamosal. It occurs in mammals only.

Cray-fish. Common name for a crustacean like *Astacus*, prawn, cambarns etc.; fresh water forms and edible animals.

Creatine Phosphate. Refers to one of the muscle phosphagens which liberates energy rich PO_4—bond to provide energy for muscle contraction.

Gribriform Plate. The part of ethmoid bone which is bearing numerous small formina for olfactory nerve.

Cricoid. A ring-like cartilage which is the larynx in vertebrates.

Cri-der-chat Syndrome. A collection of clinical features which are characterised by mental retardation, often cat-like cry of an infant. It is associated with a deletion in shortest arm of chromosome H 5.

Crista Ampullaris. A receptor structure which is lying in the expanded end or ampulla of each of membranous semi-circular canals of ear.

Cristae. Refer to the membranous formation in the inner wall of mitochondrion which has been raised into projections; projections are formed by folding of inner mitochondrial membrane into its matrix. Their thickness varies from 180-200 A^0 (two unit membranes each of thickness of 60-8-A^0 and with intermembrane space of 60-70 A^0); and possess components of respiratory chain.

Crop. A bag/sac like structure which is formed in certain animals like insects and birds. It serves to store ingested food material.

Crop-glands. A pair of glandular structures which are present in the crop of birds like, pigeon. During breeding season they secrete 'Pigeon's Milk' in both sexes under the influence of prolactin hormone from pituitary.

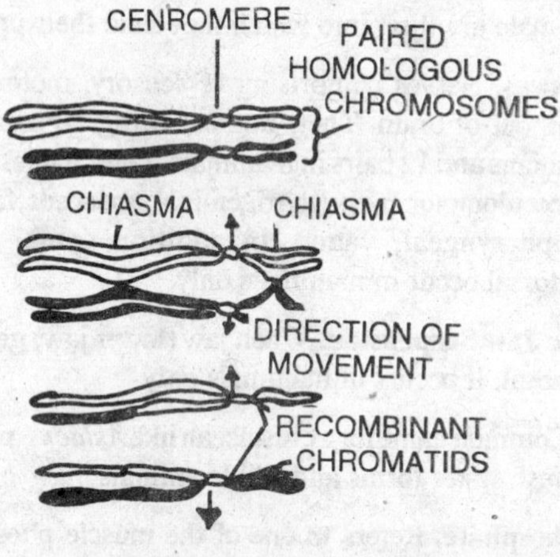

Fig. 2 : Crossing over at two chiasmata in a pair of homologous chromosomes

Crossing Over. Refers to process of the exchange of segments which takes place between members of homologous pair of chromosomes during process of meiosis. (Fig.2)

Crossoptrygii. Used for only lobe-finned fishes.

Crus Commune. A common duct which is formed by proximal parts of anterior and posterior canals of internal ear in mammals.

Cryptorchism. Used for describing a condition in which testes in a baby do not descend in scrotum and remain undescended. It is possible to treat the defect with gonadrotropic hormones or surgery.

Crypts of Leiberkuhn. Throughout the small intestine simple tubular intestinal glands called crypts of Leiberkuhn are present. They are lined by columnar and goblet cells. Former produce enzymes and latter produce mucous.

CSF (Cerebro Spinal Fluid). The fluid filling cerebral ventricles and subarchnoid space. It is formed by choroid plexus. It is a complex filtrate of blood plasma and is modified by brain cells before it returns to venous system.

Ctenoid Scale. Fish scale the posterior toothed edge of which extends out from under the preceding scale.

Ctenophora. Radiate phylum of animals having comb plates e.g., comb-jellies.

Cumulus Oophorous. Follicular cells present in the Graffian follicle of mammals which surround the oocyte.

Cupula. Gelatinous mass present on tips of sensory hairs of lateral line sense organ in fishes.

Cushing Syndrome. The clinical picture which is resulted in by excess of glucocorticoids. Persons suffering from this have been characterised by protein depletion, poor wound healing, poor muscles development, thin skin, redish-purple striae, red cheeks, moon face fat pad (buffalo hump) and pendulous abdomen.

Cutaneous. Refers to a thin non-cellular outermost covering of an organism.

Cuticula. More modern form of cuticle preferred because cuticle means inactivity.

Cuticulin. Substance making up epicuticle.

Cyanacobalamine. Also called vitamin-B_{12}. It is a water soluable vitamin which is synthesised by bacterium *Strptomyces olivaceous* under effect of Co^{60} radiations. It is present in kidney, liver and meat. It is an essential factor for growth and proper erythropoisis (RBCs formation). Its deficiency causes pernicious anaemia. Its daily requirement is 2.3 μg.

Cybernetics. Used for describing the science of information communication and control in animals and machines.

Cyclic-AMP (Cyclic-Adenosine Monophosphate). Refers to a compound which is derived from ATP by adenylate cyclase. It serves as a second messenger for hormonal action. The hormone association with membrane receptor results in the synthesis of c-AMP from ATP. c-AMP then removes inhibition of a protein kinase by removing an inhabitory subunit. The active protein kinase can be able to phosphorylate another intracellular enzyme thereby either activating or repressing it depending upon enzyme. As there are various forms of kinase having different substrates that they phosphorylate, the hormone induced increase in c-AMP concentration can bring about any variety of effects.

Cyclic-Gmp (Cyclic-Guanosine Monophosphate). Refers to regulatory nucleotide which is present in cells at a concentration lower than c-AMP. It is produced from GTP by guanylate cyclase. It promotes responses which are opposite to those promoted by c-AMP.

Cyclosis. Refers to rotary slow streaming movement of protoplasm in certain cells and animals e.g. *Paramecium*.

Cyclostmates. Used for describing primitive craniates with circular mouth lacking jaws. Body long, cylindrical, skin smooth, cartilagenous endoskeleton, gills in pouches, 5.16 pairs, heart two chambered. Examples are lampreys and hag-fishes.

Cydipped. The ctenophore larva.

Cophonautes. Ectoproct larva.

Cyst. Refers to a thickened resistant wall, a sac or bladder-like structure which is surrounding or enclosing an organism.

Cystic Fibrosis. Refers to an autosomal recessive condition in which certain glands are not able to function normally resulting in a collection of symptoms among them include excess secretion of thick mucous which obstructs the air passage of lungs.

Cysticercold. Term used for larval stage of some cestodes like, *Hymenolepis nana* that is formed in the body of certain insects e.g., dung-beetle.

Cytochalasin. A drug disrupting cytoplasmic microfilaments.

Cytochrome. Iron porphyrin having hydrogen/electron carrier that is forming part of electron transport system/electron transport chain of respiration or photosynthesis. Its various types are cyt-a, cyt-c, cyt-c_1, and cyt-a_3 etc.

Cytokinesis. Cleavage of cells during telophase of mitosis and meiosis.

It is started with an invagination at equatorial plate.

Cytolysis. The destruction of cells.

Cytopenphes. Drinking of or engulfing of water by the cell.

Cytopharynx. Pharynx or gullet of a pratozoa, e.g., *Paramecium*.

Cytoplasm Meiosis. Protoplasm of cell other than the nucleus; the ground substance in which various organelles and inclusions are suspended. It serves as the seat of cellular activity.

Cytosine. Nitrogen base found in nucleic acids and nucleotides; a pyrimidine linked with purine guanine with hydrogen bond forming cross-bars in DNA molecule.

Cytostome. The moth of a one-celled animal.

D

D_2 (Vitamin-D_2). It is formed by ultraviolet irradiation of ergosterol (a substance occurring in lower plants like yeast and fungi); also known as calciferol. It is able to regulate calcium and phosphorous metabolism in the body (Also see *vitamin D_3*).

D_3 (Vitamin-D_3). A naturally occurring substance present in yolk, butter, and fish liver oils. In the body skin it is produced from 7-dehydrocholesterol present at the surface by activation through ultraviolet light. It is also known as cholcalciferol (Also see *vitamin D_2*).

D-600. An organic compound (drug) that is able to block calcium influx through cell membranes. Chemically the substance is alphaisopropyl-alpha [(N-methyl-N-homoveratryl)- alpha-amino-propyl] -3, 4, 5 trimethoxyphenyl acetonitrile.

Dactyl. Refers to the finger of hand, toe of foot or like part.

Dactylozooid. Sensory polyp which is found in hydrozoan coelenterates such as *Obelia*.

Dance of the Bees. Refers to a precise pattern of movements which is carried out by returning forager (worker) honeybees in order to reveal the direction and distance of a food source to other workers in the hive. The dance is carried out on the vertical surface of the comb and two types are known; the round dance, carried out when food is nearby and the waggle dance, when food is further away. In the waggle dance the been are rasing in the angle between the 8 and the vertical indicates direction.

Dark Current. Steady sodium current which leaks into a visual receptor cell at outer segment. The sodium is pumped out actively at inner segment thereby ccmpleting the circuit. It gets reduced by photoexcitation.

Darwinism. Refers to the theory put forward by Charles Darwin in 1858 which stated that the species of living organisms are originating and evolving through the process of natural selection in their struggle for existence in the particular environment conditions of their habitat. It further stated that out of a large variety of members in a population only those having the traits suitable for maximising the success of survival are choosen by involving natural selection during evolutionary course.

Darwin's Finches. Refer to the thirteen types of finches which were first described by Darwin. They were unique to the Galapagos Islands of the South Pacific, but related to finches on the South American continent from which they evolved. Of special interest is the amount of adaptive radiation they display, and the fact that they occupy ecological niches that, in other parts of the world, are occupied by different birds, e.g. woodpeckers.

Dating Techniques. The methods used in determining the age of rocks, fossils or archaeological remains. Two main methods are known; *Relative dating* assesses the age of a specimen in comparison to other specimens. *Absolute dating* involves assessing the actual age of a specimen by involving some reliable measure of time.

D-Cells. In human beings 1-8% cells of islets of Langerhans are D-cells which have somstostatin (substance acting on anterior pituitary and also inhibiting secretion of insulin, glycogen and gastrointestinal hormones) also contain gastrin.

Deamination. Term used for describing the removal of NH_2 group from an amino acid and resulting in the formation of ammonia. It takes place in the liver of vertebrates during catabolism of amino acids. e.g., Alanine + $1/2O_2 \rightarrow$ Pyruvic acid+Ammonia. The process has been catalysed by specific amino acid oxidase.

Decacanth Embryo. Ten hooked embryo present in the tapeworm *Taenia saginata* and infectious to secondary host.

Decapon. Ten footed animals e.g., crustaceans and arachnids.

Decerebration. Refers to the experimental elimination of cerebral activity by section of brain stem by interrupting blood supply to the brain stem by interrupting blood supply to the brain region.

Decibel. Unit of a relative scale expressing the intensity of sound i.e., equal to 0.1 bel, where

$$\text{bel} = 2 \log \frac{\text{Pressure of sound}}{\text{Pressure of standard}}$$

The standard sound reference level adopted by Acoustical Society of America is 0 decibel at a pressure level of 0.000204 Dyne/ sq. cm- A value which is just at auditory threshold of average human.

Decibles

160	120	80	40	0
Jet plane with after-burner Pain	Discomfort subway	Heavy traffic Normal conversation	Whisper	Threshold of hearing (0.0002 Dyne Per sq. cm.)

Deciduate Placenta. On higher animals, the part of the inside skin of the uterus touching the placenta which is changed while the development of off spring is taking place and pushed out after it at birth.

Deciduous Forests. Forests having deciduous trees (leaves fall off annually). Found in moderate climate.

Deciduous Teeth. The teeth falling off during early life e.g., milk teeth of mammals, these include incisors, canines and premolars. Also known as diphyodont teeth.

Decomposer. An organisms that is feeding upon dead organisms thereby breaking them down into simpler substances. Decomposers recycle nutrients so as to make them available to producer organisms. Bacteria and fungi belong to the decomposer group of an ecosystem.

Decussation. Two groups of nerves going across one another.

Dedifferentiation. During the process of regeneration it refers to the process where cells proximal to site of injury become unspecialised they will later get differentiated into structure to be formed.

Defaecation. Refers to the passage of undigested food material from digestive tract of an animal.

Degeneration (1) Means evolution to an apparently simpler structural form. It is seen in the wings of the flightless birds (e.g. emus). Such organs are termed as vestigial. It is also common in parasites; for example parasitic protozoa have few organelles compared with their free-living relatives.

(2) Death and deterioration of cells, nerve fibres, etc.

Degenerate Code. Refers to the genetic code in which two or more codons may be representing the same amino acids.

Deglutition. Act of taking food down throat.

Dehydrogenase. An enzyme catalysing the removal of certain hydrogen atoms from specific substances in biological systems. Dehydrogenases catalyse the transfer of two hydrogen atoms from substrates of NAD and NADP.

Dehydroretinol (Vitamin-A_2). Form of vitamin-A which is present in the liver and retina of fresh water fishes, some invertebrates and amphibians.

Delayed Lethal Gene. A lethal gene which affects the carrier after attainment of reproductive age. Thus, it is the only type of lethal gene that may be transmitted.

Deletion. A chromosome anamoly in which there occurs a loss of a segment of a chromosome.

Deletion Mutation. Genetic mutation in which one or more codons are knocked; a shorter protein.

Deme. An interbreeding population.

Dementia. Disease of mind in which there is a more or less complete loss of reasoning power, common sense and generally of interest and the power of feeling.

Denaturation. A process making proteins to undergo unfolding of the characteristic structure of the peptide chain. This change may be carried out by a variety of physical factors: change in pH, temperature, violent shaking and radiation. The primary structure remains same because no covalent bonds in the backbone of the proteins are usually insoluble in solvents in which they were originally soluble.

Dendrite. One of several slender branching projections arising from or near the cell body of a nerve cell (neurone) to make contact with other nerve cells. The entire array of dendrites is sometimes known as the *dendritic tree*.

Denervation. Used for describing the destruction of nerve to any part of body which results in atrophy of the organ. In skeletal muscles it brings about muscular atrophy and abnormal excitability of muscle and increases its sensitivity to acetylcholine. Fine irregular contractions of individual fibres (fibrillations) become evident.

Dental Formula. A formula showing the number of teeth of each type on the upper and lower jaws of a mammal. The initial letter of each type of tooth—i for incisor, c for canine, pm (or p) for premolar, and m for molar—is being followed by the number of teeth of that type on one side of the jaw. The number on the upper jaw is written above a horizontal line, with the number on the lower jaw below it. The total number of teeth is obtained by adding all the numbers together and multiplying by two. The general form of dental formula is:

i 3/3, c 1/1, p 4/4, m 3/3 × 2 = 44

However, it is subjected to a great variation in mammals and their classification forms basis on the dental formula. The dental formulae of some common mammals are as follows:

Kangaroo	i 3/1, c 1/0, p 2/2, m 4/4 × 2 = 34.
Man	i 2/2, c 1/1, p 2/2, m 3/3 × 9 = 32.
Rabbit	i 2/1, c 0/0, p 3/2, m 3/3 × 2 = 28.
Rat	i 1/1, c 0/0, p 0/0, m 3/3 × 2 = 16.
Dog	i 3/3, c 1/1, p 4/4, m 2/3 × 2 = 42.
Pig, Horse	i 3/3, c 1/1, p 4/4, m 3/3 × 2 = 44.
Cow, Sheep, Goat etc.	i 0/3, c 1/1, p 3/3, m 3/3 × 2 = 32.
Elephant	i 1/0, c 0/0, p 0/0, m 3/3 × 2 = 14.
Squirrel	i 1/1, c 0/0, p 2/1, m 3/3 × 2 = 22.

Dentry. A tooth-bearing membrane bone. In mammals it is the lower jaw bone, having a single membrane bone on each side, fused together in front.

Denticle (Placoid scale). A small toothlike structure present in the skin of cartilaginous fish. e.g. dogfish or shark. Each possesses a spine that projects backwards through the skin and has an outer layer of enamel covering dentine round a central pulp cavity. The spine gets attached to a flat base of bonelike substance beneath the skin. The teeth of these fishes are strong denticles having five pointed spines. Denticles are similar in structure and origin to the teeth of higher vertebrates and are homologous with them.

Dentine. A hard substance, which is closely resembling bone. It makes up the bulk of a tooth. It possesses a higher mineral content than bone and is perforated by fine canals (canaliculae) having cytoplasmic processes of the cells (odontoblasts) that line the pulp cavity. Unlike bone, dentine does not have system of Haversians canals, lacunae, or cells within its substance.

Dentition. Refers to the number, type, arrangement, and physiology of teeth in any given species.

Deoxycorticosterone (deoxycortone). A steroid hormone, produced by the adernal cortex. It is having mineralocorticoid activity.

Deoxyhaemoglobin. Haemoglobin in which oxygen is not combined to Fe^{3+} of haeme-moeity.

Deoxyribose. A pentose sugar which is a component of nucleotides and present in DNA i.e., Adenosine, cytosine, guanidine and thimidine phosphates (AMP, CMP, GMP and TMP).

Deplasmolysis. Refers to increase in intracellular pressure which is usually accompanied by cell membrane bursting. It takes place when cell is kept in hypotonic solution.

Depolarisation. Term used for describing reduction in or reversal of potential difference existing across cell membrane during resting stage.

Depressed. Flattened vertically from above.

Dermal. Pertaining to the skin especially inner connective tissue layers of vertebrate skin.

Dermal Bone. Refers to a bone which develops from mesenchyme of dermis. Such bones are usually present in dermatocranium of tetrapods skull.

Dermal Plate. Underneath the dermis, many reptiles are having specialised scales which are forming the plates called dermal plates. Such plates are present over the head of lizards and in the armature of turtles (carapace and plastron).

Dermatitis. Inflamatory development/skin troubles, which are usually caused by deficiency of biotin, pantothenic acid (vitamin B complex).

Dermatocranium. In bony fishes and tetrapods, a part of brain box is not roofed over by cartilages in embryo, but is covered by dermal bones forming the dermatocranium. It includes, ethmoid region nasal capsules etc.

Dermatome. One of the skin divisions in the development of embryo.

Dermis. The under-coat of the skin in which are nerve-endings and blood vassels.

Desiccation. Drying dehydration.

Deterimental Gene. A gene imparting a disadvantageous effect to individual that is having it. The carrier gets hampered in some way by the gene so that chances of reproduction or survival are less than those individual which are free from the gene.

Deuterostomia. Animals in which blastopore is forming anus while the second embryonic opening is forming mouth opposite the anus.

Dextral. Term used for clockwise coiling.

Diabetes Insipidus. Refers to the physiological abnormality which is characterised by passing of large amounts of dilute urine (polyuria) and drinking of large amounts of water/fluids (polydipsia). Insufficient secretion of ADH (antidiuretic hormone) is the main cause for less absorption of water from glomerular filtrate which passes out in urine.

Diabetes Mellitus. An abnormal condition in which over-much sugar is present in the blood and caused by urine. It is caused by not enough insulin produced by the pancreas, and marked by serious loss of weight and a great desire for food and drink.

Diakinesis. Refers to the last stage of the prophase in the first division of meiosis. Chiasmata appear during the stage, and by the end of diakinesis the nucleoli and nuclear membrane have disappeared.

Dialysis. Refers to a technique for separating compounds with small molecules from compounds with large molecules by selective diffusion through a semipermeable membrane. Dialysis is carried out naturally by the kidneys to extract wastes from the blood, or, in the case of lost or damaged kidneys, artificially by machine.

Diapause. A term used for describing period of dormancy in the life-cycle of some insects during which growth and development almost stop and metabolism is greatly reduced. It is often seasonal, e.g. hibernation, or in some cases may last for several years. It takes place as a result of change in hormone production and makes the insect to survive unfavourable environmental conditions.

Diapedesis. A term used for describing the process by which many of the neutrophils enter the tissues that are insinuating themselves through the walls of capillaries. Both the plasma proteins and leucocytes pass through endothelial system and in most cases injury brings about leakages of plasma proteins long before emigration of leucocytes occurs. The escape of plasma proteins is a purely passive affair and is partly true for leucocytes which, however, undergo active amoeboid movements in this process.

Diephragm. Used for describing a sheet-like muscle which is forming a partition between the thoracic and abdominal cavities of mammals; dividing membrane.

Diaphysis. Refers to the long main body of a bone as differentiated from epiphysis.

Diapohysis. Mean transverse process which arises from neural arch but often from centrum; an articular surface which serves for attachment of tuberculum of biheaded ribs.

Diarrhoea. Refers to the condition in which faeces are liquid and sent over frequently.

Diarthrosis. Used for describing perfect and movable joints between bones. They are also known as movable joints. They may be of following types:

Ball and Socket. It is also known as enarthroses: In this one bone has rounded end and the other a cavity to receive it. It shows greatest degree of movement, e.g., joints of humerus with glenoid cavity and that of femur with acetabulum.

Hinge. It is also known as ginglymusi; it is same as ball and socket but movement is somewhat restricted e.g., elbow joint.

Pivotal. It is also known as rotaria. In this one bone is shaped like fixed pivot while other rotates on it e.g., one odantoid process of axis rotates the atlas together with head.

Sliding. It is also known as arthrodium. In this two articulating bones can slide over one another e.g., joints in zygapophyses of vertebrae and in pronation (palm facing downwards) radius is on thumb-side ulna on little finger side.

Saddle. It is similar to ball and socket but since both ball and socket do not develop, the movement of joint may take place in many planes e.g. joint between matacarpals of our thumb and wrist bones.

Diastom. Space in jaw without teeth, generally between two different sorts of teeth.

Diastole. The expansion of any part, specially the heart, in regular rhythm.

Diencephalon. Used for the region of vertebrate brain which is just posterior to cerebrum adult. In embryo it refers to the second part of two divisions of prosencephalon (fore brain).

Diet. Defined as the optimum amount of substance intake of which is required for the normal functioning (optimum) of an organism. In the organisms it means the inorganic and organic food materials which supply the requisite energy for the functioning of body and the necessary substances required for body metabolism.

Differentiation. Refers to the process in which cells and tissues become specialised for carrying out specific functions during the process of development and growth.

Diffusion. Refers to the movement of molecules from region of higher concentration to lower concentration. This movement is brought about as a result of their kinetic energy.

Digestion. Means the breakdown of complex organic foodstuffs by enzymes into simpler soluble substances, which could be absorbed and assimilated by the tissues. In most animals (e.g. vertebrates and arthropods) it is extracellular, taking place in an alimentary canal or gut into which the enzymes get secreted. In simpler animals (e.g. protozoans, coelenterates, and some other invertebrates) it is intracellular, in which solid particles being engulfed and digested by amoeboid cells.

Digit. A finger or toe. There are typically five digits which are terminating the limbs of tetrapods. Each is made up of a series of small bones (phalanges). However, reduction and modification to this general plan occur. Some species have digit which are bearing nails, claws, or hooves distally.

Digitigrade. Used for describing the mode of progression in some mammals in which only digits (fingers or toes) remain in contact with the ground. It is present in dogs, cats and most fast-running animals.

Dihybrid. Refers to a hybrid heterozygous at two loci. It is obtained by crossing homozygous parents with different alleles at two given loci.

Dihybrid Crosses. Crosses taking place between parent which are having pairs of contrasted characters.

Dihybrid Ratio. Refers to 9 : 3 : 3 : 1 ratio phenotypes among off-spring of a single cross. This ratio is only obtained when two genes involved are at different loci and immediate parents of these offspring are heterozygous for both of genes. These heterozygotes are selfed, genes are not linked, no gene interaction, fertilization is random and all the gametes, zygotes and offsprings are having an equal chance of survival.

1, 2, 5-Dihyroxycholcalciferol. A physiologically active metabolite which is produced in kidney from 2, 5-hydroxy-calciferol (produced by conversion of vitamin D_3 in the cholliver). It acts as a hormone on the nuclei of interstitial epithelial cells for initiating the formation of a protein that increases the activity of calcium transport mechanism.

Di-Iodotyrosine. Refers to a precursor in the synthesis of thyroxine. Two molecules of this undergo oxidative condensation thereby forming thyroxine.

Dikinesis. Used for describing a stage during prophase of meiosis when chromosomes get thickened and chiasmata move to their ends.

Dimorphism. Refers to the difference in size structure, form, colour etc. between two types of individuals of same species.

Dinitrophenol (DNP). Refers to any of a group of six isomers that are acting as aerobic metabolic inhibitors due to their ability to uncouple oxidation from phosphorylation in mitochondrial electron transport. The chemical formula is $C_6H_3(OH)(NO_2)_2$

Dinosaurs. A general name assigned to prehistoric giant lizards that existed during remote past. Now they are extinct, e.g., *Brontosaurus, Steigosaurus, Dimetrodon* etc.

Dioecious. Having male and female reproductive organs in separate individuals.

Diopter. The unit of refractive power of lens. The number of diopter is the reciprocal of principal focal distance in meters e.g. a lens with focal distance of 20 cm has refractive power of I/0.20=5 diopters. Human eye has a refractive power of appximately 66.7 diopters at rest.

Dipeptidase. An enzyme hydrolysing dipeptides to amino acids.

Dipeptide. Refers to union/combination of two amino acids linked together by peptide bond (CO-NH).

Diphyodont Dentition. Having two growth of teeth, the first group coming out and a second taking their place.

Dipleurula. Refers to the developmental stage of echinoderms and hemichordates, hypothetical ancestor of most duterosomes.

Diploblastic. Having only two separate germ layers in the embryo the endoderm and the ectoderm, a condition seen only in coelenterata.

Diploid. Having twice the germ cell number of chromosomes as most body cells.

Diplotene. Refers to the stage in prophase of first meiotic division when mutual attraction of bivalents gets replaced by repulsion commencing at centromeres. The chromatids at this stage are clearly seen and when the chromosomes separate it becomes evident that they are held at various points called chiasmata.

Disaccharides. Refers to the union of two monosaccharides with elimination of a water molecule

e.g. Maltose........Glucose+Glucose
Sucrose.......... Glucose+Fructose
Lactose..........Glucose+Galactose

On hydrolysis they yield monosaccharides.

Discoidal Cleavage. Incomplete cleavage which is also called meroblastic cleavage; takes place in eggs having large amount of yolk.

Discontinuous Variation. Means that phenotypic variation in some characteristics which could be recognised by two or more distinct traits or clear cut difference.

Discordant. A term used for describing that two members of a twin pair are different with respect to expression of a given trait.

Disjunction. Refers to the normal separation of chromosomes taking place at anaphase of mitosis/meiosis.

Disruptive Selection. Refers to the selection against the distribution means for extremes resulting in diversing traits.

Dissimilation. Means disintegration of protoplasm mainly by oxidation.

Distal. Means away from point of attachment e.g., hand is distal part of arm. *Opposite of* proximal.

Distal Convoluted Tubule. Refers to the renal tubules of mammals that are located in renal cottex and leading from ascending limb of loop of Henle to collecting duct.

Distalia. The carpals (five in number) of distal row in the hand which are articulating with metacarpals.

Diuresis. Means increased urine secretion which is caused by insufficient secretion of ADH from pituitary.

Diuretic. Any substance increasing urine secretion.

Diurnal Rhythm. Means time of daylight. Opposite to nocturnal.

Divergence. Means evolution of dissimilar characteristics in descendents from the same ancestor.

Diverticulum. Refers to a blind tubular or saclike outgrowth from a tube or cavity. For example, the appendix and caecum of a rabbit form a diverticulum.

Dizygotic Twins. The twins that have been derived from two separate eggs. Each fertilized by a separate sperm. They may be of same sex or different sexes. Also called fraternal twins.

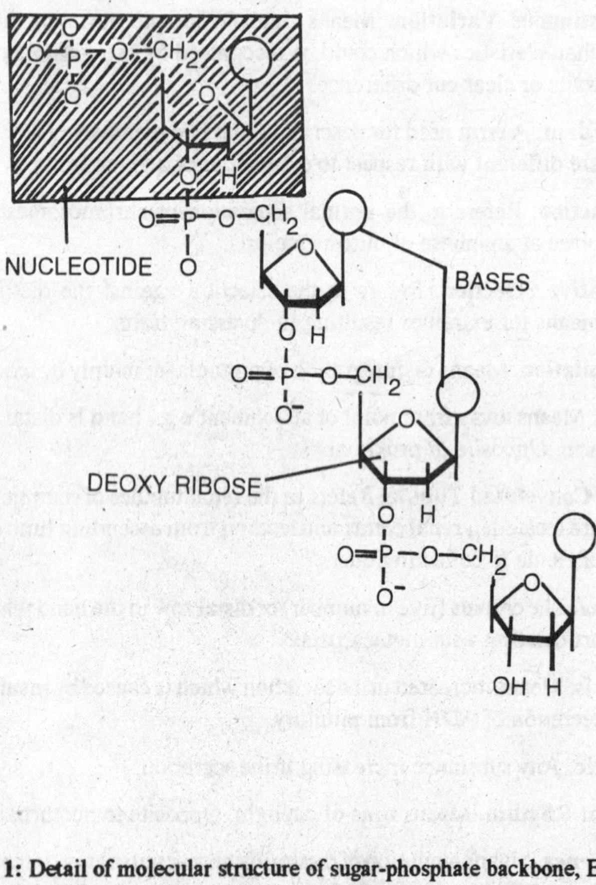

Fig. 1: Detail of molecular structure of sugar-phosphate backbone, Each deoxyribose unit is attached to a phoshate group and a base, forming a nucleotide

DNA (Deoxyribose Nucleic Acid). A substance of which genes are made up of. It determines inherited characters and consists of two polynucleotide chains which are forming a right-handed helix with one coil present more or less underneath. Four types of nucleotides are present in the polynucleotide chains although each chain may be thousands of nucleotides long. Each nucleotide is having deoxyribose sugar, phosphoric, acid and a base which may be purine (adenine, guanine) and pyrimindine (cytosine, thymine). The sequence of their pairing is adenine with thymine and guanine with cytosine. The two phynucleotide chains are antiparallel i.e. one chain sugar phosphate is composed of 3'-5' phosphodiester bonds while the other of 3'-5' phosphodiester bonds. Bases are present on the inner side of helix. By formation of hydrogen bonds two helcis are held together. DNA is able to produce copies of itself with such an accuracy so that genes (polynucleotide sequences pass unaltered from one generation to other.

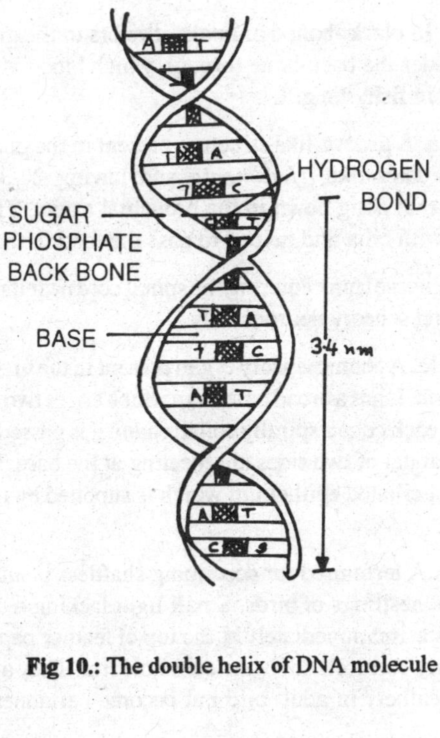

Fig 10.: The double helix of DNA molecule

DNA Hybridization. A laboratory technique which is used for measuring the similarity of one of the two species. DNA from each species on heating is denatured into single strand and a mixture of DNA strands is incubated to allow them to recombine into hydrid DNA-a process called annealing. The amount of annealing has been directly proportional to similarity of DNA strands and could be measured by heating the hybrid DNA and recording the temperature at which the strands become separate.

Dominant Trait. A trait appearing as the result of either a single dose/ double dose of a particular gene as contrasted with recessive trait which is developed only when both members of a pair of allelic genes are alike.

DOPA (Dihydroxyphenylalanine). Intermediate substance in biosynthesis, an immediate precursor of nor-epinephrine.

Dorsal. Refers to back; means opposite to ventral in bilaterally symmetrical animal.

Dorsal Aorta. In black -boned animals, it refers to the great branching artery under the back-bone through which blood is sent to all the body but in fish, the gill.

Dorsal Lamina. A groove-like structure present in the pharynx of many protochordates like, *Herdmania* and having 20-30 tongue-like languets that hang down in the bronchial cavity. They have been covered with cilia and helped to pass the food in the oesophagus.

Dorsal Root. A nerve trunk entering the spinal cord near its dorsal surface and having sensory neurons only.

Dorsal Tubercle. A chemosensory organ present in the urochordates like *Herdmania*. It has a broad base from which arises two spirally coiled cones. In each cone a spirally coiled channel is passed making three coils. Channel of two sides are meeting at the base. Entire surface is made of ciliated epithelium which is supplied by nerve from the ganglion.

Down Feather. A term used for describing shaftless insulating feathers present in nestlings of birds. Small light lacking a distinct rachis and barbs arise immediately at the top of feather papilla. Barbules do not carry barbicels so a vane does not form. They are replaced by contour feathers in adult bird but become permanent feathers in swan where they are known as powder-downs.

Downs Syndrome. On average a child among 700 children is having three autosomes on twenty-first instead of two chromosomes i.e. 2n + 1 (2I) = 47 chromosomes. The bearer is having small body size, retarded/poor mental development, thick tongue, blunt toes, open mouth, oblique eyes and folded eye lids. Also termed Mongolism.

Drum. The middle ear or the tympanic membrane.

Drumstick. Sex chromatin in polymorph nuclear leucocytes. It appears as an appendage on the lobe of nuclei.

Duck-Bill Mole. Common name for describing Australian egg laying mammal *platypus.*

Duct. A tube other than a lymphatic/blood vessel through which a liquid or other secretion of metabolism from some gland is transported.

Ductless Glands. The endocrine glands which do not have ducts to carry out their secretions. Many such glands e.g., pituitary, thyroid, thymus, adrenals, etc. occur in the vertebrates.

Ductus Botali. A connection existing between 6th arotic arch and the portion of dorsal aorta; present usually in embryonic stage. Also known as ductus arteriosus.

Ductus Caroticus. Portion of dorsal aorta which is present between 3rd and 4th arches; present usually in embryonic stage.

Ductus Cholidochus. Used for bile duct.

Ductus Communicans. In body fishes having an air bladder e.g., *Labeo* it connects the anterior and posterior chambers of air bladder and it is controlled by a sphincter.

Ductus Cuveri. In cartilagenous fishes e.g., sharks the sinus venosus is able to receive blood by a pair of large lateral veins which are ductus cuveri.

Duodenum. In man and higher animals, the part of the widest and shorter part of the small intestine going from the stomach to the jejunum and into which the bile and the pancreatic ducts have their opening.

Dura Mater. Strong outer membrane covering the brain and spinal cord in the vertebrates.

Dwarfism. Abnormally small size in humans. It occurs due to insufficient secretion of pituitary hormone during childhood and adolescence, *Opposite* to gigantism.

Dynin. A ciliary protein having Mg-activated AT Pase activity.

Dynin Arms. Refers to the projection of tubule-A of one microtubule doublet toward tubule-B of next; composed of protein having ATPase activity.

Dysarthia. Means the disturbance of speech caused by the excitomotor areas in cortex and their connections (i.e. pyramidal traits spanial nuclei, cranial nerves, muscles disturbance of tone-like regidity from lack of cerebellar control of postured and voluntary activity.

Dyschezia. A type of constipation in which feaces arrive normally at polyic colon but final evacuation is not adequately carried out. If the gall to defaecation is not undertaken the rectal wall relaxes and desire passes off. The rectum becomes in time excessively distended and contractibility of its walls gets impaired and defaecation is difficult even when more/great effort is made. Fragmentary defaecation may take place but faeces are left behind in rectum to continue distending it.

Dysgamms Globulinemia. Refers to the disturbance in the level of gamma-globulins of blood which occurs due to the abnormality of fat transport in individual lymphatics.

Dysmenorrhea. Means an abnormality of menstruation in human female which is characterised by painful menstruation.

Dysmetria. Also known as past pointing. When a person in drunken state attempts to touch an object with his finger, his motion towards the object overshoots to one side or the other. Thus, the finger oscillates back and forth and further it travels greater the exertions becomes.

Dyspnea. Refers to the laboured/difficult breathing. In this condition, the subject is conscious of shortness of breath. Dyspnea on exertion occurs due to failure of left ventricular output to rise during exercise leading to increase venous pressure.

E

Ear. One of a pair of sense organs that are located on either side of the head in vertebrates. They are concerned with hearing and balance. In mammals, the ear has an outer ear, tympanum (eardrum), middle ear, oval window (fenestra ovalis), and round window (fenestra rotunda), and inner ear. In fish only the inner ear is mainly concerned with balance.

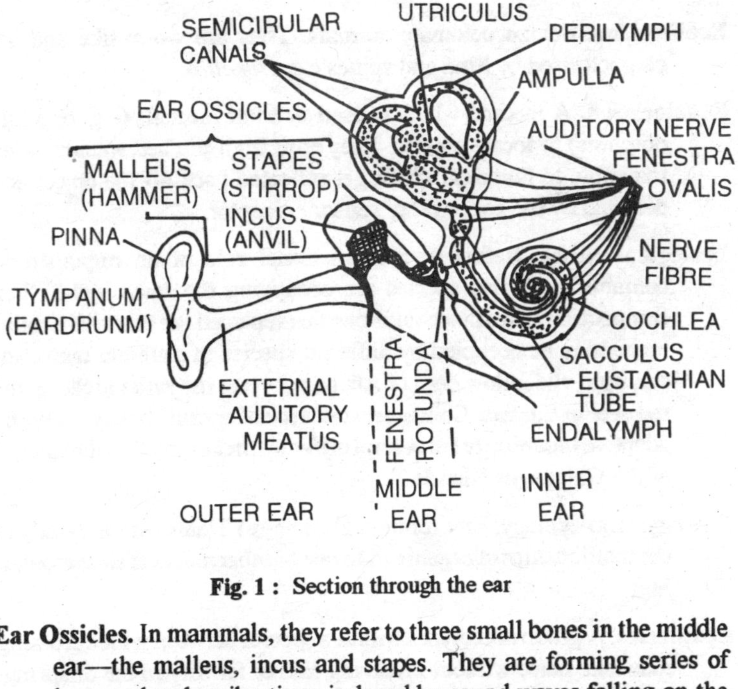

Fig. 1 : Section through the ear

Ear Ossicles. In mammals, they refer to three small bones in the middle ear—the malleus, incus and stapes. They are forming series of levers whereby vibrations induced by sound waves falling on the eardrum (tympanum) get transmitted to the oval window (fenestra ovalis) and so to the inner ear. They are homologous with certain jaw bones of lower vertebrates.

Ecad. A form within a species that is arising in response to a particular set of environmental conditions, and whose characteristics are not heritable. Differences between ecads occur due to phenotypic plasticity rather than genotypic changes.

Eccrine Gland. A gland in which the products collect at the outer part of cells and the cytoplasm is decomposed before the products gets discharged e.g., sweat glands.

Ecdyosone. Insect hormone which is initiating metamorphosis.

Echinodermata. Group of animals having rough skin due to the presence to spines, plates, ossicles etc. Strictly marine animals and non-parasitic. It includes animals such as star-fishes, sea-stars, feather-stars, brittle-stars and sea-lillies.

Echiuroidea. Schizocoelomate animals. They are worm-like and are characterised by hind end spines e.g. *Bonellia*.

Echolocation. A method which is used by some animals (e.g. bats and dolphins) to locate objects. They emit high-pitched sounds, often inaudible to humans, which get reflected back off the object and detected by the ear or other sensory receptor.

Ecological Niche. Refers to the functional role of an organism in community. If two species are occupying the same niche, then competition takes place until one has replaced the other. A similar niche may be occupied by different species in different areas, for example, the fallow deer of Africa occupies the same niche as the red deer of Eurasia. Conversely one type of organism may evolve by adaptive radiation to fill several different niches, such as the finches of the Galapagos Islands.

Ecology. (Bio-ecology, Environmental Biology). Deals with the study of the relationships of organisms to one another and to their surroundings.

Ecospecies. A group of ecotypes within a species between which crossing may take place without involving loss of fertility in the offspring.

Ecosystem. Refers to the conglometate of all life in a given area plus that life's interactions taking place with the non-living portions of the area. It comprises of biotic and abiotic components.

Ectodermal. Outer layer of cells in the gastrula which gives rise to epidermis, sense organs and the nervous system.

Ectohormone. A pheromone; a substance which is secreted externally by an organism to influence the behaviour of other organism; an ectocrine gland's secretion substance.

Ectoparasite. A parasite of external body surface e.g. tick, louse, bed bug etc. a parasite that is living outside the body of its host. *Opposite* to endoparasite.

Ectopic Focus. A portion of myocardium/AV node/bundle of His which is usually discharging independently. Such foci become generally common after an injury such as myocardial infraction.

Ectopic Pacemaker. A pacemaker which is situated outside the area where it occurs normally.

Ectoprocta. Coelomate, sessile animals having 'U' shaped intestine, mouth surrounded by lophophore and anal opening outside the lophophore.

Ectosarc. Structure in protozoan which is composed of ectoplasm.

Ectotherm. Individual getting heat directly from environment. Majority of animals except birds and mammals are ectotherms.

Edema. A term used for describing the accumulation of interstitial fluid in abnormally large amounts. It takes place due to the increased filtration pressure, decreased osmotic pressure, gradient across capillary, increased capillary permeability and inadequate lymph flow.

EDTA. Calcium and magnesium chelating agent.

Eel. Fishes having snake-like body, nacked skin and lacking pelvic fins. The dorsal causal and ventral fins are confluent. They occur is fresh water and go to sea for spawning e.g., *Anguilla*.

Effector. An organ or cell responding in a particular way to a nervous impulse. Effectors include muscles, glands and cilia.

Efferent. Carrying or conducting away.

(1) An efferent blood vessel that is able to collect blood from a capillary network, e.g., the arteries of the aortic aches of fish that lead from the gills to the dorsal aorta.

(2) An efferent nerve or neurone that is able to conduct impulses from the central nervous system to the periphery, e.g. to a muscle.

Effector Organ. Any part of body e.g., muscle or gland which transforms motor impulses into motor action.

Efferent Blood Vessel. Blood vessel which carries blood out of a structure such as gill or glomerulus of kidney.

Efferent Nerve. Nerve which is able to convey impulses outward or away from brain/spinal cord. *Opposite* to afferent nerve.

Egest. Means pass out/discharge undigested food/residue from the alimentary canal. Also known as defaecation.

Egg. A non-motile female gemete. Term limited generally to shelled structure having yolk, albumen and a cellular structure called nucleus/germinal disc.

Egg Tooth. A small tooth-like process over the tip of snout/beak which is used for breaking the egg shell. It occurs commonly in the crocodiles, birds and egg-laying mammals.

EGTA (Ethylene Glycol-Bis (β -Aminoethylether) N, N-Tetra Acetic Acid). A calcium chelating agent.

Ejaculation. The sending out suddenly and with force to liquid from part of body especially semen.

Ejaculatory Duct. A duct conveying semen to urethra, especially in mammals.

Elastase. An enzyme occurring in succus entericus (intestinal juice). It acts upon elastin and digests it.

Elastic Cartilage. Used for describing a cartilage in which yellow and elastic fibres are found in the matrix. It occurs in the pinna, epiglottis and trachial rings.

Elastin. Refers to a substance of the elastic fibres. It is a scleroprotein which is insoluable upon boiling but dissolves in proteolytic enzymes. Elastin fibres occurs in the wall of large arterise and between vertebrae (intervertebral discs of mammals).

Electroencephalogram (EEG). A recording of the changes taking place in electrical potential that are produced by activity of nerve cells in the brain, mainly the cerebral cortex. These changes have been detected by electrodes placed against the scalp and have been recorded by an oscilloscope or similar device (an *electroencephalograph*) as a series of waves ('brain waves'). Variations in wave pattern could be correlated with different mental states and neurological conditions, such as epilepsy.

Electromyogrphy. Refers to the process of recording the electrical activity of muscle on cathode ray oscilloscope. It can be carried out in unanaesthesized human beings by employing small metal discs on skin overlying muscle as the pick up electrodes or by employing hypodermic needle electrodes.

Eleutherozoa. Refers to a group of echinoderms that are composed of all extinct classes except crinoids (to which sea-lillies or sea-feathers belong).

Eldin. The cells of stratum lucidum of mammalian epidermis get flattened and transparent due to conversion of keratohyaline to substance called Pledin. One of the intermediate products obtained during keratinziation of cornification (See *keratinization*).

Elephantiasis. Human lymphatic disease which is caused by nematode *Wucheraria broncofti*. Lymph ducts are clogged new ducts are produced alongwith new connective tissue.

Elevers. Miniature cells that are about 50-55 mm long.

Embden-Mayerhof Pathway. Refers to glycolysis (glycolytic pathway) which was named after two German scientists who investigated the details of the process in 1930.

Emboli From Nuclei. Nuclei in cerebellar cortex that receive axons from lateral portions of the cerebellar hemispheres.

Embolism. Refers to blood clot forming tendency within a blood vessel.

Embolomerous Vertebra. A vertebra having double centrum as occurring in most primitive amphibia; an anterior or hypocentrum is considered to be derived from basiventral and posterior pleurocentrum is considered to be derived from basiventrals and interventrals.

Embolus. A blood clot taking place within a blood vessel.

Emboly. Invagination during gastrulation.

Embryo. Multicellular structure that is developed from zygote and is giving rise to new offsprings generally present inside female tissue or egg membrane.

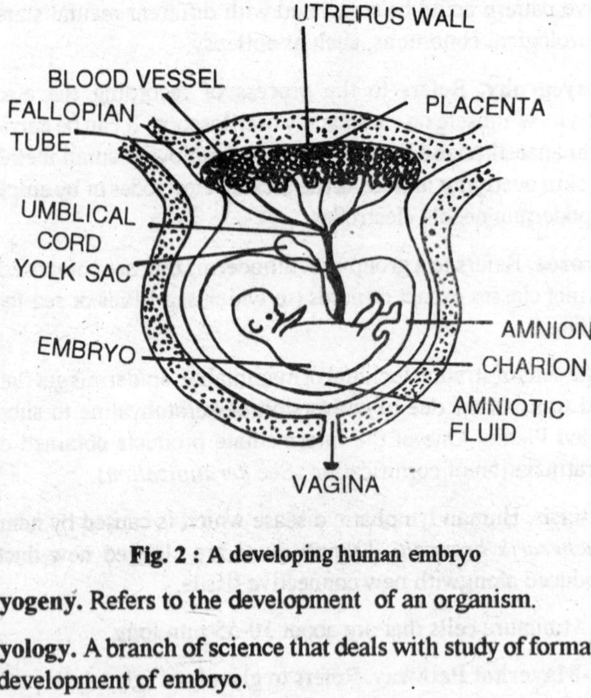

Fig. 2 : A developing human embryo

Embryogeny. Refers to the development of an organism.

Embryology. A branch of science that deals with study of formation and development of embryo.

Embryonal. Referring to an embryo.

Emeiocytosis. Also known as cell vomiting. The cell membrane around a vacuole or secretion granule gets fused with cell membrane and the region of fusion breaks down thereby leaving contents of vacuole or secretion granule outside the cell membrane intact. It needs calcium and energy. It is also termed as exocytosis or reverse pinacytosis.

EMG. Recording of activation of motor unit can be done by electromyography (process of recording electrical activity of muscle on a cathode ray oscillosope). Also see *electromyography.*

Emission. Refers to the movement of semen into urethra. The sympathetic response is integrated in the upper lumber segments of spinal cords and influenced by contraction of smooth muscles of vasa efferentia and seminal vesicels in response to stimuli in the hypogastric nerves.

Emmetropia. When ciliary muscle has been relaxed, parallel light rays striking optically normal eye focus on the retina. As long as this relaxation has been maintained, rays from the observer get focus behind retina and consequently the object appears blurred. This defect is known as commetropia.

Emphysema. Refers to a degenerative disease of lungs in which lungs loose their elasticity and their walls between alveoli break down so that the alveoli have been replaced by air sacs. The physiologic dead space is greatly increased and due to inadequate and uneven alveolar ventilation and perfusion of underventilated alveoli, severe hypoxia is developed. Inspiration and expiration are laboured and work of breathing gets greatly increased.

Encephalin. Two pentapeptides which act as natural pain killers of body by binding to opiate receptors. They are formed in the brain. Its two types are recognised—Met-encephalin and Leu-encephalin. Suffix Met-and Leu stand for 'C' terminal amino acids methionine and leucine respectively.

Encephalisation. Means relatively greater role of cerebral cortex in various functions. An important phenomenon in species in which cerebral cortex has been highly developed.

Encephalitis. Refers to the inflamation of the brain which may be caused by virus, microbes or may be secondarily due to general infections such as that of measles pneumonia etc.

Endergonic. A process or reaction which is characterised by concommitant absorption of energy e.g., protein synthesis. *Opposite* to exergonic.

Endocardium. Innermost layer of heart which is derived from mesenchymal cells between endoderm of gut and splanchinc/visceral mesoderm. Over this myocardium and epicardium are formed.

Endocrine Gland (ductless gland). A gland having no duct or opening to the exterior. It produces, hormones, which pass directly into the bloodstream. Then the circulatory system transmits them to other body tissues or organs, where activity gets modified. Thus an alternative method of communication to the nervous system is obtained.

Endocrinology. Deals with the study of the endocrine glands and their secretions (hormones).

Endocrine Gland. Glands having no ducts. Their secretion is known as hormone and is carried by blood. A number of such glands are present in the body like pituitary, thyroid, parathyroids, adrenals, interstitial cells of gonads, islets of Langerhans and certain glands in the alimentary canal.

Endocytosis. Refers to the engulfing of substances by cell due to membrane invagination proximal to particle and subsequent closure of membrane over the particle. Also see *phagocytosis.*

Endogenous. Some substance/section which is coming from within the body.

Endogenous Insulin. Insulin which is secreted inside the body when it enters the portal vein, the liver is normally exposed to concentrations of insulin that are 3-10 times greater than those in peripheral tissues. Liver is able to bind about half a dose of insulin injected into the portal vein but only 25% of a peripherally injected dose.

Endolymph. Aqueous liquid having high potassium concentration and present in scale media of cochlea of mammalian ear.

Endometrium. An epithelium lining the uterus (Also called *uterine mucosa*).

Endomitosis. Means the duplication of chromosomes without involving division of the nucleus. Endomitosis may take place in two ways: the chromatids may get separated causing endopolyploidy, e.g. in the macronculeus of ciliates, or the chromatids may remain joined multistranded chromosomes or *polyteny* e.g. during larval development of dipteran flies. Both processes result in an increase in nuclear and cytoplasmic volume.

Endoparasite. An organism living inside the body of another organism (the host) and benefitting from the relationship at the expense of the host, such as the malarial parasite.

Endopetidase. A proteolytic enzyme which cleaves peptide bond from (within) a peptide chain.

Endoplasm. The cytoplasm which is surrounded by ectoplasm within the cell.

Endoplasmic Reticulum (ER). A system of flattened membrane-bounded sacs (cisternae) which are running through the cytoplasm of all eukaryotic cells and continuous with the nuclear envelope. Although often extensive, it was only discovered with the advent of electron microscopy. Its surface is usually covered with ribosomes, forming *rough ER*.

Endorphins. Two larger peptides bindings to opiate receptors and having analgesic properties/activities. One is having 17 amino acid residues (α-endorphin) and other 17+16 additional amino acid residues (β-endrophin). Both have amino acid sequence of met-enkaphalin. Amino acid sequence present in β-endorphin in turn occurs in a polypepitde called β-lipoprotein which has been isolated from anterior pituitary.

Endoskeleton. A skeleton occurring in side the animal body, such as the bony or cartilaginous skeleton of vertebrates. It comprises of a skull, vertebral column, pectoral and pelvic girdles, ribs, and limb or fin elements. It provides shape and support to the body, protects vital organs, and gives a system of rigid levers to which muscles can attach and produce movement.

Endosmosis. Refers to the movement of solvent which takes place through semipermeable membrane from exterior to interior of the cell e.g., keeping grapes in water may bring about their bursting due to inward movement of water molecules inside them through their walls.

Endosome. Central mass having mainly chromatin material. It is present in the nucleus of certain protozoa like, *Amoeba, Euglena* etc.

Endostyle. Ciliated and glandular tract which is present in the ventral surface of pharynx of protochordates such as *Amphioxus* and *Herdmania.*

Endotherm. An animal which by producing heat within its body by oxidative metabolism is able to maintain constant body temperature like bird or mammal. *Opposite* to ectotherm.

Endothermic. A chemical reaction requiring heat/energy e.g., protein synthesis. Also called endergonic.

End Plate. A common name for vertebrate neuromuscular junction where motor axon gets remifies into fine terminal branches over a specialised system of folds in postsynaptic membrane of a muscle cell.

End-plate Potential. A brief localized depolarization or potential change which takes place across the membrane in the motor end-plate region of a muscle fibre, at a neuromuscular junction. A neurotransmitter substance gets released from the presynaptic nerve endings on stimulation by an impulse and increases the permeability of the postsynaptic (muscle) membrane to ions causing the EPP. The size of the EPP has been bound to depend on the amount of neurotransmitter released but normally it is sufficiently large to cross the threshold level for response and set off an action potential which gets propagated along the length of the muscle fibre.

Enteropeptidase (Enterokinase). A peptidase enzyme converting trypsinogen to trypsin.

End Product (Feed Back) Inhibition. Inhibition of a biochemical pathway by the end product which is formed during the reaction.

Energy. Refers to the exertion or capacity for a particular type of work.

Energy Pyramid. Refers to the energy relationships within ecosystem. When it is moving towards apex, amount of energy decereaes. For example, base may posses X kcal, assimilated by producers and apex may posses Y kcal assimilated by consumers, where Y is much lesser than X.

Energy Rich Phosphate Bond. Phosphate bond which is released by hydrolysis of ATP. The release of first bond evolves 3000 cals and the release of second bond evolves 1200 cals. The energy so released finds use in various body activities.

Enteric. Means gut or enteron.

Enterogasterone. An hormone, which is secreted by duodenal mucosa in response to fat digestion. It suppresses gastric motality and secretion.

Enterokinase. An intestinal enzyme in vetebrates which is able to covert inactive trypsinogen of pancreatic juice to active trypsin.

Enterokrinin. A hormone stimulating secretion of intestinal juice/succus enтericus in the vertebrates.

Enteropneust. Member of a hemichordate class, acorn worm.

Enteropneusta. Member of a group of chordata to which acorn worm and *Balcnoglossus* belong.

Enteroprocta. Refer to pseudocoelomate animals. They are sessile having 'U' shaped intestine. Their mouth is surrounded by a circle of ciliated tentacles and anal opening within circle.

Entoderm. Inner layer of embryo which forms mucosa of gut.

Entomology. Branch of science that deals with the study of insects.

Entropy. Measure of that portion of energy which cannot be converted into useful work in closed system; measure of molecular raddomness. It tends to increase with time in all irreversible processes.

Enurosis Nocturnal. Refers to one of the sleep disorders which leads to 'bed wetting' especially common among children.

Environment. Means the place where a species of animals is found in nature and the conditions that are present there. It means the total sum of physical, chemical and biological conditions which are surrounding an organism.

Enzyme. A substance produced within living cells that are able to catalyse specific chemical changes such as hydrolysis, oxidation, reduction, isomerisation, synthesis, lysis. It does not undergo a specific change. It consists of apoenzyme and co-enzyme.

Enzyme Kinetics. The rate which a reaction takes place has been found to depend upon concentration of substrate, product and active enzyme. If product is removed as fast as it is formed then the rate of reaction will be decided either by concentration of substrate or by concentration of enzyme. If enzyme is present in excess then concentration of substrate A decides the rate at which A gets converted to product P. Rate of conversion can be put as follows:

$$\frac{-d[A]}{dt} = k[A]$$

in which A represent concentration (instantaneous) of substrate, k the rate constant and -d[A]/dt rate of conversion of A to P. The rate of reaction does not depend upon substrate concentration when enzyme is present in limited concentration and all of the molecules of enzyme get saturated. Such reactions follows zero-order kinetics.

Eosinophilia. Refers to an increase in the number of eosinophils from their normal percentage. The disease is generally accompanied by allergy reactions.

Eosinophils. Granulocytic WBCs mammalian blood which has bilobed nucleus. The two lobes are joined by a bridge-like structure. They are non-phagocytic and can be stained with acidic stains like eosin. 1-4% normally in number. Their number increases during infections and allergy.

Ephyra. Refers to free-living larva of scyphozoan coelentrate like *Aurelia.*

Epiandrosterone. One of the 17-ketosteroid. It is formed by conversion of testesterone within the liver and gets excreted in the urine.

Epiboly. Means posterior growth of blastodermal fold over the surface of an embryo in the process of enteron formation during the gastrulation.

Epicardium. Visceral peritonium which is closely applied over the heart. It is derived from splanchnic mesoderm that is surrounding the endocardial tube during the development of heart.

Epricranium. Refers to the largest sclerite of head in a grass-hopper and related forms.

Epicuticle. During moulting it refers to those epidermal cells which are secreted under old skeleton (exoskeleton). It is made of cuticulin. In some arthopods there is a layer of some chemical substance like wax or calcium which is present over the cuticle in some arthropods and prevents waterloss.

Epidermis. Refers to outer cellular layer which covers external surface of a metazoan. It is derived from embryonic ectoderm and often secretes cuticle in some animals.

Epididymis. Coiled structure which is having efferent tubules of mammalian testis. The embryonic mesonephros in mammals which after formation of metanephros joins into the service of testis. It has three parts caput epididymis which is receiving various vasa efferntia, *corpus epididymis* which forms the middle part and *cauda epididy-*

mis which is opening into vas deferns or spermatic duct.

Epiglottis. Flap of cartilagenous tissue present in the larynx of mammals. It closes the air passage to lungs during swallowing of food.

Epilepsy. Means a nervous disorder in which a patient is able to lose consciousness and fails down. A characteristic cry may often be heard. There may occur mood disorders such as sudden melancholy, anger and less frequently gaiety. Patient's memory gets impaired.

Epimere. Refers to mesoderm during embryonic development which gets differentiated in to dorsal part called epimere. It is differentiated into dermatome dermis and sclerotome forming vertebral column.

Epinephrine. Hormone secreted by medulla of adrenal gland. Also called *adrenaline*.

Epyphyseal Plate. Refers to the end of long bone; ossifies independently of diaphysis.

Epiphysis. Refers to the end of long bone-like femur; also the pineal body.

Epipodite. A long slender structure which gets fastened to protopodite of walking leg of a crustacean.

Epistasis. A type of gene interaction in which one gene masks the expression of other gene which is having different locus and is thus not its allele.

Episternum. Means the upper part of sternum in fog which is connected to omosternum joining with clavicles and epicoracoid. It is not formed in toads.

Epithalamus. One of the divisions of thalamus possessing connections to olfactory system.

Epithelia. Means external/internal covering in the body. Also see *epithelium*.

Epithelium. Usually refers to sheet of cells which are covering either external or internal body surfaces. Several types are known:

Simple. Unicellular which is usually composed of flat cells. It is *called* squamous epithelium. It is present in Bowman's capsule.

Cuboidal. It is composed of cuboidal cells and formed in glands and ducts.

Columnar. It is composed of relatively tall cells which look like columns. It is found in gastrointestinal canal.

Stratifed. It is composed of more than one layer. It is present usually in skin.

Pseudostratified. It is composed of more than one layer. Layers are not even because the cells are of various shapes. It is present in trachea.

Ciliated. It is composed of cells which are either cuboidal or columns but cells bear cilia. It is found in trachea and some ducts.

Equatorial. Means plate-like arrangement of chromosomes in plane of equator of the spindle during the mitosis.

Equilibrium. Balance it may be static in which nothing changes or it may be dynamic in which changes in positive directions cancel each other.

Equilibrium Constant. 'K' indicative or relative concentration of atoms or others (whatever) at equilibrium.

Erepsin. Enzyme mixture of secretion of intestinal mucosa which consist of various peptidases. The peptidases split peptones and proteoses into simpler products. It does not have effect on intact protein molecules.

Ergastoplasm. Means granulated endoplasmic reticulum. When small ribosomes are present along membranes of endoplasmic reticulum, it is known as ergastoplasm.

ERP (Early Receptor Potential). An almost instantaneous potential change which is recorded from retina in response to a short flash of light that is probably corresponding to a movement of charge which takes place as the photopigment undergoes conformational change.

Erythroblastosis Foetalis. Refers to condition in foetus in which due to mother-father Rh factor incompatability, the infant's blood (Rh^+) may clot when it comes in contact with mother's blood (Rh^-). The father is having Rh^+-blood.

Erythrocurorin. Used for the haemoglobin present in the body of invertebrates like annelids, molluscs and few arthropods such as *Daphnia* and chironomous larvae.

Erythrocytes. RBC's of blood. Haemoglobin having cells which are characteristic of all the vertebrates and chordates. They are able to transport oxygen in the body. They are rounded oval, biconcave or biconvex and mostly have a nucleus except in mammals. They are formed within lymph nodes and spleen in non-mammals and in the red bone marrow in mammals. Their life period is about 100 days in non-mammals like frog and 120 days in mammals. The number is 5 million/ml of blood of man. In women the number is more. Process of their formation is known as erythropoisis.

Erythropoisis. Refers to the process of RBCs formation. It gets inhibited by a rise of RBCs in blood to supernormal value and stimulated by anaemia. This process is normally controlled by a circulating glycoprotein hormone which is called erythropoitin. Its steps are as follows:

Uncommited stem cell→Commited stem cells→Proerythroblasts→Early nomoblasts→Intermediate nomoblasts→Late nomoblasts→Reticulocytes→Erythrocytes.

Erythropoitin. A glycoprotein hormone. It is circulating in the blood and is formed by a renal factor on a globulin in plasma. It promotes differentiation of commited stem cells (erythropoitin sensitive cells) into proerythroblasts.

Estradiol. Female hormone which stimulates uterine hypertrophy. It is secreted by mature Graffian follicle.

Estrogens. Vertebrate female sex hormones which are produced in the follicles. In mammals they affect secondary sex al characteristics and make periodical changes in the walls of vagina and uterus for reception of fertilized ovum and its development.

Estrus Cycle. Refers to mammalian egg production and fertilisability. It also refers to reproductive cycle period in mammals when breeding takes place.

ETC (Electron Transport Chain). A series of enzymes which involve transfer of electrons from substrate molecules to molecular oxygen. It consists of NAD, FAD, Coenzyme-Q, Cytochrome-b, Cytochrome-c and Cytochrome-a_3.

Ethology. The branch of science that deals with the study of animal behaviour based on comparisons, the observations of patterns and consideration of environmental adaptation.

Etiology. Assignment of cause of disease or the science which deals with the study of causation.

Eucoelomata. Group of animals in which there is true coelom. (i.e., a cavity lined with peritonium is found). It incorporates all the major groups of animals from annelids to chordates.

Eugenics. Branches of science that deals with the application of knowledge of heredity to improve the human species. This term was coined by Sir Francis Galton in 1883. It includes two types of measures—one discourages the reproduction in inferior or abnormal human beings (negative eugenics) while the other encourages the production of superior human beings (positive eugenics).

Eumetazoa. Multicellular animals that have true tissue in their body and a digestive cavity. Include most groups of animals.

Eupnea. Means abnormal breathing which takes place in an individual.

Eurypterid. Used for describing extinct chelicerate arthropod of palaeozoic era.

Eustachian Tube. This tube has been named after Italian physician Eustachi. It is a tube which is leading from middle ear to phyarynx in the higher vertebrates.

Euthenics. Branch of science that deals with the study of improving human species by providing the best possible environment.

Eutheria. A group of mammals whose young ones get retained for considerable time in the uterus of mother and nourished by placenta. Also known as placentalia. In this all the mammals except egg-laying ones and marsupials have been included.

Euthyroid. Term used for describing patients who have abnormal thyroxine binding globuin level and do not exhibit signs of hyper- or hypothyroidism.

Evagination. Refers to outpocketing from any surface by growth (specially a cavity).

Eviscerate. Means removal of internal organs/viscera from an animal.

Excitability. Refers to the property of altered membrane conductance (and often membrane potential) which takes place in response to stimulation.

Excitable Tissue. A tissue that could be excited by chemical, electrical or mechanical means and produces an action potential which gets transmitted along its membrane. It includes muscles and nerves.

Excitatory Junction Potential. In smooth muscles, in which adrenergic nerve gives rise to discrete depolarisations that appear like small end plate potentials.

Excitatory Postsynaptic Potential. When single stimulus is applied to sensory nerve in experimental situation, it does not produce a propagated action potential in postsynaptic nerve cell (neuron) but instead, produces a transient partial hyperpolarisation or transient partial depolarisation. The potential due to depolarisations is termed as excitatory postsynaptic potential and during this excitability of neuron to other stimili gets increased. It is attributed to depolarisation of potsynaptic membrane which is immediately under active synaptic knob.

Excretion. A term used for the process of removal of metabolic wastes and also the substance discharged. In strict sense it is used for removal of nitrogenous wastes in the body.

Exergonic. Refers to a chemical reaction which yield energy e.g., respiration. Also known as *exothermic reaction. Opposite* to endothermic/endergonic.

Exhaution Atrophy. Like other endocrine glands β-cells of islets of Langerhans respond to stimulation with hypertrophy they get exhausted and stop secreting when stimulation is marked or prolonged. It is known as exhaution atrophy if stimulation terminates soon after this they can recover but if stimulation is continued, the cells eventually die and disappear.

Exocrine Glands. Type of glands opening through a duct e.g., liver, pancreas, sweat glands etc.

Exocytosis. Refers to the process in which vesicle membrane is fused to the surface membrane followed by the expulsion of the vesicle to cell exterior.

Exogenous Insulin. The insulin which is injected during the treatment of diabetes. It enters peripheral rather than portal circulation. However, it could not be confirmed whether the resultant unphysiologic distribution of insulin is having any consequences in the control of disease or its complications.

Exopeptidase. A proteolytic enzyme which is able to cleave only the bond joining the terminal amino acid to a peptide chain.

Exophthalmic Goitre. A disease which is resulted due to hyper-secretion of thyroid. The symptoms of this disease include large bulging eyes due to high rate of fat metabolism and their deposition in the orbits, high metabolic rate, high pulse rate, warm skin and raised nervous excitability.

Exoskeleton. A supporting structure which is present outside the body e.g., hair, feathers, horns, nails etc. *Opposite* to endoskeleton.

Exosmosis. A term used for the phenomenon in which there occurs passage of solvent molecules through the semipermeable membrane from interior of cell to its exterior e.g., if grapes are kept in concentrated sugar or salt solution, there occurs passage of solvent from interior of exterior of grapes and results in their shrinking.

Expiration. Refers to the expulsion of water or air from the gills or lungs of a vertebrate.

Explosive Decompression. Refers to the breaching of wall of pressurised cabin of an aeroplane or rocket at a high altitude which causes air embolism in persons due to rapid expansion of gases in lungs when external pressure gets suddenly reduced from atmospheric to subatmospheric.

Expressivity. Refers to the variability in phenotypic expression which is associated with a particular genotype when that genetic constitution is penetrant.

Extensor. A muscle straightening a part of body as foot. *Opposite* to flexor.

External Auditory Meatus. The opening of external ear in mammals which is surrounded by pinna.

External Respiration. Biophysical process which involves breathing in which air and blood are brought together in the lungs as well as there occurs transport of oxygen and carbondioxide between lungs and cells. It is opposite to internal respiration in which biochemical reactions occur within the cells.

External Urethral Sphincter. Refers to sphincter of skeletal muscles along the urethra. During micturition this gets relaxed so that urination becomes an involuntary act under voluntary control.

Exteroreceptors. Sense organs responding to external stimuli e.g., receptors for touch, smell, taste, sight and hearing.

Extertional Rhabdomyolysis. In potassium depleted individuals muscles' blood flow increases to a degree during excercise and there occurs a greater tendency for severe disintigration of muscles which is known as extertional rhabdomyolysis. Generally potassium ions are among most important vasodialator metabolites which are able to dialate arterioles in exercising muscles.

Extracellular Fluid. The fluid present outside the cell or cells.

Extra Embryonic Membranes. Cellular membranes which are present outside the body of embryo during the development of reptiles, birds, mammals (amniotes). These various membranes are amnion, chorion, allantois and yolk sac. The amnion and chorion play protective role and also furnish optimum moisture for developing embryo, allantois carries out respiration and excretion while yolk sac plays nutritive role.

Extra-Fusal Fibres. Fibres that costitute bulk of muscle. Inervated by alpha-motoneurons and make muscle to develop tension and shorten. *Opposite* to intra-fusal fibres.

Extra Medullary Haematopoises. Used for describing the formation of RBCs in the liver, spleen of adult man. It takes place during a disease in which the bone marrow is destroyed or fibrosed. (It is a normal mechanism in the foetus).

Extra Pyramidal System (Mechanisms). Refers to the system which is made up of those areas in central nervous system, other than pyramidal and cerebellar system, which are concerned with movement of body. This system includes a whole series of nuclei and portions of many structures such as cerebral cortex, basal ganglia, mid brain, medulla and spinal cord. This system is concerned with programming and initiation of movement as well as posture.

Extrasystole. A portion of myocardium (or AV node or bundle of His) often discharges independently (ectopic focus) and such irritable foci are usually attributed to some defect in mycardium following some injury or less blood supply. If the focus is discharged once, the beat takes place before expected next normal rhythm. This premature beat is known as extrasystole and during this there is no feeling of pulse.

Extrinsic. Means external factor, cause or mechanism which are not originating/taking place internally e.g., food supply is an important extrinsic factor that affects growth, extrinsic coagulation mechanism takes place when body tissue gets injured and thromboplastin gets liberated.

Exumberella. Used for describing the convex, dorsal or aboral surface in some animals like a jelly fish or medusa. *Opposite* to oral surface/ subumbrellar surface.

Eye. An organ of sight, or light perception. Invertebrates are having eyes that are simple photoreceptor (ocelli), sensitive to the direction and intensity of light. The higher molluscs and arthropods have compound eyes that form images. The vertebrate eye is a complicated spherical structure, which is connected to the brain by the optic nerve. It has an outer white sclerotic coat with a transparent front,

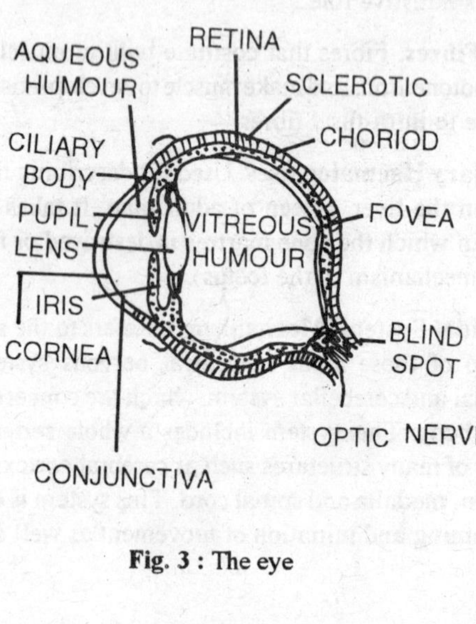

Fig. 3 : The eye

called the cornea. It is lined by the vascular pigmented choroid, continuous with the ciliary body and the iris in front. In the centre of the iris is a hole, the pupil, through which the innermost layer and has light sensitive cells (rods and cones).

Eyespot. A light-sensitive pigmented spot present in the cells of some primitive animals, including Protozoa, jellyfish, and flatworms, e.g., the miracidium larva of liver Fluke.

F

F_1. The first filial generation; i.e., the first generation which is resulted from a particular cross.

F_2. The second filial generation which is by obtained crossing within the F_1 generation. It is in the F_2 that the characteristic monohybrid and dihybrid ratios become apparent.

Facial Nerve (Cranial Nerve VII). One of the pair of nerves arising from the anterior hindbrain in vertebrates for supplying the muscles of the face. It is also carrying sensory nerve fibres from taste buds and autonomic nerve fibres to salivary glands and the lacrimal gland.

Facilitated Diffusion. A passive transport of molecules which takes place across a cell membrane along a concentration gradient. It is mediated by carrier molecules or complexes. No energy is expended in this process.

Facilitation. Used for describing the phenomenon in which passage of an impulse across a synapse makes the synapse more sensitive to successive impulses so increasing the postsynaptic response. Eventually one stimulus will evoke a response which is large enough to trigger an impulse.

Factor. Means a specific germinal cause of hereditary characteristic i.e., a gene.

FAD (Flavin Adenine Dinucleotide). A coenzyme which is formed by condensation of riboflavin phosphate with adenylic acid. It carries out important function in electron transport and is prosthetic group for some enzymes.

Faeces. Solid or semisolid material, which consists of undigested food, bacteria, mucus, bile, and other secretions. It is expelled from the alimentary canal through the anus.

Fainting. Means sudden loss of consciousness which occurs due to cerebral ischemia (reduced blood supply). Also known as syncope. Its main cause is hypotension due to sudden vasodialation of blood vessels and bradycardia (decreased heart rate) is often associated with strong emotion. The attack is short-lived and consicousness is resorted in few minutes. It is called vasovagal syncope (fainting). Pooling of blood in dependent parts of body on standing may bring about postural syncope. Persons with orthostatic hypotension possess miturition syncope. Cough syncope takes place when there occurs an increase in intrathoracic pressure, during coughing, enough to block venous return. Effort syncope refers to fainting on exertion due to inability of cardiac output to meet increased tissue demand which takes place in persons having aortic/pulmonary stenosis.

Fallopian Tube. One of a pair of ducts in female mammals also called oviducts. It conveys ova from the ovary to the womb (uterus) by the help of muscular and ciliary action. It opens anteriorly into a ciliated funnel, which is lying close to the ovary.

Family. Means principal subdivision of an order of animals/plants.

Fangs. The specialised grooved teeth present over maxillae of poisonous snakes like, cobra, viper, kraits etc., through which a pair of poison glands open.

Fascia. A band connective tissue which covers and supports or binds parts together.

Fasciculation. Jerky visible contractions of muscle fibres which take place due to pathogenic or pathological discharge of spinal motor neurons.

Fasciculi. Means a pathway structure which is present in brain stem to which second order neurons pass down the spinal cord from vestibular nuclei in the vestibulospinal tracts and ascent through these to motor nuclei of cranial nerves which are concerned with control of eye movement.

Fatigue. The build up of waste products e.g. CO_2 and lactic acid, which prevent muscular contraction in a muscular tissue (ability of muscle to contract gets reduced or obliterated).

Fat. Lipids serving as storage material in higher animals and some plants. Fats are mainly triglycerides of long-chain carboxylic acids (fatty acids).

Fat Body. A mass of fatty (adipose) tissue which forms a definite structure within the body cavity of some animals. The amphibians and reptiles are having a pair of solid fat bodies which are attached to the kidneys or near the rectum and act as a food store for use during hibernation and breeding. In insects the fat forms a more diffuse tissue around the gut and reproductive organs and stores protein and glycogen as well as fat.

Fate Map. A map of the probable fates of the various regions of the blastula surface of animals having indeterminant development, usually shown as colours: colours have been assigned to the various tissues and structures of the formed embryo (or of any later stage) and, by imagining the developmental process reversed, these coloured tissue follow back to the blasula. The three major germ layers are occupying large discrete areas of the fate maps of vertebrates. Mosaic eggs possess cell lineages instead of fate maps.

Fat Metabolism. After digestion fats are converted to glycerol and fatty acids and these substances get metabolised in two ways:

Metabolism of Glycerol. The glycerol gets Phosphorylated to form 3-phospho glyceraldehyde which is made to undergo glycolytic pathway.

Metabolism of fatty acids. Fatty acids are subjected to β-oxidation thereby yielding two molecules of acetyl coenzyme-A. This substance is involved in Kreb's cycle and also in the synthesis of cholesterol from which steroids, cholic acid etc. are obtained.

Fats. Organic compounds having C, H and O. They are esters of fatty acids and alcohol. If the alcohol is glycerol, they are tremed as natural fats but if it is of higher grade then they are called waxes. They liberate double the amount of energy (9.3 cals/gm) as compared to equivalent weight of carbohydate/protein.

Fauna. Means animal life of a given period of region.

Feathers. Structures present on skin of birds. They make body insulated from heat, cold, electricity. They make body light and help in balancing the body when attached to tail. They help in flight when attached to wings and make general contour of bird. Their many types include contour feathers, down feathers, filoplumes and bristles.

Fecundity. Refers to the capacity of an organism to produce offspring. Most organisms show such enormous fecundity that the size of their population would rapidly increase if all of them survived. In actual practice most offspring do not survive.

Feed Back. Means the return of output to the input part of a system. In negative feed-back sign of output undergoes inversion before being fed back to the input so as to stabilize the output. In positive feedback the output is not because it gets returned to the input without involving a sign of inversion and thus becomes self-reinforcing or regenerative.

Feed Back Control. Means the control which is involved through return of the output to the input part of the system. Also refers to the mechanism by which activity of a system is governed on the basis of information that is obtained from the output system's activity. In other words, it means that a system in which the accumulation of the product results a decrease in its rate of production.

Feedback Inhibition. Refers to the inhibition of the activity of an enzyme (often the first) in a reaction sequence by the product of that sequence. If the product is accumulated beyond an optimal amount, it undergoes binding to a site (allosteric site) on the enzyme, changing the shape so that is can no longer be able to react with its substrate. However, once the product is used and its concentration decreases again, the enzyme is no longer inhibited and further formation of product takes place. The mechanism finds use in regulating the concentration of certain substance within a cell.

Fehling's Solution. A freshly mixed solution which is used for testing the presence of reducing sugars and aldehydes in solution. When boiled with equal amount of Fehling's A (copper (II) sulphate solution) and Fehling's B (sodium potassium tartrate and sodium hydroxide solution) reducing sugars and aldehydes yield a brick red precipitate of copper (I) oxide.

Femur. **(1)** The long bone which forms the upper bone or thigh bone of the hindleg in tetrapods, and extends from the hip to the knee. Its upper end bears a projecting round head, which articulate with the pelvic girdle at the acetabulum in a ball-and-socket joint. The lower end has two articular surfaces (condyles) for the tibia, to form the hinge joint of the knee.

(2) The third segment of an insect leg, present between the trochanter and tibia.

Fenestra Ovalis. A gap which is left in bony labyrinth of mammals. It connects membranous labyrinth with middle ear. It is covered with vibratile membrane and oveı this is fitted stapes.

Fercula. Also known as 'wish-bone' or 'marry-thought bone'. It is formed by fusion of clavicles and interclavicle in the birds. It is a fork-shaped structure which articulates anteriorly between two pectoral girdles and posteriorly articulated with sternum.

Fermentation. Means enzymatic decomposition; anaerobic transformation of nutrients without involving net oxidation of electron transfer.

Ferritin. Means a large protein molecule which is opapue to electrons and used as a marker in electron microscopy; normally present in spleen as a storage protein for iron.

Fertile Period. The 72 hour period (14, 15, 16th day) during which there occurs release of ovum into the upper part of oviduct and fertilisation is possible. The hormonal level during this period is at maximum.

Fertilisation. Refers to the union of a mature ovum and mature sperm to form zygote.

Fever. The most universally known disease during which the thermoregulatory mechanism behave as if they had been adjusted to maintain high body temperature level than 37^{0}C. A shivering often in fever occurs due to activation of temperature rising mechanism. Drugs like aspirin are able to stimulate thermoregulatory centres to relieve it by adjusting set point to back to normal.

FFA (Free Fatty Acids). Mean extremely labile circulating fat components that are having half life period of only few minutes. They are main source of energy for many organs particularly heart. They undergo oxidation in all tissues including brain to form carbon dioxide and water. In blood they are bound to albumin.

Fibre. Thread or filament which is produced by cells but external to them.

Fibril. Thread or filament which is produced by cells and located within them.

Fibrillar Muscles. Oscillatory insect flight muscle; also known as asynchronous muscle because contractions could not individually be controlled by motor impulses.

Fibrillations. Refers to fine irregular contractions of individual fibres in animals in which destruction of nerve supply takes place.

Fibrin. A blood protein and major component of blood which clots in the vertebrates.

Fibrinogen. Soluable form of fibrin which when comes in contact with thrombin yields fibrin, a component of blood clot.

Fibrinolytic System. A system in the blood which is able to limit clotting and also able to decompose clot to fibrinogen degradation products. Its component is fibrolysin or plasmin. This substance is able to lyse fibrin and fibringoen with the production of fibrinogen degradation products that inhibit thrombin. Plasmin is obtained from plasminogen. There is an activator circulating normally as proactivator which gets activated by proteolytic fragment of active factor-XII formed by action of plasmin on active factor-XII. This positive feedback of plasmin on its own formation gets limited by available store of factor-XII and by various limiting factors.

Fibroblast. A connective tissue cell which is able to differentiate into chondroblasts, collagenoblasts and osteoblasts.

Fibrocartilage. A cartilage which is having fibres in the matrix. These fibres are made of collagen. It is present in the intervertebral discs of mammals.

Fibrous Joint. A joint lacking a cavity; bones are directly apposed.

Fibula. One of two long bones of the lower hindlimb of tetrapods. In man, it is a slender bone which bears little weight. Its upper end gets articulated with the tibia just behind and below the outer side of the knee and it extends down, lateral to the tibia, to its lower end (the lateral malleolus)—seen as a prominence on the lateral side of the ankle joint. In some species it gets reduced and partly or wholly fused with the tibia.

Fighting Reaction. When an animal has been threatened it makes an attempt to flee but if cornered, it fights. During this state spitting, growling, piloecrection, pupilary dialation, often bitting and clawing etc., occur. It generally happens during sexual selection.

Filial Generations (F_1 & F_2). Refer to the first and second filial generations with respect to successive generations following cross-breeding (the hybrids produced by cross between original parents form F_1 generation and the products formed by selfing the F_1 hybrids form F_2 generation).

Filiform Papillae. Small conical projections which are on the greater part of tongue of man. They lack taste buds.

Filoplume. A hair-like feather having few barbs at the tip.

Filopodium. Filamentous pseudopodium of sarcodine protozoan.

Filtration. Refers to a process by which fluid has been forced through membrane or the other barrier due to a difference in hydrostatic pressure on the two sides. The amount of filtrate is proportional to the pressure difference and the surface area of the membrane. Molecules of smaller size than the pores of membrane had to pass along with the filtrate while larger molecules are retained.

Filum Terminale. Refers to the thread-like terminal part of spinal cord which is non-nervous. In mammals this part terminates in lumber region even if there exists a tail.

Fin. Means locomotor or starting structure and extension of body in an aquatic animal. These may be paired/unpaired. The paired fins have pectoral and pelvic fins while ventral, anal and crudal fins have unpaired ones.

Final Common Pathway. As motor neurons supplying the extrafusal fibres in skeletal muscles are different side of reflex arch, it means that all neural influences affecting muscle contraction are finally funnelled through them and due to this reason they are said to constitute final common pathway.

Fin-Fold-Theory. The theory which postulates that ancestral vertebrate possessed a pair of lateral fin fold one on each side of trunk which got fussed behind the anus with median ventral fin which extended along the tail to mid dorsal side of the body. Persistence of some parts and disappearance of others in fin fold and median fin had resulted in the formation of paired an median fins respectively in fishes. The tetrapod limbs then evolved form paired fins of fishes.

Finger Printing. A procedure which is used to separate the component parts of a protein following enzyme cleavage of protein into fragments and their separation by involving a combination of electrophoresis and paper chromatography.

Fins. See *Fin*.

Firing Level. First manifestation of approaching impulse takes place at the start of depolarisation. After an initial 15 mV depolarisation its rate increases. The firing level is defined as the point at which this range in rate takes place.

First Order Enzyme Kinetics. See *enzyme kinetics*.

Fission. A type of asexual reproduction in which a parent cell undergoes division into two (binary fission) or more (multiple fission) similar daughter cells. Binary fission takes place in many unicellular organisms (protozoans, bacteria); multiple fission takes place in Sporozoa. Fission starts with division of the nucleus by mitosis, followed by cytoplasmic division and sometimes sporulation.

Fitness. In an evolutionary context, it refer to the ability of an organism to produce a large number of offspring that are able to survive to a reproductive age.

Facidity. Refers to the state in which the cells are not turgid and consequently the tissues remain loose. When motor nerve is cut from the muscle the metabolism gets impaired and turgidity is lost. The muscle offers very little resistance.

Flagellum. Used for describing whip-like structure in some animals like, *Euglena* having 9+2 arrangement of inner filaments.

Flame Cell (Solenocyte). A cup-shaped cell that has a group of cilia and present in large numbers within the tissues of various invertebrates (e.g. Platyhelminthes, Rotifera, and some Annelida).

Flare. Redness which speads out from injury and occurs due to arteriolar dialation. The pain in this area occurs due to chemical agents which are liberated by antidromic impulses in sensory nerve fibres. The agent is a proteolytic enzyme and brings about liberation of vasodialator kinins. The kinins decrease threshold of the pain fibres in the area and they may themselves produce pain.

Flat-Worms. A common name for platyhelminthes e.g., *Fasciola Taenia* etc.

Flatus. When large amount of air is swallowed along with food (as an case of nervous persons who hyperventilate sometimes), some air passes out by regurgitation but much of the air is passed to colon where some oxygen gets absorbed and H_2S, CO_2 and methane formed by colonic bacteria are added to it and is finally removed as flatus.

Flavoprotein. A conjugated protein in which a flavin (FAD or FMN) gets joined to a protein component. They act enzymes in the electron-transport chain.

Flexor. A muscle whose function is to bend or flex a joint thereby decreasing the angle of two component parts. *Opposite* to extensor.

Flight or Fight. When an animal is threatened, its first attempt is to flee but on being cornered in such situation it attempts to counteract. These responses are termed as 'flight or fight' respectively.

Floating Ribs. The two most caudal ribs which lack ventral attachment.

Flocculus. Refers to lateral portions of cerebellum which are known as flocculonodular lobe. Its connections are essentially all vestibular. When it has been damaged or destroyed, animals walk in staggering fashion on the broad base. They tend to fall and are reluctant to move without a support.

Fluorescence. Refers to the property of emitting light upon molecular excitation by an incident light. The energy of the emitted light is always less (of short wave length) than the light producing excitation.

Fluorescent Strains. Refers to those strains which on combination with chromosomes render them luminous due to ability to emit light some of which gets absorbed.

Flux. Means the rate of flow matter or energy across a unit area.

Flying Fish. Common name for bony-fish *Exocoetus.* The fish could fly due to the presence of large pectoral fin which acts as patagium-like structure.

Flying Fox. Used as a common name for flying mammal *Pteropus medius* (large frugivorous bat).

Flying Lizard. Used as a common name for lizard *Draco*. The animal possesses a patagium between hind and fore limbs and is capable of flight.

Foetus (Fetus). Used for the embryo of mammal especially a human embryo, when its external features resemble those of the mammal after birth, i.e. after it has developed limbs, eyelids, etc.

Foliaceous. Refers to the leaf-like appendage which may also be multi-lobed.

Foliate Papillae. Broad, leaf-like papillae which lie near the base of tongue in parellel rows. They bear taste buds on their sides. They are present in man.

Folic Acid. One of the water soluable vitamins (member of vitamin B complex group) which is essential coenzme for 'one carbon' transfer involved in methylation reactions. It occurs in leafy green vegetables. Its deficiency causes sprue, anaemin etc.

Folium Lobules. Refers to the lobules in cerebellum where auditory and visualstimuli are able to evoke electrical responses. Each stimulus brings about a stimulation and the stimulation is responsible for turning of head and eyes to ipsilateral side.

Follicle. A cellular sac or covering. The term is generally resticted to a mass of cells surrounding the developing ovum (Graffian follicle), mass of cells surrounding hair shaft (hair follicle).

Follicular Phase. The term used for that part of estrus and menstrual cycle which is characterised by formation of and secretion by Graffian follicle.

Food Chain. Means linear interactions of organisms in an ecosystem in terms of food consumption e.g. plant to herbivore to carnivore to decomposer.

Food Vacuole. The intracellular digestive apparatus as present in *Amoeba, Paramecium*.

Food Web. Mean complex nutritional interactions which take place within an ecosystem. It includes alternatives e.g., an animal may be able to eat different foods if primary energy source gets deleted from environment.

Foramen. A natural opening in an animal organ or other structure, especially in a bone or cartilage.

Formen Magnum. The opening in the skull through which is spinal cord is passing.

Foramen of Megendie. Refers to the median foramen in the roof of medulla which communicates with the cerebrospinal fluid in subarachnoid space.

Foramen of Monro. Used for the passage between lateral ventricles and third ventricles of brain (also see *ventricles of brain*).

Formina of Loschka. Refers to two lateral foramina in the roof of medulla which communicates with cerebrospinal fluid in sub-arachnoid space.

Foraminifera. Used for Saracodine protozoa that have fragile calcareous shells with pores through which pseudopodia come out.

Fornix. Refers to a band of longitudinal fibre present below the corpus callosum in the mammalian brain. The two strands of fibres called taeniae hippocampi, form the fornix. The ridges get separated both in front and behind by the main body of fornix forming anterior or posterior pillars. Through posterior pillars, the splanium gets united. The anterior pillar passes ventrally thereby connecting hippocampal lobe with hypothalamus and mammillary body.

Forward Failure. Cardiac output which is not adequate to perfuse muscles, specially failure of output to rise with the excercise. Such persons cannot undertake excercise.

Fossa. A pit/depression present in a bone.

Fossil. The remains of, or impressions which are left in rocks by long dead animals and plants. Most fossils have hard skeletal material because soft tissues and organs rot away very quickly. Mineral salts from surrounding rocks gradually replace the hard organic material, to give a cast in a process called *petrification*. Alternatively the organic material dissolves away leaving an impression or mould in the surrounding rocks. Trace fossils (e.g. of dinosaur footprints) give indirect evidence of prehistoric life forms.

Fossil Hominid. Any of the manlike fossils belonging to the family hominidae. It includes genera like *Australopithecus*. Fossil hominids occur in great number and variety but none of the lines identified are regarded as the direct ancestors of modern man, the only existing hominid.

Founder Effect. Refers to a type of genetic drift where new population is founded by relatively few individuals, hence their limited gene pool forms basis for all future generations.

Fourth Sound. During each cardiac cycle, besides normal first and second sounds, often third and fourth sounds are also heard in the stethoscope. It can be heard before first sound when atrial pressure is high or ventricle is stiff in conditions such as hypertrophy. It occurs due to rapid ventricular filling and is rarely heard in normal adults.

Fovea. The point of most acute vision on the retina of the vertebrate eye. The image of an object is incident on this region, which is directly opposite the centre of the pupil and lens. It consists entirely of densely packed cones for daylight vision. Some birds possess two foveas for sharp forward and lateral vision.

Fovea-Centralis. Optic centre of the mammalian retina where there are only cone cells present.

Fragmentation. A form of asexual reproduction occurring in certain metazoan animal such as some aquatic annelids (e.g. *Lumbriculus*) and sea anemones. It involves the breaking up of the body into two or more pieces that subsequently develop into complete organisms. A form of fragmentation (*strobilization*) is seen in the formation of medusae in jellyfish.

Frame Shift Mutation. Refers to genetic mutation where one, two, $(\text{three+one})_n$ $(\text{three+two})_n$ base pairs get deleted or added. It results in the formation of disturbed code and as a result different amino acids may be coded for. It may also result in premature synthesis termination.

Fraternal Twins (Dizygotic twins). Two offspring born to the same mother at the same birth, that result from the fertilzation of two eggs at the same time. They may be of unlike sex and are no more genetically similar than any two siblings.

Free Energy. The energy available for work in a system at a given temperature and pressure.

Freeliving. An organism that is neither attached nor parasitic.

Freemartin. A masculinized but genetically female calf. In some mammals (e.g. cattle) fraternal twins share a common blood circulation in the uterus and this may result in a female twin which receive male hormones from here brother. The freemartin exhibits some male characters and is sterile.

FRH (Follicle Releasing Hormone). A hormone which is secreted by hypothalamus. It acts on pituitary to secrete follicle stimulating hormone (FSH) which acts on ovarian follicle in females and testicular follicles in males.

Fringing Reef. A coral reef which is built up from bottom of ocean (Also see *coral reef*).

Frons. Refers to the anterior portion of the fore of an insect.

Frontal. Means parellel to main body axis and at right angles to sagittal plane.

Frontal Lobes. Refers to structure which are related to cerebral cortex. They lie anterior to excitomotor cortex and also extent on to medial aspect of hemispheres as back as anterior end of corpus callosum thereby including precallosal part of singular gyrus. They connect with thalamus, hypothalamus and correlate with activities of higher intellect, personality, emotional effect and forms of behaviour of social significance.

Frontoparietal. Wall of long flat which forms roof of cranium in the vertebrates like frog.

Fructose. Refers to keto-hexose sugar present (mostly in free state) mainly in the plant tissue; component of sucrose. In solution it forms furanose ring to form fructofuranose. Its polymers are known as fructosans (Most important is inulin).

FSH (Follice Stimulating Hormone). Refers to an anterior pituitary gonadotropin that is able to stimulate follicles in females and testicular spermatogenesis in males.

Fundic Glands. Refers to the glands which are situated in the main part (fundic region) of the stomach in vertebrates. These glands secrete all types of substances which are present in gastric juice.

Fundus. Means the base of a hollow structure of part which is farthest from the opening.

Fungiform Papillae. Toad-stool shaped papillae which lie on the margins of the tongue in the man. On them taste buds are present that respond to the taste of the food.

Funiculus. A cord e.g., spermatic cord or umbellical cord.

Fusimotor System. The gamma-motoneurons and the interfusal fibres that they innervate collectively to form this system.

G

G_1-Phase. Phase of cell cycle which is before chromosomal replication and after the telophase of mitosis.

G_2-Phase. Phase of the cell cycle which is after chromosomal replication and before prophase of mitosis.

Gaba (Gamma Aminobutyric Acid). Refers to an inhibitory transmitter whcih is identified in crustacean motor and in the vertebrate nervous system.

Gagging. Thing is thrusted into mouth to prevent speech or to hold it open for operation. It is reflex response which is integrated in medualla oblongata.

Galactorrhea. Dopamine like aldopa tends to inhibit prolactin secretion in mammals and during its disturbance there occurs abnormal secretion. The drugs like reserpin increase prolactin formation and stimulate this process. In this condition there occurs excessive milk secretion.

Galactose. Constituent of milk sugar lactose which is formed by action of lactase enzyme on it. In the liver it gets metabolised by its conversion to glycogen or to galactosamin which acts as an essential constituent of nervous system.

Galactosemia. Refers to an inborn error of metabolism occurring due to congenital deficiency of phosphogalactose uridyl transferase which is responsible for reaction between galactose I-phosphate and uridine diphosphoglucose so that ingested galactose gets accumulate in the blood.

Gall Bladder. Bladder-like structure present in the liver. It gets a number of cystic ducts from liver lobes and a single duct mostly and rarely two bile ducts. The main function of bile ducts is to bring the stored bile to the duodenum.

Gall-stone. Spherical masses of crystals which are formed chiefly from cholesterol and calcium salts which under certain conditions, may get accumulated in the gall bladder.

Galls. Abnormal growths of plant tissues resulted by various organisms which irritate the plant and possibly lead to the production of some type of growth hormone. The principal gall-causing insects or *gallicolae* belong to three different orders: (1) Gall-wasps and some sawflies belonging to the Hymenoptera; (2) Gall-midges belonging to the order Diptera; (3) Certain aphids and their allies among the Hemiptera. Galls are also sometimes formed by mites, eel-worms and fungi.

Galton's Law. According to the law person or animal gets 1/2 of his inherited characters from the two parents; 1/4 from his grandparents; 1/8 from great grandpaprents and so on. It is merely a statistical approximation which is independent of any biological theory of genetics.

Gamete. A male or female reproductive i.e., a cell having the hapid number of chromosomes which, when it undergoes conditions with another haploid cell at fertilization, gives rise to diploid cell or zygote from which the off spring develops. Male gametes are generally small and motile (spermatozoa) whereas female gametes are larger and have yolk (ova). In some organisms, however, the gametes may be alike (isogametes).

Gametocyte. Used for describing the potential gametes of malarial parasite (*Plasmodium*) which do not undergo further development as long as they are present in the man. However, they become active when sucked by secondary host mosquito during its feeding operation on such a person.

Gametogenesis. The process of gamete formation which is of two types; spermatogenesis involving the formation of sperms by the testes by the male and oogenesis involving the formation on ova by the ovaries of female.

Gamma-Motoneurons. Neurons which are originated in the spinal cord and synapsed with intrafusal muscle fibre.

Ganglion. A small compact mass of nerve cells and connective tissue. In a vertebrate the chief ganglia are; the dorsal root ganglia of the spinal nerves; the sympathetic ganglia which are forming chains on each side of the spinal cord; the colleteral ganlgia in the mesenteries. In many invertrates (e.g. annelids and arthropods) there is a ventral chain of ganglia which are arranged segmentally.

Ganglion Cells. Refers to some nerve cell bodies specially those which are located in ganglia of invertebrates or outside the vertebrate nervous system.

Ganoidal, Ganoid Fish. In the older system of classifcation the group included a miscellaneous collection of fossil and present-day fish all having thick dermal scales but otherwise not having a much in common.

Ganoid Scales. Ganoin. Ganoine. Refers to the fise-scales in which there is a hard shiny layer of *ganoin* on the upper surface and a bony layer of *isopedine* below. This type of scale occurs in a thin form on many modern bony fish and in a thicker form in such fish as the sturgeon and the garpike which used to termed as 'ganoids'. In the thinner scales, however, the layer of ganoin may get lost.

Gap Junctions. Used for referring to the specialisation for electrical coupling between cells where intercellular spacing is only about 20 A^0 and tubular assemblies of particles are connecting the apposed membranes.

Gastralia. Abdominal ribs or ventral ribs that are present in some extinct amphibia, living fossil lizard *Sphenodon* and the crocodiles. They have 'V' shaped dermals bones present in the ventral region of the abdomen.

Gastric Juice. A digestive fluid which is secreted by cells of gastric epithelium. It is a mixture of water, enzymes like pepsin, renin etc. and salts.

Gastrin. A protein hormone secreted by gastrin cells and pyloric glands and tends to induce the secretion of gastric juice by the stomach.

Gastrocentrous Vertebra. Vertebra present in repitles and mammals in which pleurocentrum gets enlarged to form centrum while hypocentrum gets reduced to a small piece (as in reptiles) or it get changed to form an intervertebral disc (as in mammals).

Gastrocoel. The primitive digestive cavity of metazoan embryo which is formed by gastrulation.

Gastroderm. The lining of coelentrate cavity.

Gastronemius. Large muscle that is present on the posterior side of lower leg of a vertebrate.

Gastropoda. Refers to group of mollusca which includes snails and slugs. The root gets attached below stomach and looks like modification of later.

Gastrotricha. Minute unsegmented worms swimming by means of their ciliated epidermis. They are hermaphrodite and have some affinities with the Rosifera and with the Nematoda. They are, however, sufficiently distinctive to be classed in a separate phylum.

Gastrovascular System. A system of channels for water circulation which is connected to the gastric cavity (colenteron) in larger coelenterates such as the jelly-fish.

Gastrula. Refers to the stage in an embryo in which the primary germ layers have been laid down (see *Gastrulation*).

Gastrulation (Embryology). The term used for describing the process following cleavage in which, by a complex series of cell-movements, the primitive germ layers, *ectoderm, mesoderm* and *endoderm* are laid down in definite positions which are ready to give rise to the principal organs of the later embryo.

GBG (Gonadialsteroid Binding Globulin). During metablosim of testesterone 2/3 rd testesterone is bound to plasma whereas some is bound to -globulin and albumin. This particular globulin is known as GBG.

G-Cells. Cells present iin lateral walls of glands in antral portion of gastric mucosa. They are flask-shaped with a broad base having many granules and narrow apex that reaches mucosal surface. They secrete gastrin hormone.

Gel. Jelly-like substance which is formed by a colloidal soluation while in the solid phase e.g., plasmagel of *Amoeba.*

Gemmule. Refers to a multicellular vegatative bud of certain sponges like, *Spongilla.*

Gena. Lateral portion of insect's head; feathered portion of mandible (lower jaw) of bird.

Genae. The lateral cuticular plates of an insect's head, below and behind the eyes.

Gene. A hereditary factor or unit of inherited material which is able to carry the power of transmitting some particular quality from one generation to the next. The genes are made of some type of nucleoprotein (derivatives of ribonucleic acid) and arranged in linear series along the chromosomes (see *Genetic Code*).

Gene Flow. Refers to the flow of gene pool of alleles from another population through interbreeding that alters the original allele frequencies.

Gene Frequency. The proportion of an allele in apopulation in relation to other alleles of the same gene.

Gene Mutation. A mutation (change in hereditary features) which is caused by a chemical or physical change in the composition of gene (See *DNA* and *Genetic Code*).

Gene Pool. The total number and vareity of genes which are existing within a breeding population or species at a given point in time.

Generation Time. The average time between the cell division of parent and daughter cells within a population of cells.

Generator Potential. A recorder potential which depolarises the spike initiating zone of an axon.

Genetalia. Refers to the reproductive organs of male or female individual. The internal genetalia means reproductive organs like testes, vas deferens etc., in male and ovaries uterus etc., in female. While the external genetalia means the copulatory organs like penis, scrotum in male and vulva, opening of bulbourethral chamber, clittoris etc. in female.

Genetic Code. Refers to the sequence of bases present along the DNA molecule. A group of three bases codes for one amino acid e.g., the triplet CAA (cytosine, adenine, adenine) code for valine. A string of triplets codes for a complete protein. As 64 different triplet combinations corresponding to 20 amino acids are known, thus many amino acids could be coded for by two or more triplets. The code is called *degenerate*, because the only the first two bases, and in certain cases only one base, are needed to ensure the coding of a specific amino acid. Three triplets, called 'nonsense triplets', do not code for any amino acid and have other functions, e.g. marking the beginning and end of a cistron.

Genetic Death. The term used for describing the failure of an individual to reproduce due to sterility, failure to mate or elimination from population before reproductive age.

Genetic Drift (Sewall Wright effect). Refers to the fluctuation of allele frequencies in a small population which is due entirely to chance. If the number of matings is small then the actual numbers of different types of pairing may depart significantly from the number expected on purely random basis. Genetic drift constitues one of the factos that may distrub the hardy-Weinberg equilibrium.

Genetic Engineegring. Deals with the manipulation of an organism's genetic material. Work of this nature is usually carried out with bacteria and viruses and involves the addition of new genes to the chromosome complement or the replacement of certain genes by other genes. Such research is carried out under strict control measures due to the possiblity that new virulent and potentially very dangerous self-replicating organisms might be the end product of certain procedures.

Genetics. The term introduced by Bateson to describe the study of inheritance and variation and the factors controlling them. Now the subject has three main subdivisions Mendelian genetics, population genetics, and molecular or biochemical genetics.

Geniculate Body. A projection of geneculate nucleus which is able to relay incoming sensory information to the cortex; named for its knee-like shape in cross section.

Gential. Term used for the reproductive organs (also see genetalia).

Genome. The set or sets of chromosomes carried by each cell of an organism. Haploid organisms possess one set of chromosomes, diploid organisms two sets, polyploid organisms many sets sometimes from the same ancestor (autopolyploids) and sometimes from different ancestors (allopolyploids).

Genotype. Refers to the genetic make-up of an organism. The actual appearance of an individual (the phenotype) has been found to depend on the dominance relationships between alleles in the genotype and the interaction between genotype and environment.

Genotypic Ratio. Refers to the ratio of genotypes which results from particular cross e.g. in monohybrid cross between tall and dwarf parents the genotypic ratio in F_2 geneation will be I Tall (pure homozygous): 2 Tall (heterozygous): I Dwarf (homozygous).

Genus. Refers to a collection of similar species. Genera are subdivided into a sub-genera, sections, subsections, series and subseries. Similar genera have been grouped into families.

Geological Time Scale. A system which is used to measure the history of the earth by studying the rocks of the Earth's crust. We know that new rocks are generally deposited on top of existing material, those lower down are oldest. The strata of rock have been classified according to their age, and a time scale corresponding to this can be constructed. The various main divisions (eras) are the Palaeozoic, Mesozoic and Caenozoic. These have been further subdivided into periods and epochs. See also Palaeozoic Mesozoic, Caenozoic.

Geotropism. Refers to the response in behaviour of an organism which takes place with respect to gravity.

Germ Cell Layer. Term used for one of the primary layers in an embryo-ectoderm, endoderm and/or mesoderm.

Germinal Layer. A layer capable of forming new cells; from whcih new cells can be budded off.

Germinal Variation. A type of variation/change, which occurs due to the modification in germ cells. It is opposite to somatic variation which involves only somatic cells. It is heritable.

Germ Layers. Three germ layers which are formed in the embryo-ectoderm, endoderm and mesoderm. Ectoderm gives rise to skin and nervous system, endoderm the mucosa of the gut where as mesoderm coelom, muscles, blood and blood vessels, bones, connective tissue, kidneys, gonads, etc.

Germplasm. Material basis of heredity. A term used for the herditary substance in germ cells by Weismann.

Gestation. The period between fertilization and the birth of an eutherian mammal, During this period the entire development takes place. It has been lound to vary in different mammals. The gestation periods of some common mammals are as follows:

Opposum —13 days/about 2 weeks.
Mouse, rat etc. —21 days/3 weeks.
Rabbit —32 days/about 4-5 weeks.
Dog. cat etc. —60 days/about 8-9 weeks.
Guinea pig —68 days/about 9-10 weeks.
Pig —112 days/16 weeks.
Sheep —147 days/21 weeks.
Monkey —168 days/about 6 months.
Man —270 days/about 9 months.
Cow —280 days/about 10 months.
Horse —336 days/about 12 months.
Elephant — 600 days/80 weeks/20 months.

GFR (Glomerular Filtration Rate). Defined as the amount of total glomerular filtrate which is produced per minute by all the nephrons of both the kidneys. It is equal to clearnace of a freely filtered and non-reabsorded substance like insulin.

GH (Growth Hormone). A protein hormone which is secreted by anterior lobe of pituitary. It stimulates growth by acting upon thyroid gland and other body functions such as protein, carbohydrate and fat metabolisms. It is also known as somatotropin.

Gigantism. Refers to the excessive growth of body which takes place due to higher/hypersecretion of growth hormone by anterior lobe of pituitary since birth. *Compare with* dwarfism.

Gill Arches. Walls which are adjacent to gill-slits; walls bear gills is an aquatic animal such as fish.

Gill Cletts. External branchial apertures through which the gill pouches of pharynx are opening to exterior (outside); slits between two successive gill arches. Also known as gill slits.

Gill Filament. Used for a finger like subdivision of a gill. Various gills may be including rows of gill filaments.

Gill Rackers. The spiny ossifications which are attached to first four gill araches. They serve to filter the large particles form incurrent water.

Gill Rays. Refers to small, slender and cartilagenous structures which are unbrached. They support the walls of gills and are present on epibranchials and creatobranchials (Also see *branchial arches*).

Gills. The respiratory organs for aquatice animals such as fihes. They have gill pouches which are present in the wall of pharynx. There are extra branchial apertures which lead these proches to outside. Further, there are intra branchial apertunes through which these pouches open into pharynx and gill lamellae. The latter are respiratory surfaces and formed by raised mucous membrane of the gill pouch.

Gill Slits. Refers to the series of paired openings which are present in between the wall of pharynx and the body of chordates. They form the diagnostic character of chordates. They form the diagnostic character of chordates. They are also known as gill clefts or pharyngeal clefts and pharyngeal slits (when do not carry gill). They are formed by invagination of outer wall of body and pharynx and evagination of inner wall of pharynx and their subsequent fusion.

Gizzard. Used for describing the muscular part of the digestive tract as in earthworms, cockroches and birds. It is involved in grinding the ingested food materials in the want of a grinding apparatus in the buccal cavity.

Gland. An organ which is able to synthesize a specific chemical substance. This is secreted either through a duct into a tubular organ or onto the surface of the body or directly into the bloodstream.

Gland, Anal. Any gland present in the neighbourhood of the anus or rectum in a number of different animals, e.g., the odoriferous anal glands of snakes.

Glans. Term used for the anterior most sensitive part of penis in mammals, which is highly inervated with blood vessels. It reingorces sexual stimulation.

Glaucoma. Refers to the increased intraocular pressure which is caused due to blockage of canal of Schelmn through which old eye humors (aqueous and vitreous) pass out. It is accompanied by intense pain in the eye-ball, may or may not be followed by vision loss.

Glisson's Capsules. Envelopes of fibrous connective tissue that are enclosing the lobules of the vertebrate liver.

Globin. A protein of the histone group which, when on combination with the iron-containing pigment *haem,* forms the blood pigment called haemoglobin.

Globulins. A group of proteins which an hydrolysis yield the amino-acid *glycine*. Like the albumens they yet get coagulated by heat but unlike them, they are very soluble in salt solution. They occur in many plant and animal cells as well as in blood-plasma.

Glochidium. The larva of a lamellibranch mollusc such as a mussel. It has a small bivalve shell and a tentacle or sucker or attachment to ffsh after it is relesed from the gill-chamber of the parent.

Glomerulus (Glomus). A small cluster of capillaries such as those in the Malpighian bodies of the vertebrate kidney. The term glomus is limited to the more diffuse capillary network in the early stages of development before definite Malpighian bodies are formed.

Glomus Tissue. Refers to the tissue present in carold/aortic body. It consists of islands of two types of cells surrounded by sinusoidal vessels. The cells have been naed type-I and type-II. Type-II cells are gliat cells which are surrounded by type-I cells. Unmyelinated endings of glossopharyngeal nerve fibers are present, at intervals, between types-I and II cells.

Glossopharyngeal. Refers to the ninth cranial nerve which goes to tongue and pharynx. It has mixed fibres.

Glottis. The opening from pharynx which leads into the larynx (voice chamber).

Glucagon. A protein hormone which is released by the α -cells of islets of Langerhans. The low blood sugar or by growth hormone induces the secretion of this hormone which stimulates glycogenolysis i.e., brings about conversion of glycogen to glucose to tackle the problem of hypoglycemia.

Glucocorticoids. Steroids which are synthesised in the adrenal gland cortex having a wide ranging matabolic activities. They include cortisone, cortisol, corticosterone and II-deoxycoticosterone. They are responsible for maintaining the carbohydrate metabolism in the body.

Glucogenic. Refers to the glucose forming process e.g., amino acids that after deamination metabolise like carbohydrates.

Gluconeogenesis. Refers to the synthesis of carboydrates from non-carbohydrate sources like fatty acids and amino acids.

Glucose [D (+)]. The main product of carbohydrate digestion. It is a monosaccharide (hexose) which is used as a principal body food for energy yielding processes. Normal level in bloo l is 120 mg/100 ml of blood but its excess results in hyperglycemia and deficiency results in hypoglycemia.

Glucostasis. Means the regulation of blood sugar in the body so that its normal level is always maintained. It is carried out by a combined effect of two hormones such as insulin and glucagon which are regarded to be responsible for glycogenolysis respectively. The process has been influenced by PSH (pancreas stimulating hormone) from pituitary which is secreted under the control of PSh-RF (pancreas stimulating hormone releasing factor) from hypothalamus.

Glutamic Acid. One of the products which is formed during Kerb's cycle and also formed by transfer of NH_2 group over α - ketoglutaric acid. It is able to depolarise mammalian neurous when kept on their membranes.

Glutinants. Refer to larger nematocysts of *Hydra* etc., having long armed thread. They are involved in capturing the food, anchoring the body with substratum and the locomotion. (Also see *nematocysts*).

Glycerokinase. An enzyme involving in phosphorylation of glycerol and giving rise to the formation of 3-phosphoglyceraldehyde during the metabolism of glycerol (The substance so formed enters the glycolytic pathway. It is an example of intermediary metabolism).

Glycholates. Salts (Na^+ and K^+) of glycholic acid whcih are present in bile juice. They combine with lipids to form micelles water soluble complexes from which lipids could be absorped in the intestine.

They take part in emulsification of fats and absorption in the intestine. They are able to stimulate reesterification of fatty acids and glycerol synthesis from glucose in the mucosal cells of intestine.

Glycholic Acid. Cholic acid which is formed in the liver by metablism of cholesterol and conjugated proteins within the liver to produce sodium and potassium glycholates.

Glycine. An amino acid synthesised in the body provided other nutrient elements alongwith nitrogen and carbon components are available. It is mediator which is responsible for direct inhibition in spinal cord. It occurs in neurons and almost all living cells.

Glycogen. A homopolysaccharide which is formed within the liver from excess of glucose, fructose etc., and stored in the liver, muscles and other tissue as reserve food. The process of its formation is termed as glycogenesis. When there occurs a lowering in body glucose level the substance is converted to form glucose by involving the process of glycogenolysis. Its synthesis has been controlled by hormone insulin and break-down has been controlled by glucagon.

Glycogenesis. Refers to the process by which excess glucose in the body has been converted to glycogen. It takes place within the liver. The process has been regulated by hormone insulin and useful in preventing hyperglycemia in the body. In event of the disturbance in this process the sugar passes into blood and urine causing a pathological condition called diabetes mellitus. The various steps of this process can be summarised as follow:

Glucose ——————→ Glucose-6-phosphate ——————→

phosphorylase Phosphohexo

ATP→ADP isomerase

Glucose-I-phosphate ——————————→ Uridine diphospho-

+Uridine diphospho-

galactose

glucose ————————————————→ Glycogen.

Various intermediate steps.

Glycogenolysis. Refers to the processes of conversion of glycogen to glucose when the body blood sugar falls below normal (as occurs during hunger). The process takes place within the liver and is regulated by hormone glucagon. Both glycogenesis and this process are involved in maintaining glucostasis in the body. The process can be summarised as follows:

Glycogen ——Phosphorylase, ATP→ADP——→ Glucose-1-phosphate ——Phosphoglucomutase——→ Glucose-6-phosphate ——Phosphates in the liver——→ Glucose.

Glycogen Synthetase. An enzyme catalysing the formation of glycogen from glucose.

Glycolipid. Refers to lipid which is conjugated with carbohydrate molecule (galactose in most cases). It is generally found in the nerve tissue.

Glycolysis. Means anaerobic respiration of cabohydrates which is stepwise breakdown of glucose, starch and glycogen etc., to form pyruvic acid (a substance needed to start Kreb's cycle). It occurs within the cytoplasm and outside the mitochondira. The various steps taking part in this process can be put as below:

Glucose ——Phosphorylase, ATP → ADP——→ Glucose-6-phosphate ——Phosphohexo isomerase——→ Fructose 6-phosphate ——Fructokinase, ATP→ADP——→ Fructose-1-6-diphosphate ——Aldolase——→ 3-phosphoglyceraldehyde ——Dehydrogenase——→ 3-phosphoglyceric acid ——Phosphoglyceromutase——→ 2-phosphoglyceric acid ——Enolase——→ Phosphoenol-pyruvic acid ——phosphates, ADP—ATP——→ Pyruvic acid.

Glycosuria. The excreation of excessive amount of glucose in the urine.

Gnasthostomata. Term used for a group of vertebrates that have a pair of jaws. Most of the vertebrates are included in this subdivision e.g., fishes, amphibia, reptiles, birds and mammals. *Opposite* to agnathostomata.

Goblet Cells. Epithelial cells having a flask-shaped cavity which secretes mucus; present in many parts of the alimentary canal, e.g. the duodenum of vertebrates.

Goitre. An abnormal increase in size of thyroid gland which is usually resulted by the dietary lack of the iodine.

Goitrin. An active antithyroid agent. Brassica vegetables (family Cruciferae) especially cabbage and turnip have progoitrin and a substance which converted these compounds into goitrin. It may bring about goitre and is goitrogen. Chemically the substance is L-5-vinyl 1, 2-thioxaxolidone.

Goitrogens. The substance producing goitre/antithyroid acitivity e.g., goitrin found in the cabbage etc.

Golgi Bodies. Minute coiled threads of protein and lipoid material which are found in the cytoplasm of most animals cells and few plant cells. Their function is not known with certainty.

Golgi Cells. Teh cells present in the granular layer of cerabeller cortex. Their dendrites are projecting into molecular layer and are recovering inputs from parallel fibres. Their cell bodies are receiving via collaterals from incombing fibres and Purkinjie cells.

Gonadotropic Hormones (Gonadotropins). Hormones obtained from naterior pituitary lobe that are able to influence the activity of goands. They stimulate gonands for carrying out both reproductive and hormonal functions.

Gonands. The reproductive organs in the individuals. In male they have testes and in females the ovaries. They produce sex cells/gametes by the process of gametogenesis.

Gonangium. Refers to reproductive individuals of hydrozoan colony e.g. *Obelia* colony.

Gonotheca. Used for the firm external covering of gonangium.

Gonozooid. Reproductive zooid in a hydrozoan like *physalia*.

Gorilla Rib. An extra rib which is present in some humans (normally present in gorillas and chimpanzees).

Graded Response. One that tends to increase as a function of the energy applied. A membrane response which is not all-or-none.

Graafian Follicles. Fluid-filled spherical cavities in which the oocytes are developed in the ovary of a mammal (see *Follicle*).

Granulated Endoplasmic Reticulum. See ergastroplam.

Granulocytes. WBCs which have granulated cytoplasm. Three types could be recognised — eosinophils, stained with acidic stains, basophils stained with basic dyes and neutrophils which take neutral dyes. They concern with immunity reactions and defence of body.

Grave's Disease. Thyroid has been diffusely enlarged and hyperplastic and there occurs protrusion of eye ball. It is also termed as exophthalmic goitre.

Gray Crescent. Refers to primary organiser. A portion of gastrula which induces formation of endodern and mesoderm. It becomes dorsal lip of blastopore during processes of gastrulation.

Gray Matter. Nerve tissue which is composed of non-myelinated neurons. *Compare with* white matter.

Green Glands. Excretory organs of Crustacea which are situated in front of the cephalothorax. They have opening at the base of the antennae

Each gland is having an end-sac, a spongy, 'labyrinth', a tubular duct and a dilatable bladder.

Gregarine Motion. A wriggling motion which is caused by contraction of myonemes, as in the gregarine protozoa.

Gregarious. Means living in company, flocks, herds etc., as seen in locust. *Opposite* to solitary.

Grey Nerve Fibres. Nerve fibres of the non-medullated types (i.e.. without a fatty sheath), that connect the sympathetic ganglia with various internal organs.

Grey Ramus. A branch which connects each spinal nerve to the corresponding sympathetic ganglin, having non-medullated which give it its grey colour when contrasted with the white colour of medullated fibres.

GTP (Guanosine Triphosphate). A high energy molecule which is similar to ATP. It is also an energy providing substance fibres which takes part in several energy requiring processes like bond formation during protein synthesis.

Guanine. A nitrogen base combing with deoxyribose sugar and phosphoric acid to form one of the nucleotides which make up RNA and DNA. In addition, the hydrolysis of GTP (guanoisine triphosphate) evolves energy to drive some energetically unfavourable reactions. Also *refers to nucleotides*). Chemically the substance is 2-amino-6-oxypurine.

Gubernaculum. Refers to the fibrous cord-like structure through which the testis in mammals gets attached to the wall of scrotal sacs. The contraction of this cord makes the descent of testis. It is a normal feature in most mammals except during cryptorchism.

Gunylate Cyclase. An enzyme converting guanosine triphosphate (GTP) into c-GMP (cyclic-gunanosine monophosphate).

Gustation. The sense of taste. Also, refers to chemoreception of ions and molecules in solution by specialised epithelial sensory receptors. Tongue and the taste buds are present on its surface which respond to this sense.

Gustatoreceptors. The organs of taste. They have mainly taste buds which are present over the tongue (Also see *the taste buds*).

Gynadromorph. Individual in which part of an animal body may exhibit male and part of body as female. It is not different from hermaphrodites which are having both sets of reproductive organs (male and female) in one single individual.

Gynecomastia. Refers to the development of breast in mal. It is present in mild transient form of 70% of normal boys at puberty. It occurs due to a variety of seeminlgy unrelated conditions which include enuchoidism due to primary testicular disease, hypothyroidism, hyperthyroidism and cirrhosis of liver.

Gyrodactyloid. Refers to a free-swimming larval stage of some ectoparasitic flukes (Monogenea). In *Polystoma*, for example, this larva has five rings of cilia, four eye-spots and a ring of hooks at the hind end. It lives freely for about two days before it enters the gill chamber of a tadpole and finally mature in the cloaca of an adult frog.

H

Habenular Commisure. A transverse nerve tract which runs across the roof of the third ventricle of the vertebrate brain just beneath the pineal organ.

Habitat. Term used for to the environment in which an animal lives. It is of various types like, aquati meaning habitat dominated by water, testes trial meaning habitat of normal land, subterranian meaning habitat inside earth, arboreal meaning habitat on trees.

Habituation. Refers to the progressive loss of the behavioural response probability which takes place with repetition of a stimulus.

Haemal. The blood vascular system.

Haematocrit. Refers to the percentage to blood volume which is occupied by erythrocytes. In the man the normal percentage of haematocrit is usually 40-50%.

Haeme. Iron containing red pigment. Chemically it is a porphyrin derivative. It forms a part of haemoglobin, myoglobin, methanoglobin and cytochrome-c (Also see *haemoglobin*).

Haemianopsia. Means half bindness. Leisions which affect optic chiasma such as pituitary tumours expanding out sella turica give rise to the destruction of fibres from both nasal hemiretinas and cause this pathological condition.

Haemochromatosis. Refers to a disease having large ferritin and haemsiderin deposit. It is characterised by pigmentation of skin, pancreatic damage with diabetes, cirrhosis of liver, high incidence hepatic carcinoma and the gonadial atıophy (non-functioning of gonads).

Haemocoel. Refers to the body cavity which functions as part of circulating system. It is present in insects like cockroach. It is filled with haemolymph.

Haemocyanin. A bluish respiratory pigment which is present in the insects' haemolymph. It transports oxygen to body tissue.

Haemocytoblasts. Refer to primary cells of red bone marrow from which all kinds of blood corpuscles coluld be derived. They are irregularly shaped having large deeply staining nucleus and clear cytoplasm. Haemocytoblasts giving rise to granulocytic RBCs are known as haemorythroblasts, those giving rise to granulocytic WBCs are known as myeloblasts, those giving rise to agranulocytic WBCs are known as nomoblasts and lynphoblasts those giving rise to platelets are known as megakaroblasts.

Haemoerythrin. An invertebrate respiratory pigment which is a protein but lacks haeme.

Haemoglobin. Refers to the pigment in RBCs of vertebrates which carries oxygen in the body. In invertebrates it is dissolved in plasma. It undergoes combination with oxygen to form oxyhaemoglobin and when the blood goes to tissues with this compound, it gets decomposed and the oxygen liberated finds use in the oxidative part of respiration (Kreb's cycle).

Haemolymph. Blood of invertebrates having open circulating system as present in the arthropods like cockroach.

Haemolysis. When RBCs are placed in hypotonic solution the corpuscles get disrupted and haemoglobin is dissolved in solution. It is present in liver and spleen in the mammals and the result is formation of bile pigments which enter the bile juice and a number of amino acids enter amino acid pool, where they find use in various metabolic processes in the body.

Haemophilia. Refers to the abnormal inherited condition in man in which delayed clotting of blood takes place. It is sex-linked disease and usually common among males.

Haemoposis. Refers to the process of RBCs formation in the vertebrates (found in red marrow in the adult mammals and liver and spleen in amphibia etc.). In this process the haemocytoblasts are differentiated into proerythroblasts and these cells transform into early and late normoblasts which are converted to reticulocytes and ultimately produce RBCs.

Haemopoitic Tissue. Refers to the tissue in which RBCs are produced. The term used for red bone marrow in adult mammals while spleen, kidney and liver are in embryonic stages of mammal an lower vertebrates.

Haemorrhage. The term used for bleeding from some internal organ/part of the body which is associated with defective clotting mechanism in the body (See *clotting mechanism*), difective capillary contractibility and due to combined defects.

Haemostasis. The damage in a small vessel which is able to initiate a series of events that give rise to the formation of clot and the result is stopping of bleeding.

Hair. Refers to the slender filamentous skin accessary of mammals and of exposed surfaces of some arthropods.

Hallux. Used for describing the first digit of foot. It is not present but present in some birds and other mammals like primates. It is also oppasable like thumb (Also see *pollex*).

Harem. A group of females that are kept normally for purpose of copulation.

Haversian Canals. Named after an English physcian Havers. Used for descrining the canals present in the bone that allowpassage of blood vessels and nerves through it. They form characteristic structure in the mammalian bone. They are interconnected by canal of Volkmann.

Haversian System. A system in mammalian bone which is formed by lamellae, Haversian canals and Volkmann's canals. The lamellae of bone (osteon) have osteocytes, which surround Haversian canal. The various Haversian canals are interconnected together by the Volkmann's canals.

HCI Gastric. Hydrochloric acid of normallity 0.16 N. which is secreted in the stomach by oxynitc cells. It is able to kill harmful bacteria, makes food acidic so that gastric enzymes pepsin and remin may act upon it.

Head. The specialized anterior part of an animal on which the mouth and the chief sense organs ar situated. It is often enlarged to accommodate the brain or the principal ganglia.

Head Fold (Embryology). Refers to the fold of blastoderm which bends under and is able to limit the anterior end of an amniote embryo separating it from the yolksac. By the formation of this and a corresponding tail fold, the embryo is lifted up from the underlying yolk and gets 'pinched off' from the surrounding tissues. The space enclosed by the head fold forms the beginning of the fore-gut.

Heart. The chief pumping organ whose pulsations are able to drive blood round the circulatory system. Usually it consists of an enlarged muscular part of a blood vessel and has at least two valves to prevent back flow. It may have more than one chamber; that which receives blood being an auricle and that which drives out blood being a ventricle. Fish possess one auricle and one ventricle; amphibians and most reptiles two auricles and one ventricle; birds and mammals two of each. In insects and other arthropods the heart is an enlarged dorsal vessel having a number of ostia or openings through which blood enter from the pericardial sinus.

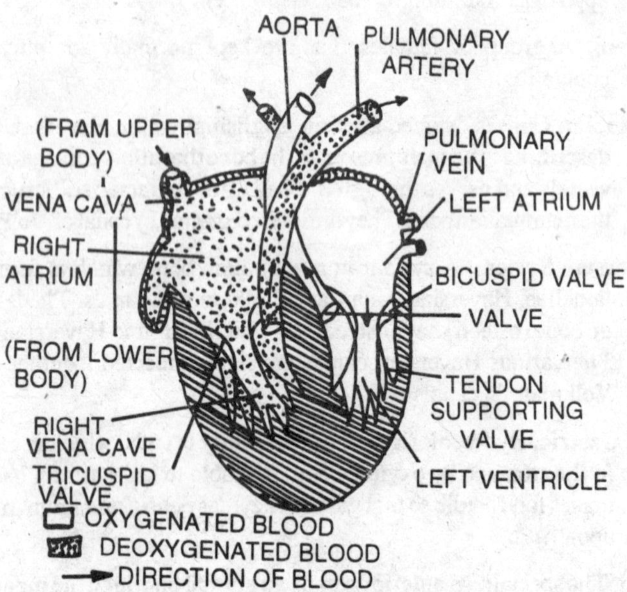

Fig. 1 : Structure of the mammalian heart

Heavy-Meromyosin-(H-Meromyosin). Refers to the segement of myosin of striped muscle beyond 850 A^0. It is possible to split into two segments H-meromyosin—S_1 and H-meromyosin—S_2.

Heavy-Meromyosin—S_1. One of the regions of heavy meromyosin which has maximum ATPase activity. It helps in cross-linking of the cross-bridges during muscle contraction.

Heavy-Meromyosin—S_2. The second portion of heavy-meromyosin which is having comparatively lesser ATPase activity.

Hectodactylus. The male cephalopod mollusc's modified arm which is used mainly for transfer of sperm to female.

Helicotremma. At apical end of cochlea, the area from where basilar partition (basilar membrane) ends to apex.

Heliotropism. Refers to the response in the organism which takes place with respect to Sun-light.

Helix. Used for any structure having spiral shape e.g., shell of snail like *Pila,* DNA helix, cochlea of mamalian ear etc.

Hemiballism. In ballism, the involuntary movements are failing, intense and violent. They appear when subthalamic nuclei get damaged and sudden onset of movements of body taking place on one side due to haemorrhage in contralateral subthalamic nuclei are called hemiballism. It is one of the most dramatic syndromes in clincal medicines.

Hemibranch. In fishes the gill lamella are present on each side of gill arch. The two such lamellae belonging to one gill arch and the interbranchial septum in between them form a complete gill. Only a single gill lamella known as half gill or hemibranch. The first gill in fishes like *Scolidon* is able to support a hemibranch.

Hemichordata. Refers to deuterostomial enterocoelomate animal phylum to which *Balanoglossus* (acorn-worm) belongs.

Hemipenes. Used for the copulatory organs presnet in the reptiles like, lizards and snakes etc. They form one pair and each gets lodged into a sac-like structure. They have a pedicel or a stalk bearing a groove along its length, also bearing a glans-like structure at its tip sometimes. They find use in transferring sperms into the reproductive tract of female.

Hemizygous. Refers to be having particular loci which are present in just a single dose as in case for all sex-linked genes of human male.

Hensen's Node. Anterior end of primitive streak in the embryo of bird (chick embryo).

Hensen's Discs. Lightly staining discs which run transversely across a fibril of striped muscle when the latter is in a relaxed state. Each Hensen's disc divides the 'A-disc' into two parallel halves, but when the muscle contracts these two halves become one.

Heparin. A protein-containing molecule which prevents blood coagulation. In the body it is secreted by Mast cells of areolar tissue (which is present around main blood vessels aortae etc., and among the visceral organs in the septa nd mesenteries). It prevents thrombosis of blood inside the blood vessels during its flow through them. (Also see *areolar tissue*).

Hepatic. A structur like duct, blood vessel etc., referring to liver e.g., hepatic portal vein, hepatic duct hepatic artery/vein etc.

Hepatic Caecum (Mesentieric Caecum. Pyloric Caecum). A branch or diverticulum from the mid-gut, in which digestion may take place.

Hepatic Portal System. The term used for describing a double capillary circulation which is associated with intestine, liver etc. The veins collect blood, digested food etc., from alimentary canal and then open into a single heptic portal vein. This vein opens into the liver and distributes whole of its blood to this part. From liver tissue another set of capillaries collect the blood and pass it to heart via hepatic vein. This system is present in all chordates and maintains the amount of sugars, amino acids, fatty acids etc., in the blood excess of which are retained in the liver for storage.

Hepatomegaly. The hepatic (liver) enlargement which occurs during congestive heart failure.

Herbivorous. An organism that feeds mainly upon plants and vegetable matters e.g., rabbit, elephant etc. In ecosystem these animals are regarded as primary cosumers. (Also see *food chain*)

Herdarian Gland. Refers to modified sebaceous glands which are placed behind the eye-balls.

Herdity. Refers to the tendency of offsprings of an organism to resemble their parents. It deals with the study of inheritance. It includes several laws and principles that deal with transfer of traits (characters) from parents to offsprings as for example Mendel's laws, crossing-over, sex-linked inheritance etc.

Hertability. Refers to that proportion of phenotypic variation in expression of characteristics which is attributed entirely to genetic component.

Hermaphroditism. A term used for describing condition of possessing both reproductive organs in one single individual as in earthworm, *Fasciola, Taenia, Hydra* etc. It is different from gynandromorphism where part of individual's body is having male appearance and part of body as female.

Heterocercal Tail. Used for descrinbing a tail which is asymmetrical internally as well as externally. In such a tail the axis (portion of vertebral column) is bending upwards and the two lobes of tail fin (epicaudal/upper lobe and hypocaudal/lower lobe) are not of equal size. Such a tail is present in sharks like *Scoliodon* and bony fishes like sturgeon.

Heterodont Dentition. The type of dentition in which different types of teeth as in mammals are present. In such dentition four types of teeth-incisor (front teeth), canines (pointed teeth following incisors), premolars and molars (both grinding cheek teeth) are present. Teeth in this dentition may be monophyodont (like molars) or diphyodont (like molars) or diphyodont (like other remaining types). The equation expressing heterodont teeth in a mammal is termed as dental formula.(Also see *dental formula*).

Hetregametic Sex. Used for defining that sex which produces two kinds of gametes with respect to sex chromosome e.g., man is heterogametic because he produces two types of gametes one X-bearing and other Y-bearing.

Heteromorphosis. Regeneration of new part that is different from the part it replaces.

Heteronomous. A condition in which metameres of an animal are not similar. Metameres are specialised in various parts of the body.

Heterophil. A type of white blood corpuscle whose granules have been found to stain differently in accordance to the species of animal from which they come, i.e. they may be basophil, acidophil or meutrophil.

Heteropoda. Pelagic molluscs of the group Taenioglossa in which the shell is reduced or absent and the foot may be modified in various ways as a swimming organ.

Heterosexual Behaviour. Refers to the sexual drive of an individual which occurs towards an individual of the opposite sex. *Opposite* to homosexual behaviour.

Heterosexuality. Refers to carrying out of sexual act by an individual of opposite sex of the same species. It is *opposite* to homosexuality.

Heterothermic. The animals who are not able to maintain constant body temperature but show endothermic regulations during their activity period e.g., protothria and metatheria among mammals.

Heterotroph. An organism which has different ways of nutrition except autotrophic one. Such ways of nutrition are holozoic, saprozoic and parasitic methods of nutrition which are common among most of free living and parasitic animals.

Heterotropic Bones. Bones not belonging to endoskeletion or dermal skletion. They include following bones!

Seasamoid bones. These are bones which are formed by ossification of tendon where tendon moves over the body surface e.g., patella.

Oscordis. They are in the interventricular septum of some ruminants.

Os-penis. It is a bone that gives rise rigidity to penis as in rodents, carnivorous aquatic mammals and lower primates.

Os-rostralis. It is present in snout of pig.

Os-falciforme. It is present in palm of moles and helps in the digging.

Sclerotic. It forms ring of eye-ball of lizards and birds.

Heterozygous. Used for an individual having two members of an allelic pair of genes which are dissimilar. An example is that tall hybirds are having genetic constitution. Tt or tT, they produce two kinds of gametes one with factor-T and other with fuctor-t. It is *opposite* to homozygous.

Hexacanth Embryo. An embryo of some tape-worms like *Taenia solium* which has six hooks and is surrounded by protective layers. It is present in the faecal matter of the primary host (man). It is infections to secondary host (pig) and mode of infection is oral. In the pig it enters the volutary muscles by changing into cysticercus larva.

Hexokinase. An enzyme catalysing phosphorylation of hexose sugars like glucose.

Hexose Monophosphate Shunt. The pathway to pyruvic acid is generally glycolysis of glucse. The pathway for this by involving pentoses is called hexose monophosphate shunt. The overall importance of this shunt acts more as a source of chemicals rather than source of energy.

Hind. The posterior region of alimentary canal; limbs (derived from legs of deer-like animal).

Hind Gut. In arthropods, it refers to the posterior portion of digestive tract. It is lined with cuticle. It is also present in the invertebrate's embryo, rudiment of posterior portion of digestive tract.

Hippocampal Commissure. The median connection which joins two hippocampal lobes in mammals. Generally it is called psalterium of lyra.

Hirschprung's Disease. A disease which affects young children. In this disease there occurs degeneration of cells of Aucrbach's plexus in the region of pelverictal junction, often great length of colon may take part. This affected protion fails to transmit peristaltic waves and consititutes a functional block. The colon above block is greatly distended. There occurs a marked constipation, great abdominal distention and attacks of intestinal obstruction may take place.

His' Bundle. A bundle of muscle and nerve fibres which are able to connect the auricles and ventricles of the heart and transmit impulses from the pacemaker.

Histamine. A derivative of the amino-acid histidine which is produced by damaged cells of vertebrates. When it is released it has the effect dilating capillaries and lowering blood pressure. It is also believed to be one of the active substance present in the stings of bees, wasps etc.

Histogenesis. Origin, development and differentiation of tissues of an organism.

Histotoxic Hypoxia. The condition of oxygen deficiency at tissue level in which amount of oxgen delivered is sufficient but due to action of toxic agent the tissue cells cannot use all the orygen supplied to them.

H-Meromyosin. Refers to the zone of myosin after 8ᴗ0 A^0. Its two regions are H-meromyosin-S_1 and H-meromyosin-S_2. The thin filaments of actin are bound to H-meromyosin-S_1 and the zonse is having ATPase activity.

Hodgkin Cycle. Refers to the regenerative or positive feed back loop which is responsible for uptake of active potential. Depolarisation results in an increase in Na^+ permeabilty allowing an increased influx of Na^+ which further tends of depolarise the membrane.

Holandric Inheritance. Refers to sex-linked inheritance where 'Y' chromosome has genes in question. e.g., hairy ears in males have been regarded as inherited holandrically in man.

Holoblastic Cleavage. Complete division of zygote which gives rise to two blastomeres. It is generally present in eggs having less amount of yolk as in case of frog. It is opposite to meroblastic cleavage.

Holocrine. A method of production of digestive juices or other fluids in which the cells of the glands undergo disintegration to form part of the fluid. Such a mode of secretion takes place in the sebaceous glands of mammals and in the intestines of certain insects.

Holocrine Gland. A gland in which the cells when filled to capacity with products die and get discharged from epithelial wall and new cells are able to replace the discarded ones e.g., oil glands or sebaceous glands or sebaceous glands in mammalian skin.

Holonephros. In primitive vertebrates the entire body segements bear one nephric tubule on each side connected with two longitudinal ducts.

which are oppening into cloaca. Myxine and larve of some limbless amphibia have the kidney which looks like this type of kidney but no living vertebrate is known to have this type of kidney. It is also known as archinephros.

Holophytic. Type of nutrition of which an animal or organism is able to prepare its own food from carbon dioxide, water etc., in the sunlight in case plants. It also takes place in some protozoa like *Euglena*. This is also known as autotrophic nutrition.

Holothuroids. A group of echinoderms to which sea-cucumbers belong.

Holotricha (Holotrichcida). Protozoa having cilia which are uniformly distributed over the whole surface of the body, but without an adoral wreath.

Holotrichous. Having cilia of equal size which are arranged all over the surface of the body, as for example in *Paramecium*.

Holozoic. Feeding completly on organic substaces and unable to do photosynthesis. This happens with all animals and with those plants which lack chlorophyll. Sometimes this is the only means of deciding whether a unicellular organism could be classed as an animal or a plant.

Homeostasis. Refers to that tendency by which the living organisms are able to create and maintain their steady state at the cost of their environment and always have a tendency to remain in optimally tuned condition. The term was introduced by Claude Bernard as 'mileu interior' which implies internal environment.

Homeotherm. Used for an individual that is able to maintain constant body temperature inspite of fluctuations in external environmental temperature. This is carried out by shivering etc., in cold and by sweating etc., in hot. The hypothalamus plays important role in maintaining the constant body temperature. Examples are birds and mammals.

Hominid. Extent or extinct man or man-like type.

Homodont. Means dentition in which all the teeth are alike. It is opposite to heterodont. This is common in amphibia, reptiles etc.

Homocercal Tail. The type of tail which is externally symmetrical but asymmetrical internally. Such a tail is present in most bony fishes (e.g. *Labeo*).

Homologous. Chromosomes which carry genes for same traits; one is paternal being transferred from father and the other being transferred from mother. These chromosomes come to unite or pair during synapsis of meiosis.

Homologous Chromosomes. Chromosomes which, in normal diploid cells, exist in pairs, are usually similar in shape and have identical sets of loci (positions of genes). In each pair, one of maternal origin, the other paternal. They may both have genes for the same or for the same or for opposite allelomorphic qualities (see *Mendelism*).

Homologous Organs. Organs in different species showing similarity of origin and development in the embryo although they may become very much modified and take on a different function in the adult. The small bones in the ear of mammal, for instance, are homologous with certain bones of the jaw in fish and reptiles (cf. *Analogous*).

Homology. Refers to the basic similarity; structural likeness of an organ/part of one animal with comparable unit in other which results from common ancestry. These organs may/may not be carrying out same function e.g., wings of birds have homology with paired fins of fishes although the wings help in flight and the fins help to position the body in the direction of locomotion.

Homosexucal Behaviour. Used for sexual drive of an animal/individual whcih take place towards another individual of same sex (in the same species). Opposite of heterosexual behaviour.

Homosexuality. Refers to sexual act by an adult with a member of same sex within same species. It is *opposite* to heterosexuality.

Homozygous. Used for an individual in which the members of a given pair of genes are alike/similar; e.g., hydrids having genetic constition TT or tt. *Opposite to* heterozygous.

Hoof. A borny casing of the toe which is formed from *keratin*, a hardened form of epidermal cuticle. Hoofs are peculiar characteristic of the ungulated group of mammals.

Hooklet. Hooled structure (generally small) which are able to hold the barbs of a feather of a bird and help in the formation of van in the contour feather. They are not present in down feathers and filplumes etc.

Hook-worm. A common name for nematode parasite of mam *Ancylostoma duodenale*.

Horizontal Cells. A nerve cell whose fibres are able to extend horizontally in outer plexiform layer of retina.

Hormone. A chemical sustance which is produced by an endocrine or ductless gland and able to pass in the blood-stream to all parts of the body and to exercise a controlling effect on various metabolic functions. Examples in the vertebrate body are *thyroxin* from the thyroid gland and *adrenlin* from the adernal glands.

Hormoproteins. Hormones which are also proteins, e.g. insulin, secretin etc.

Horn. Cornified cap-like outgrowth of skin that is present on the head region of animals like cows, sheep, goats, etc. The following types of horns are known:

Prong horn. It is the permanent projection of frontal bone which is covered with horny sheath and shed annually. It is present in *Antelope*.

Antler horn. It refers to the branched horn which is formed by branching outgrowth of frontal bone. It is covered, during growth, by hair skin which is shed annually. It is found in deer family.

Giraffe horn. It is just like antler but the velvet (soft skin covering the horn) which is never shed. Short and unbrached. It is present in giraffe.

True horn. It is formed by the growth of skin of head region and its subsequent conification. It is present in cattle like, cows, sheep, buffaloes etc.

Fibre horn. It is formed of keratinised cells of epidermis and consists of mated, keratin fibres which are bound together but fibres are not true. These are also known as hair-horns and present in rhinoceros.

Horripilation. Refers to the erection of hair that reduce heat transfer across the layer thereby heat losses are decreased.

Host. The term is applicale to an organism that provides food, shelter or other benefits to another organism without any gains.

Host Cells. The epitheliomuscular cells of cnidarians which lodge cnidotlasts.

Humoral. The term used for body fluids such as hormones or other chemical agents.

Humus. The term used for the organic part of the soil.

Hunger. The term used for physiological state of body which is stimulated by fall in glucose level of blood, a rise in free fatty acids' level. These factors may be able to promote appetite and initiate feeding. These stimuli to eating form a quick acting response to need for food and there occurs intake of food.

Huntigtion's Disease. Nervours disorder which is generally mainfested well after puberty. It is characterised by involuntary movements of face and limbs along with a gradual loss of mental faculties.

Hyaline. Means glassy or semitransparent. The term is generally applicable to clear substance of protoplasm, matrix of cartilage etc.

Hyaline Cartilage. A cartilage having clear glassy semitransparent matrix which is present in the hyoid apparatus of frog.

Hyaluronidase. An enzyme produced by sperm which makes help the sperm to break through the egg membrans and in the fertilzation by having them into the ovum.

Hybrid. Individual formed by the union of sperm and egg which is different in one or more genes; heterozygote.

Hybrid Vigour. The term used when a hybrid exhibits more vigours growth, greater yield, increased disease resistance than either parent. The effect occurs due to accumulation of dominant alleles each having additive effects and masking effects of deleterious recessive alleles.

Hydranth. Expanded end of a branch of hydroid colony (e.g., *Obelia*) which is specialised for feeding. Hydra-like zooids having mouth, tentacles, cup-like hydrotheca play nutritive role in the colony.

Hydrated State. Surrounded by water molecules, usually occurs due to polarity.

Hydrocaulus. Main stalk-like stem of a hydroid colony.

Hydrocortisone. A glucocorticoid effecting carbohydrate metabolism, protein metabolism, lipid metabolism, electrolyte and water relieving muscular weakness, correcting eosinophilia, anaemia, lymphocytosis and neutropenia etc., overcoming myocardial weakness, controlling nervous activity, pepsin production, metabolism of bone infection, inflammation and trauma etc.

Hydroid. Similar to *Hydra;* having a sac-like body with a mouth and tentacles at the top and a body-wall made of only two layers of cells (see *Coelenterate* and *Diploblastic*).

Hydroid Generation. The asexual phase in the life of a colonial coelenterate of the class Hydrozoa, consisting of a number of polyps whose body-cavities are linked by a common tubes or *coenosarc*. Alternating with this generation is the *medusoid* or sexual generation consisting of free-swimming jelly-fish which bud off from the parent colony.

Hydrops Foetalis. Due to the mismatching of Rh factor some of the infants born may be having haemolysis of blood. If the haemolysis of blood is severe, the infant may die in uterus or develop anaemia, severe jaundice and edema which is termed as hydrops foetalis.

Hydropyle. A specialized region of the shell of an insect's egg through which water can enter.

Hydrorhiza. The root-like portion of a colonial coelenterate such as *Obelia,* which consists of a much branched tube by which the colony gets attached to the surface of seaweeds or rocks. From this root-like part there arise stems or *hydrocauls* which are bearing numerous sedentary polyps whose digestive cavities are all linked by a common tube.

Hydrospires. Long internal prouches which are radiating out from the centres of certain fossil echinoderms. They opened to the exterior by spiracles present near the mouth and were probably used for respiration.

Hydrotheca. Transparent membrane extending from perisarc and surrounding the main part of hydranth.

Hydroxyapatite. Mineral in bone is mainly in form of a complex salt which resembles and is identical with hydroxyapatite—$Ca_{10}(PO_4)_6(OH)_2$. The salt forms crystals which are measuring 2.0 × 3.7 nm.

Hydroxy. Refers to OH group.

17 α-Hydroxy Pregnenalone. One of the intermediate substances which are formed during metablolism of cholesterol from which various steroids could be synthesised.

17 α-Hydroxy Progesterone. An intermediate which is formed during biosynthesis of steroids.

19-Hydroxy Testesterone. Refers to one of the intermediates which are formed by hydroxy-pregnenalone from which estrdiol, estriol and other metabolites could by synthesised.

Hydroxy Tryptamine. A substance obtained by hydroxylation of essential amino acid tryptophan which forms seratonin by involving decarboxylation.

Hydoid Apparatus. The series of bones and cartilages (see *Hyoid Arch*) to which muscules for moving the tongue and floor of the mouth are attached. The latter are best developed in fish and Amphibia whose breathing has been found to depend upon this movement.

Hyoid Arch. The second visceral skeletal arch of vertebrate, immediately behind the mandibular arch and supporting the floor of the mouth. In cartilaginous fish it is separating the spiracle from the first gill-slit and consists of a *basihyal* cartilage, two *ceratohyols* and two *hyomandibulae*. In bony fish there are also *epihyals* and *hypohyals*. In land animals in which the spiracle is converted into the Eustachian tube there is usually a flat plate at the root of the tongue and a series of small bones which are passing upwards towards the cranium.

Hyostylic Suspension. A jaw suspension present among sharks where upper jaw does not have direct attachment with cranium but is suspended by ligaments with brain box through hyomandibular which is movably articulated to otic region of cranium.

Hypapophysis. Hypocoentrum in some vertebrae of reptiles, birds and mammals linger on the ventral side of centrum as a bony nodule which is termed as hypapophysis.

Hyperaldosteronism. Refers to the aldosterone secretion enhancement in the individual. It is referred to primary when some aldosterone secreting tumours of adrenals' cortex produce excess of mineralocorticoids. It is referred to secondary when it occurs due to nephrosis, congestive heart failure etc. (due to elevated levels of renin and angiotensin-II and may cause increased aldosterone secretion).

Hyperalgesia-A. Refers to the pathological condition in which the sensitivity of the pain receptors is changed. There are two types of alterations—primary when pain in the flare (see *flare*) is felt and secondary when another aberration of sensation is marked following an injury.

Hyperdactyly. A condition when an animal has large number of digits. It is present in some aquatic dinosaurs.

Hyperglycemia. Refers to the increase in blood sugar level from normal. It occurs due to disturbed insulin production in the body. There occurs reduced ability to metabolise or convert them into fats and loss of proteins which are metabolised for energy in place of glucose. The disease can be cured by administration of insulin. Refers to an induce glycogenesis so that there occurs control of excess of blood sugar.

Hyperkalemia. Refers to an increase in K^+ ions of the blood. It is considered to be very dangerous and potentially lethal condition due to its effect on heart. It may bring about paralysis of atria and prolongation of QRS complexes. Altered repolarisation gives rise to the appearance of tall peaked 'T' waves.

Hyperopia. Means near sightedness, a condition where eyeball is relatively long, resulting in the eye's point of focus lying in front of eye instead of retina.

Hyperparasitism. A condition in which the parasites get parastised by other parasites. It may be secondary, tertiary etc.

Hyperphagia. Leisions in staity centre in hypothalamus which causes ingestion of very large amount of food material thereby leading to this condition which is opposite of hypophagia.

Hyperphalangy. Means a condition when an animal is having a large number of phalanges as seen in aquatic dinosaurs.

Hyperpnea. Increesed lung ventilation/hyperventilation in an animal.

Hypertensin. A substance which takes part in regulating blood pressure in the kidneys. It is formed by the breaking down of a substance *hypotensinogen* under the influence of a hormone, *renin*.

Hypertonic. A comparative term which is used to denote a solution of greater osmotic pressure than that of another, the latter being *hypotonic*. If two solutions of unequal concentration are separated by semipermeable membrane (e.g. a cell-well), water will flow, by osmosis, from the hypotonic to the hypertonic solution.

Hypertrichosis. Used for the presence of large hair on the external ear or pinna. It is Y-linked character which is inherited only in males of human beings (man).

Hypertrophy. Refers to the abnormal increase in size of an organ or the other part of the body.

Hypervitaminosis. A disease which is caused by taking an excessive dose of vitamin. For example hypervitaminosis of vitamin K brings about gastrogenital disturbances and anaemia. Hypervitaminosis of vitamin D brings about weight loss and calcification in many soft tissues etc.

Hypnotoxin. A poison which is found in the nematocysts of cnidarians like *Hydra*. The organism uses it to paralyse the prey during ingestion period.

Hyoblast. The germinal layer which spreads down under trophoblast to form yolk sac.

Hypocentrum. In embolomerous vertebra the anterior centrum is derived from basiventrals. It is called hypocentrum.

Hypocercal Fin. The caudal fin in which notochord or vertebral column is extending into ventral portion of tail. The ventral portion is extending further than the dorsal portion of tail.

Hyoconvertinemia. Due to congenital reason there occurs deficiency of blood clotting factor-VII (proconvertin factor) in some individuals. Due to this there is less activation of factor-X (which is responsible for conversion of prothrombin to thrombin) and blood has low capacity for coagulation.

Hypodactyly. A condition in animals like birds in which the animal is having lesser number of digits (three toes in most of the birds and only two in ostrich).

Hypodermis. A layer which is situated below dermis of skin. It is also called subcutaneous tissue. It consists of connective tissue fibres and cells. In mammals it is made up of adipose tissue and forms a leyer called panniculus adiposus. In aquatic mammals like whales it is made up of a very thick layer of fat called blubber.

Hypolossal Nerve. The twelfth cranial nerve of a reptile, bird or mammal; a motor nerve which is innervating the floor of the mouth and the tongue. In Amphibia it is the first spinal nerve which comes out immediately behind the skull; in fish it is formed from several spinal nerves which are joined together.

Hypoglycemia. Refers to a decrease in blood glucose level of body. The brain is unable to function and there often occurs 'hypoglycemic fit'. Under this condition alpha-cells of islets of Langerhans secrete glucagon which stimulates glycogenolysis that reduces conversion of glycogen to glucose so that normal blood glucose level is restored. Patients under hypoglycemic fit need immediate intravenous injection of glucose.

Hypogonadism. Refers to the low secretion of gonadotropins due to hypothalamic disease which brings about under-functioning of goands resulting in improper body development, delayed puberty, improper development of secondary sexual character, improper development of reproductive organs and disturbed reproductive cycles and disorders.

Hypomere. When mesodermal pouch during embryonic development is differentiated into two parts the ventral part of this is known as hypomere. It forms coelom in the embryo (dorsal part of the pouch forms myotome and dermis).

Hypophagia. Refers to a condition in which animal is able to reduce amount of ingested food material. It occurs due to leisions in the area nearby the satiety centre in hypothalamus.

Hypopharyngeal Groove. A narro, ciliated and glandular groove present along the ventral side of pharynx in protochordates such as *Herdmania.* Also known as endostyle.

Hypophysial Sac. A cavity present beneath the brain of some vertebrates. It is representing a duct which orginally led from the hypophysis of the pituitary body into the mouth. In higher vertebrates it reduced to a cleft between the anterior and intermediate lobes of the pituitary body.

Hypophysis. Refers to that part of the pituitary body which is formed from an upgrowth of the ectoderm in the region of the mouth. The whole pituitary body is a combination of this with the *infundibulum* or down-growth from the third ventricle of the brain. In the fully developed organ the hypophysis gives rise to the formation of anterior lobe and the infundibulum to the posterior lobe.

Hypoyhle. Posterior opening of claspers attaching with the cloaca in sharks. This opening acts as an exit through which sperms leave the claspers and pass into the cloaca of female.

Hyporthachis. The aftershaft of a feather.

Hypothalamico Hypophysial Relay System. A system in which portal veins linking the capillaries of hypothalamic medium eminance with those of adenohypophysis. These veins are able to transport hypothalamic neurosecretions directly to adenohypophysis.

Hypostome (Hypostoma). Any structure present below or immediately behind the mouth, as for instance the oral cone of a hydra.

Hypostomides. Small flying-fish having large horizontal pectoral fins and with the body entirely covered by bony plates. They inhabit sanday shallows of the Indian and Pacific Oceans.

Hypotension. Systolic blood pressure under normal condition is usually 145-150 mm of mercury. A fall in systolic blood pressure below normal range is known as hypotension. Blood pressure below 100 mm of mercury means feeble circulation and below 80 mm of mercury means impending death. The low blood pressure results in due to shocks, severe haemorrhage, burning, hypothyroidism and disturbances of adrenal gland.

Hypothalamus. The thickened floor of the third ventricle in the vertebrate brain. In mammals it has the nerve-centres which control the temprature of the body.

Hypothesis. A tentative solution of a problem which has not been proved to the either true or false.

Hypotremata. The group of cartilagenous fishes having ventral gill slits. They are commonly known as rays. Examples are electric ray, sting ray etc.

Hypoxia. Means reduced oxygen level in the body of an individual.

Hypoxica. Refers to the problem of breathing in normal individuals at high altitudes and a complication of pneumonia and number of common diseases such as paralysis of respiratory muscles, skeletal deformities decreased elastic recoiling etc.

Hypural. Haemal spines which are present in tail fin region of tail vertebrae of some animals like snakes.

H-Zone. Refers to the light region in the centre of 'A' band of stripped muscule fibre. *Also* known as Hensen's disc. Its middle portion is having a transverse line called M-line.

I

I-Band. Refers to the region between A-band and Z-disc of muscle sarcomers in a striped muscle which appears light when viewed, microscopically. It possesses actin filaments (thin filaments) without overlap from myosin filaments.

Ichthyopterygium. Used for decribing lobed fins of extinct fishes like *Sauropteurs* and *Eusthenopteron* which were paddle-like. They are the type of structures from which the tetrapod limbs are likely evolved.

ICSH (Interstitial Cell Stimulating Hornone). Hormone secreted by anterior pituitary lobe. This hormone stimulates ovulation and lutenization of ovarian follicles in female and testesterone secretion in the male individuals.

Icterus. Refers to a pathological condition in which skin, scleres, mucous membranes etc., become yellow due to accumulation of free or conjugated bile pigment bilirubin. The condition could be only detected when total plasma bilirubin becomes greater than 2 mg/dl.

Idac. Suffix representing family (term related with taxonomy) like Ascaridae (to which the round worm *Ascaris* belongs).

Identical Twins. Refers to two individuals that are developed from a single fertilized egg and therefore having identical set of genes. Also termed as 'one egg twins' or 'Monozygotic twins'.

Idiopathic Hypertrophy. Used for hyperfunctioning of any part of body without involving any distinct pathological reason.

Idioventricular Rhythm. Used for describing the beating of ventricles at a slow rate independent of auricles when conduction from auricles to ventricles gets interrupted completely or third degree of heart block is said to be present.

I-Disc. Formerly called the j-disc, a light coloured band which is running transversely across the fibres of striated muscle. A region which is occupied by filaments of the protein *actin* between two successive *myosin* filaments.

IDL (Intermediate Density Lipoproteins). Used for describing the substances which are formed during fat metabolism in the liver by conversion of very low density lipoproteins by loss by some triglycerides and protein in the complexes.

IJP (Inhibitory Junction Patentials). In tissues which are inhibited by adrenergic stimuli (adrenaline) the hyperpolarizing inhibitory junction potentials are produced by stimulation of adrenergic nerve.

Ila (Insulin Lile Activity). In plasma, insulin-like activity could be bioassayed by doing comparisons of the effects of plasma samples with those of known amount of insulin on a variety of physiological parameters. The ILA values in plasma have reported to be generally higher than the values determined by assays, and a significant ILA perisists even after pancrease has been removed. Normal concentration of insulin measured by immunoassay in peripheral venous plasma of fasting normal humans is 0-70 units/ml. The amount of insulin secreted per day in a normal man is usually to be 40 units.

Ileocolic Ring. A small rounded thickening which is present at the posterior end of mid gut in protochordates like *Amphioxus.*

Ileocolic Valve. A valve which is present between ileum an the colon in alimentary canal of various vertebrates.

Ileum. The term used for describing the hind-part of the small intestine of a vertebrate. In the mammaliam alimentary canal it comes after the jejunum or after the duodenum and ends at the *sacculus rotundus* where it meets the colon and the caecum. In some vertebrates, however, it is less distinct and the name may be used vaguely for the hind part of sometimes for the whole of the small intestine; it is also used to denote any analogous part in some invertebrates.

Ileus. Means decrease in intestinal motality as happened generally after abdominal operations due to traumatised condition of intestine or irritation of peritonial lining which brings about reflex inhibition due to increased discharge of adrenergic fibres in splanchnic nerves which reduces peristaltic activity and the contents of intestine do not propel into colon it becomes irregularly distended by pockets of gas and fluid.

Iliac Arteries. Blood-vessels that are leading from the dorsal aorta to the hindlimbs of a tetrapod vertebrate.

Iliac Process. A process corresponding in position to the ilium but not forming a separte skeletal unit, as for instance on the pelvic girdle of a fish.

Iliac Veins. Refer to the blood-vessel which return blood from the hind-limbs to the posterior vena cava in a tetrapod vertebrate.

Ilium. Refers to the anterior dorsal bone on each side of the pelvic girdle in tetrapod vertebrate which is fused to the ischium and the pubis and also to the transverse processes of one or more sacral vertebrae. This fusion of bones gives rise to strong support for the body in the region of the hind legs.

Imaginal Discs (Imaginal Buds). Refer to the clusters of cells which undergo rapid division to form the rudiments of future organs during the metamorphosis of an insect.

Imago. The adult form of an insect after metamorphosis.

Imino Acids. Used for substances like proline and 4-hydroxyproline which do not have amino group (NH_2) but have imino group (NH).

Immunoglobulins. Refer to the proteinous substances which are produced by lymphocyte-plasma cell system and their five types can be recognised-Ig-G (12, 100 μg/ml) possesses localised protection in external secretions like tears etc., Ig-M (930 μg/ml) plays role of complement of fixation, Ig-D (23 μg/ml) carries out antigen recognition by β-cells and Ig-E (0.5 μg/ml) liberates histamine from basophils and mast cells. All are related to immunity reactions taking place in the body on individual.

Immuno Sympathectomy. Refers to the destruction of sympathetic ganalia which is brought about by antiserum of nerve growth factor.

Implantation. Means the burrowing of blastodisc (blastocyst) into the endometrium of uterus as happening in eutherian mammals. It occurs due to erosion of endometrium by syncytiotrophoblast of embryo and takes place on the dorsal side of wall of uterus. A placenta is then developed and tropphoblast is remaining associated with it.

Impulse. A physiological disturbance created by the stimulus tramsmitted normally along the axon to its termination.

Inborn Metabloic Disorders. The term used for describing the heritable diseases because of some malfunctioning in a gene coding for some specific enzyme which is involved in a particular step during metabolic pathway thereby resulting physiological disorder. These are more than 200 such deseases e.g., alkaptoneuria in which defective functioning of homogenistic acid oxidese brings about the blockage of conversion of homogenistic acid into melylacetoacetate due to which homogenistic acid passes into urine and urine of such person becomes black in contact with air.

Inbreeding. Means the crossing of closely related individuals e.g., first cousins or brothers and sisters.

Incisors. The term used for decribing teeth in the frontal part of mouth cavity in mammals. They are single rooted having flat crown and possess sharp edge. They are used for cutting food material. Their number in a generalised mammals is $3/3 \times 2$ but most mammals are having $2/2 \times 2$. (See *dental formula*).

Incomplete Dominance. Refers to the failure of one allelomorphic character to dominate another in a heterozygous (hybrid) individual, e.g. in certain cases when a black animal is crossed with a white one the resultant offspring may be grey, blue-grey or variegated. A completely black individual will be only produced by the presence of two genes for blackness.

Incubation Period. (1) Implies the time between the laying of an egg an the hatching of the young bird, reptile etc., if kept at the necessary temperature. (2) Also refers to the interval between an infection and the onset of disease.

Independent Event. Mean two ro more alternatives whose separate occurrences are not influencing one another in any way.

Inductor. Used for describing an organizer or other tissue which is able to influence production of a specific morphogenetic efffect in one tissue of developing embryo.

Inflammation. Refers to the condition in which normally blood vessels are dilated. Leucocyte conglomeration and edema become evident.

Normally it is a response to injured tissue or tissues.

Infra. Suffix which means below-like infraorbital infraintestinal etc.

Infranodal Block. Refers to a disease in conducting system of heart below the node.

Infundibular Organ. Refers to a depression in the cerebral vesicle of *Amphiroxus*. It is lined with ciliated cells. It is able to detect changes in the pressure of fluids within the neural tube.

Infundibulum. Refers to that part of the pituitary body which is formed from a down-growth of the thalamencephalon in the vertebrate brain (see also *Hypophysis*).

Ingest. The term used for taking a substance from outside specially food intake in the alimentary canal by an individual having holozoic nutrition. It is carried out with mouth.

Inguinal Canal. Refers to the communication existing between space of scrotal sac and abdominal coeloms in early stage of mammals. When the decent of testes from abdomen to scrotum takes place the canal is shut up by partition which is called perinium. Any rupture occurring in this partition makes intestine to descend into scrotal sacs causing one of the hernias.

Inguinal Glands. Scent glands present between anus and penis or vulva of mammal.

Inguinal Region. Refers to the ventral part of body between lower abdomen and anus in a mammal.

Inheritance. Defined as the sum of all characteristics which are being transmitted by germ cells from parent to offspring.

Inhibin. A substance which is secreted by some elements in seminiferous tubules. It is able to control FSH secretion independently by feed back.

Inhibitory. In neurophysiology the term is used for decribing a reduction in probability of generating an action potenital.

Inner Segment. Means the portion of vertebrate photoreceptor cell that has the cell organelles and synaptic contacts.

Innominate. Bifurcation of ventral aorta in fihses; each half of pelvic girdle and the union of common carotids and subclvian in mammals.

Inosine Triphosphate. One of the phosphorylated compounds which have pyrimidine or purine other than adenine. It is having inosine.

Insemination. Any device by which sperm cells are introduced into the body of a female prior to fertilization of the ovum.

Insertion. Means the placing of attachment of a muscle to a movable part in contrast to the origin which forms the point of attachment to a relatively immovable part.

Insertion Mutation. Genetic mutation in which an extra amino acid has been coded for because an additional base triplet has been included in the chromosome.

Insessorial. Feet or claws which are specially adapted for perching.

Inspiration. Means the taking of air into to lungs of other breathing organs.

Instar. Refers to a stage in the development of an insect between two moults. Generally there are a number of larval instars, varying from three to about thirty, followed by the pupa and then by the adult or imago. The pupal stage may not be present.

Instinct. An innate disposition which helps an animal to perceive an object or situation and to respond to it by a specific pattern of behaviour. The main difference between instinct and intelligence is that in the former the behaviour pattern is relatively immutable whereas the latter makes an animal to 'learn', i.e., to modify its actions with every changing situation.

Insulin. The term used for a hormone produced by the *islets of Langerhans* that are small glands in the pancreas of vertebrates. It has the effect of accelerating the change from glucose to glycogen in the liver. Deficiency of this hormone makes excess of blood-sugar or *diabetes mellius*. Chemically insulin is protein which is built up from chains of amino-acids linked by sulphur atoms.

Insulinoma. An insulin secreting tumour which is responsible for chronic elevation of insulin level. It is bale to affect glucotasis in the body.

Integration (Neural). Synthesis of an output which is based on the sum inputs to a neural networm.

Integument. The term used for outer covering specially the skin of a vertebrate alongwith its derivatives. It is made up of epidermis, derived from ecotoderm, and dermis, mesodermal in origin. Besides it is also including various structures such as hair, feather, scales, plates etc., present on the skin.

Interacuricular Septum. A partition which is present between two auricles in the heart.

Intercalated Discs. Cardiac muscles' fibres are able to be divided into shorter cylindrical units by transverse partitions which are called intercalated discs. Electron microscopy reveals that there occurs no continuity of protoplasm in this region but they are formed by interdigitation of cell proceses which are taking place from either sides. They are actually relay stations which transmit wave of contraction from one fibre unit to the other.

Intercartilagenous Bone. Bone formed from a cartilage which was originally formed of mesenchyme.

Intercellular. Used for describing some structure, substance present in between cells.

Intercolumnar Tubercle. Refers to one of the area present outside the brain whcih gets stained when acidic dye is injected into animal. There are fenestrated capillaries and many relatively large quantities of hypothalamic hormone. It functions as chemoreceptor.

Intercostal Muscle. Refer to the muscle that makes the ribs to operate in mammals and also enables lowering of sternum during inspiration and expiration.

Intermediary Metabolism. The interrelationship among carbohydrate, protein and fat metabolisms. This provides required metabolite in the body immediately irrespective of its dietary intake by the individual.

Intermedin. A hormone which is secreted by intermediate lobe of pituitary in some vertebrates like fishes and amphibia. Its main activity is to control skin pigmentation by carrying out the control of the chromatophores.

Internal Secretion. Used for a hormone, neurosecretion etc. Which takes place from some ductless gland or tissue.

Interorbital Septum. The bony septum present in between two orbits., It is found only in birds and mammals. Such a skull having it is called tropibasic skull in contrast to platybasic skull in which it is not present.

Interoeceptors. Receptors/sense organs found in the body which are able to detect temprarature, pain, hunger, thirst, fatigue and muscular position. It is *opposite* to exteroreceptors.

Interphase. A resting phase sometimes taking place between the first and second meiotic divisions of cell. Normally each cell during meiosis gives two haploid cells and these immediately undergo division again giving four cells in all. If the second process gets delayed, however, an interphase may take place during which the chromosomes disappear.

Interphase Cell. Cell that is not able to undergo mitosis; a metabolising cell; stage betweeen two mitotic divisions.

Inter-Renal Bodies. Endocrine organs present between the kidneys in a fish. They are homologous with the adrenal cortex in mammals (see *Suprarental* and *Adrenal*).

Intersex. An individual which is intermediate between male and female, either due to an abnormality of the sex chromosomes or of the sex hormones or both. They are usually sterile (see also *Gynandromorph*).

Interspinous Bones. An alternative name for the *somactids* or basal skeletal elements of the median fins of a bony fish.

Interstitial Cell. Cells which occupy the interstices between other cells, as for instance those in the testis or ovary which secrete secondary sex hormones; those in the body of a Hydra or similar animal which develop into nematoblasts or other types of cell etc.

Interventrals. During development of vertebral column, three ventral arcualia developing from less dense caudal part sclerotome and from posterior part of vertebra.

Interventricular Septum. Refers to the partition present between two ventricles in the heart. Generally it is an extension of interauricular septum into the ventricle and its later on fusion with the apex of ventricle.

Intervertebral Discs. Pads of fibro-cartilage present between the centres of adjacent vertebrae in many animals. In the centre of each disc is a soft area, the *nucleus pulposus*, which represents the vestige of the embryonic notochord.

Intestines. The parts of the alimentary canal following the stomach. In mammals these are the duodenum, jejunum and ileum which comprise the small intestine, and the colon, caecum and rectum which comprise the large intestine.

Intracellular Digestion. Refers to the digestion of small particles of food taking place in vacuoles within a cell after having been engulfed by means of pseudopodia. Such digestion occurs not only in protozoa such as *Amoeba*, but also in certain amoeboid cells in the endoderm lining the digestive cavity in coellenterates and Platyhelminthes.

Intrafusal Muscle Fibres. Muscle fibres in the form of a small spindle-shaped bundle which are surrounded by a thick sheath of connective tissue. They are generally present in striped muscle close to the insertion of tendon. These fibres are often shorter and narrower having distinct striations than those of normal stripped muscle.

Intrinsic Factor. Any factor influencing the functioning of a system and having its origin within the same system. For exmaple gastric mucosa secretes an intrinsic factor which is essential for absorption of vitamin B_{12} from small intestine.

Intromittent Organ. Means penis in the male mammal. It is a copulatory organ which is made of three vertical columns of erectile tissue. On the ventral side of it, there is corpus spongiosum that surrounds urethra and above which paired corpora cavernosa are present which are attached proximally to ischia. The anterior end of corpus spongiosum is able to form glans, covered by a loose fold of skin which is called prepuce.

Introvert. A part of an animal which can either be projected outwards (everted) to tucked in like the finger of a glove (inverted), e.g., the mouthparts of certain annelids and molluscs.

Intussusception. Growth of living organisms which takes place by adding new materials within the protoplasm.

Inulin. An indigestible starch which finds use in studies of kidney function because it is freely filtered and not actively transported.

Invaginate. Means unfold. The folding of a layer of cells into a cavity as that of blastula to form a gastrula.

Invagination. Refers to an in-pushing of the outer layer of an organism. At the mouth and anus, for instance, the epidermal layer is often folded in to form a covering for part of the oesophagus and the rectum. These parts are called the *stomodaeum* and the *proctodaeum* respectively. A similar process occurs in the early development of many embryos when a blastula (hollow ball of cells) is converted into a gastrula (sac-like structure) by invagination of one side to form the endoderm.

Invagination Canal. Means endolymphatic duct which arises from sacculus and opens externally above the head.

Inverse Stretch Reflex. Upto a certain point the harder a muscle is stretched the stronger will be the reflex contraction. However, if tension becomes sufficiently great the contraction ceases suddenly and muscle relaxes. This stretch is known as inverse stretch reflex. It is also known as autogenic inhibition. The receptor for the reflex is Golgi tendon organ.

Inversion (Genetics). A from of mutation in which part of a chromosome gets detached from the rest and then joins up again the opposite way round.

Invertase. An enzyme which is acting upon sucrose and hydrolysing it to glucose and fructose.

Invertebrate. An animal that is lacking a backbone or vertebral column e.g., an annelid.

Investing Bones. The bones that are developed in connective tissue also know as membrane bone e.g., Frontoparietal, parasphenoid, nasals, vomars, septomaxillary, palatine, pterygoid, maxillae, premaxillae, qudratojugal, angulosplenial, dentary, mentomeckalian in the skull of frog.

In 'Vitro'. Means a glass; in an artificial environment it means outside the body. *Opposite to* 'in vivo'.

In 'Vivo'. Means within the living organism or tissue.

Involuntary Muscle (Unstripped Muscle). Muscular tissue which consists of narrow spindle-shaped cells each having a centrally placed nucleus and with no transverse striations. Such muscles have the power of slow contraction which is carried out automatically and unconsciously; movements of the alimentary canal are an example.

Involution. Means the process of rolling or turning of cells over a rim.

Iodine. Essential elements needed by body. It is involved in biosynthesis of throxine. Its deficiency in diet is responsible for goitre.

Iodoacetic Acid. An agent which is able to poison glycolysis by inhibiting glyceraldehyde phosphate dehydrogenase.

Iodopsin. A violet-coloured photochemical substance which is present in the retinal cones of some animals. It is not so sensitive as *rhodopsin* but possibly similar in action.

Iodotyrosine Dehydrogenase. A microsomal enzyme which is able to deiodinate the iodianted tyrosines. The iodine liberated in this way is reutilised.

Ionophore. Molecules that are able to carry certain ions or molecules across the membranes.

Ipsilateral. Pertaining to the same side of an organism.

I.Q. (Intelligence Quotient). Defined as the ratio of mental age and chronological age of a persion multiplied by 100. It is useful in deciding the intelligence of persons. The I.Q. of some type of human beings are summarised as follows:

140—or more	—Genius.
120—139	—Most superior.
110—119	—Superior.
90—109	—Average.
80—89	—Ordinary.
70—79	—Dull.
50—69	—Moron.

25—49 —Imbecile.
0—24 —Idiot.

Iridiocytes. Also called guanophores and have crystals of guanine which refract light. They are type of chromatophores.

Iris. A pigmented disc which is present between the cornea and the lens in the eyes of vertebrates and of certain molluscs such as the squid and cuttlefish. A certain hole, the pupil, permits light to pass through to the retain. Radial and circular muscles in the iris are able to adjust the size of the opening to be varied so as to regulate the amount of light entering the eye.

Iris Cells (of insects, Crustacea etc). Pigmented cells that surround each ommatidum or optical unit of a compound eye. In bright light they spread round each crystalline cone, isolating it from its neighbours, so that vision becomes *mosaic*; in dim light they may retract so that the light entering each ommatidium merges with that of its neighbours causing continuous but rather ill-defined vision.

Irridiscence. Re-emission of absorbed energy by an organism/substance in the form of radiation in the ware length of visible lights.

Irritable Foci. A portion of myocardium is able to discharge independently and such areas are known as irritable foci. These are common especially after an injury such as myocardial infraction.

Irritability. Sensitivity; the ability of an animal or plant to respond to any change in its environment, e.g., to light or to chemical ro tactile stimuli. In unicellular organisms the whole cell will respond but in higher animals there usually occurs stimulation of specialized cell or *receptors* followed by transmission of the impulse through a nervours system and reaction by *effectors* (muscles etc).

Ischemia. Used for the absence of blood flow to an organ or tissue.

Ischemic Hypoxia. A condition in which blood flow to a tissue has become to much low that adequate oxygen is not supplied to it despite a normal p_{02} (partial pressure of oxygen) and haemoglobin concentration.

Ischiatic Symppysis. A fusion of the two ischia, as for example in the pelvic girdles of some birds.

Ischium. One of the component bone in the pelvic girdle which often forms ischial symphysis. It is separated from pubis by an ischiopubic formen.

Islets of Langerhans. The microscopic endorine structure which are disposed throughout the pancreas. They have three types of cells—α–cells, β-cells, and δ-cells. The α-cells secrete glucagon, β-cells insulin and the δ-cells somatostatin and may also have gastrin.

Isocitrate Dehydrogenase. An enzyme that is able to convert isocitric acid to oxalosuccine acid during Kreb's cycle by dehydrogenation.

Isoenzymes. Different molecular forms of proteins which have enzymatic activity.

Isogammetes (Homogametes). Gametes in which there occurs no obvious distinction between male and female.

Isogamy. (1) The ability to form isogametes.

(2) The union of two isogametes.

Isogenic. (1) Animals of the same sex from the same litter.

(2) Embryonic tissues having a similar origin.

Isolation. Refers to the segragation of populations into smaller units so that new species may be evolved. It may happen due to geographical by physical separation in distance or ecological in different types of environments in same region or seasonal where two populations are breeding at different times or physiological when there occurs incapablity in mating and fertilization of gametes so that the animals of two groups will not be able to mate with each other.

Isolicithal Egg. Eggs having yolk distributed evenly throughout the cytoplasm. Two types of such eggs are known:

Allecithal. These eggs are having either no yolk or extremely small amount of it e.g., eutherian eggs (viviparous mammals).

Meiolecithal. These eggs have some yolk e.g., eggs of Amphioxus.

Isomerase. Enzyme which is able to catalyse conversion of one isomer to the other.

Isometric Contraction. The muscle contraction in which length of contracting muscle is remaining same/unchanged e.g., pressing hand against wall or table top.

Isorenin. The renin like the enzymes which are extracted from uterus, placenta, foetal membranes, aminiotic fluid, pituitary gland, blood vessels and brain. These enzymes do not have renal origin hence called isorenins instead of renin.

Isorhizas. Type of nematocysts of *Hydra* that are used for adhering the body with substratum or for offence. There are two types—holotirchous isorhizas which are having oval body and long thread with rows of spines and atrichous isorhizas which are having oval body and long thread with rows of spines and atrichous isorhizas which are little small but the thread is having no spines.

Isotonic Concentration. In terms of living cell, it refers to the condition in which the concentration of water molecules ramains same inside and outside the cell. As a result of this water molecules pass in and out in equal numbers thus resulting in no change in cell size/shape.

Isotonic Contraction. Refers to the contraction in which the tone of muscle is not changed e.g., holding of weight by hand.

Itch. A sensation in the body which is produced by repetitive low frequency of 'C' fibres (fibres in dorsal root respond to pain reflex response). The itch spots are present in the region of body where there exist many nacked endings of unmyelinated fibres. It is found in skin, eyes and certain mucous membranes only and not in deep tissue or viscera.

Iter. The full name in its Latin form is *Iter a tertio ad quartum ventriculum*: a passage which connects the third and the fourth ventricles of the vertebrate brain. It is also known as the *aqueduct of Sylvius.*

Itanidae. Gall-midges: an alternative name for the *Cecidomyidae (q.v).*

J

Jacanidae. Lotus-birds; tropical and sub-tropical birds that are characterized by a light body and enormously elongated unwebbed feet which enable them to walk easily on the floating leaves of water lilies.

Jacobson's Anastomosis. A network of nerves that connect the seventh cranial (facial) nerve with the ninth (glosspharyngeal) in many vertebrates.

Jacobson's Organs. A pair of pouches communicating with the nose and with the mouth in many vertebrates; their function is probably connected with smelling of food in the mouth. They are very highly developed in snakes.

Jaundice. Yellowishness of shin, scleres, mucous membranes due to excess production of bilirubin but decreased uptake of bilirubin, disturbed intracullular protein binding or conjugation intrahepatic or extrahepatic bile duct obstruction etc. Also known as *icterus*.

Jaws. Used for describing the structures which are normally used for holding or chewing the food materials. They occur in gnothastomates like fishes, amphibia, reptiles, birds and mammals.

Jaw Suspension. Splanchnocranium in ganthostomates (vertebrates with jaws) suspends the lower jaw with cranium and forms jaw suspension. Various types of jaw suspension are as follows :

Amphistylic. Upper jaw gets attached directly to otic region and hyomandibular gets movably attached to otic region and is slinging jaw with crauium. It is present in primitive sharks.

Hyostylic. Upper jaw does not have direct attachment with cranium but is suspended by means of ligaments to it through hyomandibular which gets movably attached to otic region. It is present is sharks.

Amphidiastylic. Upper jaw gets attached (by ligaments) with cranium and the hyoid arch is not involved in the action of jaws. It is pursuit in acanthdi fishes.

Autostylic. Upper jaw gets fused with the cranium and lower jaw gets articulated with quadrate. It is present in amphibia, reptiles, birds etc. Two subtypes of this are known—monimostylic when quadrate is immovable as in most lizards and *streptostylic* when quadrate is movable as in snakes.

—*Craniostylic.* When lower jaw is articulated with squamosal. Found in mammals.

Jejunum. Part of intestine which extend from duodenum (the ileum in a vertebrate).

Jelly-Fish. Common name for a scyphozoan coelenterate *Aurelia.*

Jenus Green. A vital dye with which mitochondria get stained.

Jog-Cells. Cells in kidneys which produce renin.

Jogging. Means slow walk or trot.

Joints. Refers to the junction between two separate bones or other hard structures. Joints fall into three classes that differ in the degree of

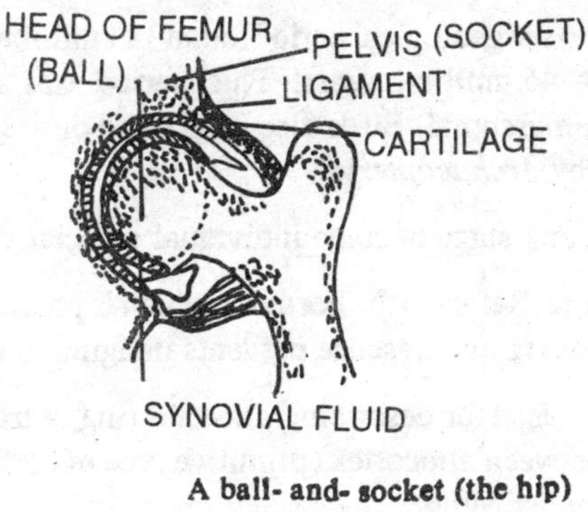

A ball- and- socket (the hip)

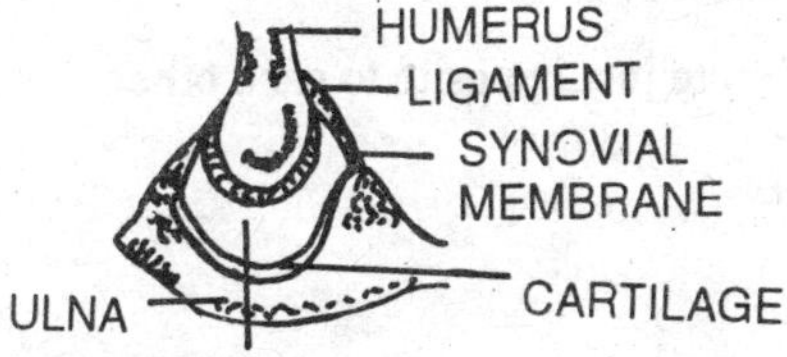

Fig. 1. Types of freely movable joints

freedom of movement they allow: (i) Immovable joints, (ii) Slightly movable points, (iii) Freely movable or synovial joints. Synovial joints include the ball-and-socket joints and the hinge joints (Fig. 1).

Jugal. Refers to one of the investing bones of skull forming part of upper jaw in frog. It forms secondly upper jaw with premaxillae, maxillae in mammals and forms slender bar with maxilla and quadratic jugal joining quadrate posteriorly in reptiles.

Jugular Artery. An artery referring to throat.

Jugular Pulse. Refers to the pulse which is felt in large jugular vein (in the neck region). It can be recorded by employing suitable tambour applied to neck skin. The pules is having three waves—a *wave* occurs due to atrial systole, e-wave gets transmitted manifestation of rise in atrial pressure produced by the bulging of tricuspid valve into atria forming isometric ventricular contraction and the-v *wave* mirrors reveal rise in atrial pressure before tricuspid value opens during diastole. Jugular pulse is having clinical importance.

Jugular Vein. Vine which pertains to throat.

Jurassic. Refers to the geological period found 181 million years ago and existed for 46 million years. This period was dominated by Dinosaurs, marsupial. Birds also originated during last phase as evidenced by *Archaeopteryx*.

Juvenile. The young stage of some individual especially insects etc.

Juvenile Hormone. Refers to the hormone which is produced by corpora allata in insects. Its presence prevents imaginal moults.

Juxthallocortex. Used for describing a second ring of transitional type of cortex between allocortex (primitive type of cortex) and rest of cerebral hemispheres.

Juxtraposed. Refers to just opposite to each other.

K

Kallidin. Used for describing one of the kinins which is formed in plasma. Its name is lasylbradkinin which is a decapeptide.

Killikrein. Refers to the proteolytic enzyme which converts kiniogens to kinins.

Kappa-Particles. In some strains of *Paramecium aurelia* individuals possess DNA-containing particles which on coming in contact with individuals, not having them, brings about their death. These particles are not having a killing effect on the bearer. The individuals bearing them are known as killers and the poison produced be them is paramecin.

Karyokinesis. The sequence of events resulting in division of nucleus through mitosis/meiosis. It may / may not be followed by cytokinesis.

Karyotype. Refers to the pictorial representation of chromosomal complement of an individual. Usually chromosomes have been separated into identifiable groups having common morphological characteristics.

Katabolism. Refers to the breaking down of complex substances into simpler ones in a living organism with liberation of energy.

Keel. Distinct median ridge-like structure which is along the ventral part of avian sternum.

Keber's Organ. Refers to the pericardial gland of a mollusc such as *Anodonta*. It is a large glandular excretory organ which is situated in the anterior of the body and discharging into the pericardial cavity.

Keratin. A hard horny protein-like substance from which claws, nails,hoofs etc., are formed. Refers to the protein which is formed by certain epidermal tissue such as vertebrate skin. The process of its formation is called keratinisation.

Kernicterus. Refers to a neurological syndrome in which there occurs deposition of bile pigments in basal ganglia. It may also develop if birth is complicated by a period of hypoxia.

Ketone. Organic compounds having C=O group e.g. acetone. It may be formed in the body during metabolic activities and regarded as important source of energy in some condition. It gets discharged in urine (in form of ketone body) and also through expired air.

Ketone Bodies. In many tissues acetyl coenzyme-A units undergo condsenation to form aceto-acetyl coenzyme-A. In liver, which is having deacylase, free acetoacetic acid gets liberated. This β-keto acid is converted to β-hydroxybutyric acid and acetone. As these compounds undergo metabolisation in liver with difficulty, they diffuse into circulation and come in urine. The ketone bodies are formed generally during starvation when there occurs gluconeogenesis.

Ketosis. Refers to the accumulation of ketone bodies in blood steams due to the ability of tissues to oxidise ketones. It is exceeded due to metabolic disorder.

17-ketosteroids. Refers to the derivatives of cholesterol and hormone of adrenal cortex (steroids). Most of steroids are having keto group at 17-C position and thus called 17-ketosteroids.

Ketosterone. An intermediate compound which is formed during biosynthesis of estriol from estrone.

Kidney. Usually used for describing the main organ for excretion of liquid nitrogen wastes in vertebrates; also often loosely used for describing analogous organs in certain other animals as in crayfish, snails, mussels etc. (Fig. on p. 195).

Killer-T-Cells. Cellular immunity gets mediated by T-lymphocytes. When they encounter the antigens on cells from other individuals they become activated. They enlarge, undergo division and release lymphokinins. T-lymphocytes responsible for lysis of homograft and tumour cells independent of complement are known as killer T-cells.

Kinase. A enzyme which functions as an activator to another enzyme, e.g , enterokinase and thrombokinase.

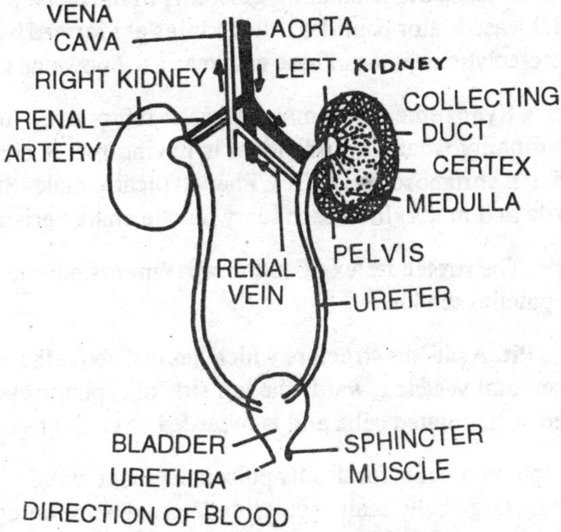

Fig. 1. The Kidneys of a mammal

Kindred. Pedigree. A chart which shows mating relationship within a given family.

Kinesis. Refers to the movement of an animal or part of an animal in response to a stimulus but having no relationship to the direction of the stimulus. Some animals, for instance, get stimulated to burrow by dry conditions in the soil and come to rest when conditions become sufficiently moist.

Kinetin. A synthetic form of kinin which is able to stimulate the formation of DNA and hence to promote cell-division.

Kinetochore. Region of chromosome to which the spindle fibers are attached at nuclear division. Often used synonmously with centromere, although some regard it s separate entity which is located in the region of centromere.

Kininases. Two enzymes converting kinins to inactive peptides—one is kininase-I which is carboxypeptidase and removes C-terminal amino acid residue and the other *kinanase-II* which is able to convert angiotensin-I to angiotensin-II.

Kinino Cilium. A true (9+2) cilium which present in sensory hair cells.

Kininogens. Means the circulating globulins in the blood plasma from which vasodilator peptides called kinins are formed by the action of proteolytic enzyme. These enzymes are known as kallikreins.

Klinfelter's Syndrome. In human beings it refers to a chromosomal abnormality where an individual is having two X chromosomes and a Y chromosome (XXY). Phenotypically male, the bearer is sterile and may exhibit secondary female characteristics.

Keen Jerk. The stretch reflex of qudreiceps femoris muscle which tops the patellar tendon.

Kolliker's Pit. A pit-like structure which lies just above the anterior end of cerebral vesicle towards the left side of cephalic eye-spot. It is lined with ciliated cells and is regarded as chemoreceptor.

Krait. A common name for deadly poisonous snake *Bungarus* which is having large belly scales (crossing laternal extremities), rounded tail, largest 4th infralabial scale and hexagonal dorsals.

Krause's End Bulb. Thermoreceptor in skin where skin is hairless.

Krause's Membrane. Disc by which I-band in striped muscle fiber gets crossed. Also known as Z-disc or Dobie's line.

Kerb's Cycle. Involves the final oxidation of acetyl coenzyme-A (formed by oxidative conversation of pyruvic acid obtained during glycolysis) to CO_2 and water. The energy liberated finds use in the reduction of certain co-factors mainly NAD and flavoproteins (Also see *glycolysis and ETC*).

Krypts of Leiberkuhn. Refers to the depression in between villi of intestine lined with columnar and goblet cell's into these Brunner's glands open which secrete mucous and enzymes.

Kupffer Cells. Stellate phagocytic cells in the sinuses of the liver. They from part of the reticulo-endothelial system and apparently have as their main function the ingestion of dead red corpuscles.

Kymogram. The recordings of muscle contraction which are obtained by a kymograph.

Kymography. An apparatus which is used for recording muscle contraction.

Kynotryncha. Refers to the pseudocoelomate class of Aschel-minthes.

L

Labial. Any structure which pertains to lips like labial artery, labial muscles, labial glands etc.

Labial Cartilages. Small cartilaginous skeletal elements present at the corners of the mouth in some fish.

Labial Glands. Refer to the salivary glands which are arranged in a row in the upper and lower jaws of some reptiles. In poisonous snakes those of the upper jaw are modified so as to form the poison-glands which usually have ducts which lead to the bases of the fangs : the latter are grooved or perforated making channels down which the poison flows.

Labial Hooks. Refer to the hooks or spines on the labial palps of the nymphs of dragon flies (Odonata). In these insets the whole labium forms an enlarged prehensile organ called the *mask* which is capable of being shot out rapidly to catch tadpoles or other creatures.

Labrun. Pertains to upper lip of insect's mouth. Dorsal to mouth's opening.

Labyrinth. The part of internal ear-membranous labyrinth which is found in all vertebrates and its hearing portion is known as lagena. In mammals the periotic bone also forms bony labyrinth around membranous labyrinth.

Labyrithine Organs. Refers to the special air breathing organs which occur in climbing fish *Anabas*. They are kept in dorsals enlargements of gill chamber. They have three or more cocentrically arranged bony plats with frilled margins attached to bony base which in turn gets attached to upper end of 4th gill arch.

Lachrimal. Refers to tears.

Lachrymal. In most of vertebrates such as birds, mammals, etc. large orbit has been bounded by a ridge to which the bone attached is termed as lachrymal. It is also called sometimes supraorbital.

Lachrymal Glands. The tear glands found in the eyes of vertebrates.

Lactation. Refers to the production of milk by the mammary glands.

Lactation Anoestrus. A phenomenon common to most mammals in which ovulation and sexual desire get inhibited for a time during lactation i.e., immediately after giving birth to the young.

Lacteals. Used for the central lymphatic cavities in the villi of the small intestine, into digested fats get absorbed. The fluid in them is a milky emulsion having fat globules suspended in lymph. From the lacteals, fats are carried to all parts of the body.

Lactic Acid. The prolonged anaerobic oxidation of glucose during glycolysis in which pyruvic acid is converted to this substance under the influence of lactic acid dehydrogenase. Accumulation of this substance in muscles is responsible for muscles fatigue. On getting oxygen in the system it may be disposed to CO_2 and water.

Lactogen. A hormone or group of hormones from the pituitary gland which stimulate the production of milk in mammals.

Loctogenic. Refers to the milk producing/milk secretion stimulating hormone/substance such as prolactin from pituitary gland.

Lactose. A disaccharide present in milk only. It is also called milk sugar. When hydrolysed by lactase, it produces glucose and galactose.

Lacuna. Refers to any space between cells or a space having a matrix such as bone.

Lacunae of Trophoblast. Spaces or cavities in the trophoblastic part of a mammalian placenta which become filled with maternal blood diffusing through from the wall of the uterus.

Lagena. An outgrowth from the sacculus of the ear in fish and Amphibia; a forerunner of the cochlea of mammals but smaller and not coiled.

Lamarckism. The evolutionary theory populated by Jean Lamarck (1744-1829) but now largely discredited. Its main principles were: (1) That frequent use of an organ caused it to become larger and better developed, whereas its disuse caused it to become small or degenerate.

(2) Such changes acquired during life could be inherited and eventually resulted in the formation of new species (see also *Neo-Lamarckism*).

Lamella. Refers to thin leaf-like layer as gill lamella.

Lamina Propria. Middle layer of mucosa of intestine having thin layer of peculiar connective tissue supporting epithelial layer over a basement membrane and has in it small nodules of lymphatic tissue, blood vessels and nerves.

Lappet. Flap-like projection which is present on each side of tentaculocyst of jelly-fishes etc., or at the head of a planarian.

Larva. The immature living stage in the life cycle of various animals which gives rise to the adult stage by undergoing metamorphosis.

Larynx. Refers to the organ which is situated between trachea and base of tongue. Its opening is known as glottie and it has vocal cords. It occurs in all vertebrates having lungs except fishes.

Latent Period. Refers to the period between occurrence of infection of host by a pathogen and appearance of identifiable symptoms of disease. During this period the activity of pathogen is low. With reference to nerve tissue it means isopotential interval following the stimulus artifact (brief irregular deflection of baseline) when stimulus is applied.

Lateral. Means the side of the body. At each side of the median line.

Lateral Line. Sense organs in fishes and few amphibains that are extending bilaterally and longitudinally. Also presents in nematodes at hypodermal longitudinal thickenings.

Lateral Line System. Refers to the series of hair cells on medial walls of canals which are running the length of head and body of fishes and many amphibians. These channels are having opening to outside and respond to water movements.

Lateral Plate Mesoderm. Unsegmented mesoderm in the ventrolateral region of a vertebrate embryo.

Lateral Ventricles. The cavities in the two cerebral hemisphere. Also called *paracoels*.

Latitudinal Migration. Refers to the migration taking place in a north-south plane.

L-Dopa (Levo-Dopa). During the biosynthesis of catecholamines there occurs the formation of dopamine. L-dopa takes part in this. It is used in the treatment of Parkinson's disease and involved in the control of prolactin secretion in animals and human being.

Learning. Refers to the ability of nervous system of an individual which takes place toward an external stimulus in a particular way based upon some past experience having similar situation.

Lecithality. Means yolk concentration in oocyte/egg.

Lecithin. A phospholipaid substance present in the yolk of most substances.

Lemnisci. Glandular organs present on the pharynx of Acantho cephala spiny-headed worms.

Lens. A transparent structure present in the eye that has refractive power and derived from ectoderm of head.

Leptotene. In meiosis it refers to the appearance of thread-like chromosomes within the nucleus during the first meiotic division.

Lethal Gene. A mutant allele that is able to kill the bearer (individual possessing it). The mutation commonly takes the form of a small chromosomal deletion so that genetic code for a key position becomes missing or gets disrupted e.g., yellow lethal in mice.

Leucine. Neutral amino acid ; an amino acid having unsubstituted chains. It is an essential amino acid. Its isomer is isoleucine which is also an essential-amino acid.

Leukemia. Refers to the conditions in which there occurs abnormal increase in WBCs number from 7,000 to 700,000 (or even more). Regarded as a type of blood cancer.

Levator. A muscle which raises a body part.

Leydig Cells. Intestitional cells of testes that get stimulated by luteinising hormone to secrete testesterone.

LH (Luteinsing Hormone). A hormone which is secreted by anterior pituitary. It induces Graffian follicle to brust open and release the egg.

Ligament. Though, fibrous connective tissue band which binds bones of supporting viscera.

Light-Meromyosin. Refers to the rod-like fragment of myosin molecule which constitutes molecule's backbone.

Limbic System. In mammals and man there is the small portion of rhinencephalon which is concerned with olfactory power. However, there is the remaining portion which is concerned with emotional responses, instincts and complex neuroregulatory functions and is known a limbic system or limbic lobe. Each limbic lobe is having a rim of critical tissue around hilus of cerebral hemisphere and a group of associated structures—the amygdala, hippocampus and septal nuclei.

Limbs. Used for the paired locomotory structures of tetrapods. There are two pairs of limbs—fore limbs and hind limbs. Each limb is having three parts—brachium or stylopodium which is forming upper arm in fore limb and thigh in hind limb, *antebrachium* or zeugopodium which is forming fore arm is fore limb and shank in hind limb, *autopodigm* forming hand in fore limb and foot in hind limb. The autopodium is further divisible into basipodium which is forming wrist in fore limb and ankle in hind limb, metapodium which is forming palm in fore limb, acropodium which is forming fingers in hand and toes in foot.

Limy. Containing Ca^{2+} salts especially $CaCO_3$.

Lingua. Refers to the tongue or any analogous organ, e.g., the hypopharynx of an insect or the radula of a mollusc.

Lingual Artery and Vein. Refer to vessel supplying blood to the tongue; in tetrapods the artery is a branch of the carotid while the vein returns to the external jugular.

Lingual Ribbon. A name sometimes used to describe the narrow ribbon-like radula of certain molluscs.

Linguatulida. An alternative name for *Pentastomida (q.v).*

Linin. A viscoid substance which forms an apparent mesh throughout the nucleus of a cell.

Linkage. Refers to the carrying of several different genes on one chromosome so that the qualities associated with these genes are generally inherited together as if they are becoming a single unit. Occasionally such linkages break down due to the phenomenon of *chiasma or crossing over (q.v).*

Linked Genes. Refers to genes which are associated on same chromosome and thus tend to be transmitted together in a cross. As linked genes are present at different loci, they are not allelic.

Lipase. Refers to the fat digesting enzyme which is present in intestinal juice. Also present in gastric secretion, it converts fats to fatty acids and glycerol.

Lipemia. Means the milky appearance of plasma after meals due to presence of so many large particles in blood which are known as chylomicrons. These are formed in the intestinal mucosa during absorption of long chain fatty acids into lymphatic ducts. The lipemia becomes clear by action of lipoprotein lipase.

Lipids. Refers to fatty acids, neutral fats, waxes, steroids and phospholipids. They are hydrophobic and look/feel greasy.

Lipogenesis. Used for describing the formation of fat from non-fatty sources in the body of individual.

Lipoio. Means substance of fatty nature.

Lipophilic. Describing for an affinity for lipids.

Lipoproteins. Used for conjugated fats and proteins. They form characteristic substances of animal and plant tissues. They are also formed in the body by action of lipoprotein lipase on chylomicrons.

Lipotropins. (β & γ). Refers to the peptides having fat-metabolising activity. These are isolated from pituitary gland of mammals e.g., sheep. γ-Lipotropin is having 58 amino acid residues while β-lipotropin may be precursor of peptides that bind to opiate receptors (enkepkalins). Both lipotropins are having MSH (melanophore stimulating hormone) activity in the body.

Liquor Folliculi. Used for describing the liquid/fluid present in the autumn folliculi of Graffian follicle of mammals.

Lithosphere. Refers to the solid, rocky component of earth, Earth's crest.

Litoral. Used for describing floor of sea from shore to the edge of continental shelf.

Litoral Fauna and Flora. Animals and plants that are inhabiting the sea-shore and the shallow sea near the shore.

Liver. Refers to a digestive gland whose duct opens into the intestine; in a vertebrate it is the largest gland of the body, producing bile, making and storing glycogen and helping to break down proteins and fats.

Lobi Inferiores. Refers to lateral branches and thickenings of the pituitary in fundibulum generally present in fish.

Lobopodia. The blunt type of pseudopodia such as those present in *Amoeba*.

Locus. Refers to the specific position which is occupied by a particular gene of a chromosome. Any one of the alternative forms of genes may be formed at a given locus.

Loop of Henle. Kidney structure which is present in the medulla in case of mammals where most of water reabsorption is occurring. Its activity has been controlled by ADH (antidiuretic hormone). It deals with diuresis (urine formation).

Lophodont. Molar and premolar teeth with transverse ridges of enamel running across the crown: the typical grinding teeth of ruminants and many other herbivores.

Lophophore. A circle or horse-shoe of ciliated tentacles which are surrounding the mouth in Polyzoa and also in Brachiopoda.

Lorica. Protective covering which is secreted by certain organisms e.g. ciliate protozoa.

Lown-Ganong-Levine Syndrome. Refers to a pathological condition which leads to attacks of atrial arrhythmias. In this condition depolarisation presumably is passing from atria to ventricles via an aberrant bundle and then entering the interventricular conducting system distal to AV node.

LSD (Lysergic Acid Diethylamide). Refers to the psychomimetic agent and serotonin antagonist. The compound is able to produce transient hallucinations and other mental aberrations.

Luciferase. Enxyme contributing to light production in certain animals like-flies.

Lumber Vertebrae. Vertebrae of the abdominal region to which are attached the muscles of the back.

Lumen. Refers to the internal cavity within body or a structure such as lumen of intestine, gland to a blood vessel.

Luminiscence. The production of light due to the chemical reactions in the cells. An example is the production of bluish-green light by *Noctiluca* (a marine protozoa) during night hours.

Ling-Fishes. Another name for fishes of group dipnoi. They are having developed lung and by which they engulf air to carry out respiration.

Luteolysis. The regression of corpus luteum so that next menstrual cycle may start.

Luteotropin. Means LTH (luteotropic hormone) which is also known as prolactin hormone. It stimulates milk secretion and maternal behaviour.

Luxation. Used for describing easy passage of faecal matter out of alimentary canal. Gastrointestinal motality, mucous, secretion by large intestine play vital role in the process.

Lymph. Means blood minus RBCs. It is produced by lymph nodes and mainly concerned with body defence mechanism.

Lymphagogues. Agents increasing lymph flow. The include a variety of agents that increase capillary permeability and the agents bringing contraction of smooth muscles.

Lymphatic System. Refers to a system of vessel and nodes in vertebrates which is leading from tissue spaces to large-veins entering the heart. It is a part of circulatory system.

Lymph Hearts. Refers to the localised enlargements of lymph vessels with valves and contractile walls. They occur in some vertebrates like frog. They are present in the body generally at places where lymph vessels fall into veins.

Lymph Nodes. Area of lymphatic system which produces antibodies.

Lymphocytes. Agranuclocytes (a type of WBCs) that have round of kidney shaped nucleus. They constitute about 28% of WBCs and produce antibodies and opsonins to destroy foreign matter.

Lymphokinines. Lymphocyte products which are mediating cellular immunity. They are responsible for delayed allergic reactions, rejection of transplanted organs and rejection of tumour cells.

Lyon-Hypothesis. According to this hypothesis, in any given cell of a mammalian female one X-chromosome is entirely active while the other being inactive completely or in parts. The inactivation takes place at random so that the paternally derived X-Chromosome becomes active in some cells and maternally derived one active in others.

Lyse. Means breaking up usually by chemical method (breaking chemically).

Lysine. One of the essential amino acids. It is basic amino acid which has diaminomonocarboxylic group.

Lysis. Means the break-down of the organised structure of biological matter. This term is usually restricted to break-down of living cells.

Lysolecithin. The lispase-A in the pancreatic juice takes away a fatty acid from lecithin to form lysolecithin. This substance damages cell membranes. In acute panceratities, a severe disease, there occurs activation of phospholipase with the formation of lysolecithin from lecithin. This brings about disruption of pancreatic tissue and surrounding fat.

Lysosome. Refers to subcellular, digestive structure which contains enzymes of various types.

Lysylbradykinin. See *kallidin and kinins.*

M

M-Curve. Also known as summation. It is used for describing a condition if a contracting muscles is imparted one more stimulation during its condition phase and then the second contraction is added to first one yielding 'M-curve' on kymogram.

m-RNA (Messenger -RNA). A part of RNA which is responsible for transmission of informational base sequence of DNA to ribosomes.

MSH (Melanocyte/Melanophore Stimulating Hormone). A peptide hormone which is released by the adenohypophysis. It affect melanin distribution in mammals and creates skin-colour changes in lower vertebrates like fished, amphibia and reptiles.

Macromere. Used for describing the relatively large yolk laden cell which is present during cleavage stages of certain animals like frog etc.

Macronucleus. Refers to the large nucleus in ciliated protozoa as in *Paramecium*. It control all body activities except reproduction.

Macrophage. A large amoeboid cell that is able to engulf, ingest, and destroy bacteria damaged cells, and worn-out red blood cells. This process is termed as *phagocytosis* and constitutes an important part of the body's defence against disease.

Macrophage System. Refers to the body defence system which is formed by macrophages, lymph nodes, spleen, Kupffer's cell. This system by involving active phagocytosis engulfs the foreign bodies and damaged tissue parts. Also known as *reticulo-endothelial system*.

Macula (1) A small area of sensory epithelium present in the sacculus and utriculus of the inner ear. It has sensory hairs which are embedded in an otolith- a gelatinous mass has crystals of calcium carbonate having.

(2) A region rich in cones surrounding the fovea in the vertebrate eye.

Macula Densa. At the point where afferent renal arteiole is entering glomerulus and the efferent renal arteriole is leaving it, the tubule of nephron touches the arteriole and glomerulus. At this poin' a modified region of tubular epithelium which makes the start of distal convolution is known as macula densa and it lies in close approximity to J G-cells.

Macula Lutea. Yellowish pigmented spot at the posterior pole of eye which makes location of fovea centrails.

Maggot. Used for describing worm-like legless larva of fly.

Magnum. Refers to part of oviduct in birds just behind the funnel and between isthmus. It secrets albumen around the egg.

Malaria. Any one type of the several fevers each produced by specific sporozoan parasite belonging to genus *Plasmodium*. This parasite invades liver and RBCs of various mammals causing chill and high fever. This fever can be cured by the drugs like quinine, paludrin etc.

Malignant Transformation. Pertaining to a change from non-cancerous to cancerous to cancerous state.

Malleus (Hammer). The hammer-shaped bone which gets attached to the tympanic membranes (eardrum). It forms the first and largest of the ear ossicles in mammals. It is homologous with Meckel's cartilage.

Malpighian Body (Malpighian corpuscle, renal corpuscle). Refers to the part of the excretory unit (nephron) within the cortex of the vertebrate kideny. It consists of a blood capillaries (glomerulus) and a surrounding cupshaped Bowman's capsule.

Malpighian Layer. A layer of dividing cells at the base of the epidermis of vertebrates. It cells are having granules of the pigment melanin, which protects the body against ultraviolet radiation.

Malpighian Tubules. Slender blind-ending tubes which are opening into the anterior end of the hindgut of insects, spiders, millipedes and centipedes. They act as excretory organs.

Maltese Cross. Centrum in cartilagenous fishes is not purely cartilagenous, calcification occurs and fibrocartilage is developed in whole centrum except at 4 wedge-shaped places which are known as maltese cross.

Maltose. A sugar present in germinating cereal seed. It is a disaccharide which is composed of two glucose units.

Mammae. Used for the structure in skin of mammals where various mammary glands are opening through their ducts. The characteristics shape and bulk of the structure has been formed by adipose tissue surrounding mammary gland. The mammae open at nipples that are formed by erectile tissue. The distribution is usually pectoral (as in woman) or inguinal (as in cattles) or from pectoral to inguinal (as in pig). See *breasts.*

Mammalia. The class of vertebrates including the most successful tetrapods. They are homoiothermic, having an insulting body covering of hair and usually with sweat and sebaceous gland in the skin. Typically, the young are born alive and are suckled on milk which is secreted by the mammary glands. It has three subclasses, the Monotermata, the Metatheria (marsupials), and Eutheria (placental mammals).

Mammary Glands. The modified sweat glands of skin in mammals. They secret milk. Each is having many lobules and excretory ducts from lobules combine to form large ducts which meet similar ducts and pour into main ducts which finally open out separately at the nipple.

Mammilary Processes. Refer to the additional processes which are associated with lumber vertebrae of mammals. Two pairs—one is dorsolateral to prezygapophyses which is called metapophyses and other is between postzygapophyses and transverse processes which is called anapophyses.

Mandible. (1) Refers to one of a pair of feeding appendages (mouthparts) of various arthropods, like crustaceans, insects, millipedes and centipedes.

(2) Also refers to the extended upper and lower jaws of birds; they form a beak.

(3) Used for the lower jaw of vertebrates. In many species it mainly consists of a paired membrane bone, the dentary which in mammals

is bearing the lower set of teeth.

Mandibular Arch. The first visceral arch; it is modified to form the upper and lower jaws in Chondrichthyes (cartilaginous fish).

Mentle. Refers to a tissue covering most of the body of molluscs. It secretes the shell(s) and in shell-less molluscs it is tough and protective. The mantle gets folded to enclose the *mantle cavity*, which possesses the respiratory organs.

Map Unit. A figure which is used to measure distance between two linked loci as known from the amount of crossing over which takes place between them. For loci lying close to each other, the map units are equal to the number of new combinations that could be detected so that 1 map unit=1% recombinations. For loci which are more distantly situated with reference to each other, the amount of multiple cross even reduces the number of new combinations which could be detected to that the map units and 1% recombinations are not equal.

Marine. Of sea water; forms or bodies of salt water.

Marker Gene. Used for describing any specific gene or genetic trait which can by employed for following the transmission of a specific chromosome or chromosomes region in mating.

Marrow. Used for the tissue which is filled in the spaces of long bones like femur etc. The marrow of vertebrates in general is having yellow bone marrow which stores fat but in mammals it is having red bone marrow besides yellow bone marrow. The red bone marrow possesses four types of cells—haemocytoblasts that losm RBCs, myeloblasts that form WBCs (agranulated) nomoblasts that form WBCs (agranulated) and megakaryoblasts that form marrow platelets.

Marsupials. Mammals that are bearing pouch or marsupium.

Marsupial Bones. A pair of cartilage bones that extend forward from pubes in the ventral wall of abdomen. Also known as *epipubic bones*.

Marsupium. Used for describing a pouch on the abdomen of marsupials and some monotermes. It consists of a fold of skin supported by the epipubic bones of the hip girdle. It covers the mammary glands and protects the young, which migrate there after birth for completing development.

Masseters. Refers to the muscles which are behind temporalis (arising from otic capsule and inserted on coracoid process of lower jaw) and arise from quadratojugal to get inserted on outer surface of lower jaw.

Mass Pyramid. A mode of expressing mass relationships within an ecosystem. When moving towards apex, mass decreases e.g. base may be composed of producers whereas apex may be having secondary or tertiary consumers.

Mastax. Toothed chewing apparatus which is found in rotifer plarynx.

Mast Cell. A cell occurring in the ground substance (matrix) of areolar (loose) connective tissue, which it is secreting. Mast cells also occur in the lining of blood vessels, and some occur in the blood, as basophils. They possess lobed nuclei and cytoplasmic granules that are stained by basic dyes. Mast cells secrete heparin (an anticoagulant) and histamine.

Mastication. Used for describing chewing food with teeth or other mouth parts as found in many insects and other animals like mammals.

Mastigonemes. Small delicate processes that arise from one side of flagellum in *Euglena* or similar type of animal.

Mastigophora. A group of protozoa bearing flagellum e.g., *Euglena* is autotrophic bearing chlorophyll.

Maternal. Pertains to mother's side/mother.

Matrix. A term used in histology for intercellular surface. It is also used for non-cellular substance of connective tissue as in bone or cartilage.

Maxilla. 1. Used for one of a pair or two pairs of feeding appendages (mouth parts) of various arthropods, like crustaceans, insects, millipedes and centipedes.

2. Also used for a pair of large bones of the upper jaw of vertebrates. In mammals they bear the molar and premolar teeth.

Maxilliped. Refers to one of the first three paris of thoracic appedages in crustaceans such as cray-fish.

Meatus. Refers to a passage or channels in the body, such as the *external auditory meatus* in birds, mammals, and some reptiles, which leads from the external opening of the ear to the eardrum.

Mechanoreceptors. Sense receptors to mechanical distortion or pressure.

Meckel's Cartilage. Refers to the ventral division of mandibular arch in vertebrates that makes lower jaw.

Medial Longitudinal Fasciculi. Refers to the never tract through which second order neurons, after passing down the spinal cord, form vestibular nuclie in the vestibulospinal tract ascend to the motor nuclei of cranial nerves that control eye ball movement.

Medial Olfactory Stria. Used for the nerve tract through which band of fibers (mostly axons of mitral cells) are passing from glomeruli to cortical and medial portion of ipsilateral amygdaloid nucleus and to the perpiriform and periamygdaloid cortex in the brain.

Median. Used for describing the mid-line or near the middle of the body.

Median Eye. An eye present in the middle of the head. It is found in some crustaceans, such as the microscopic pond animal *cyclops*. It is a simple light receptor (ocellus).

Mediastinum. 1. A membranous septum in the midline of the thorax of mammals which separates the two pleural cavities ventrally.

2. The space between the pleural membranes surrounding the lungs, that is having the heart, thymus, oesophagus and trachea.

Medulla. Used for the central region of an animal organ, when it differs in structure or function from the outer regions. An example is the medulla of kidney.

Medulla Oblongata. A region of the hindbrain concerning with the functioning of the visceral organs, e.g., the stomach, lungs and heart.

Medullary Centres. A group of neurons in the medulla oblongata that are involved in integration of information as in regulation of respiration (respiraticn centres) and circulation (cardiovascular centers).

Medullary Groove/Tube. This term is synonymous with neural plate, groove or tube. It refers to the successive stages in the embroynic development of vertebrate central nervous system.

Medullary Sacretions. Used for epinephrine and norepinephrine which are commonly known as catecholamines.

Medullary Sheath. Refers to a sheath which is present around the medullated fibres (greater than I in diameter in mammals). It is made up of fatty substance between which thinner layers of protein are present. Also known as *myelin sheath.*

Medullated. Mainly used for certain nerve fibers that are covered with myeline sheath.

Medullated Nerve Fiber (myelinated nerve fiber). A nerve fiber surrounded by a fatty (myelin) sheath. Most nerves of vertebrates have thousands of medullated fibers, which appear white due to the fatty sheaths.

Medusa. A stage in the life cycle of coelenterates in which the body is having the shape of a bell or inverted saucer with a fringe of tentacles around the rim and a mouth beneath. The medusa represents the free swimming dispersal stage of the life cycle, in scyphozoans (jellyfish) it is the only form. Medusae possess sex organs (male or female); sperm swim from the male or fertilize eggs in the female. The fertilized egg is then developed into a planula. See also *polyp.*

Megakaryoblasts. Refers to the cells of red bone marrow that are destined to form platelets.

Megalecithal Egg. Any cell which is loaded heavily with yolk as birds and egg-laying mammals. Also termed as *polylecithal egg.*

Meiosis. The process of cell division which leads to the production of daughter nuclei having half the genetic complement of the parent cell. Cells formed by meiosis produces gametes and fertilization restores the correct chromosome complement. Meiosis involves two divisions during which the chromosomes replicates only once. Like mitosis it also involves the stages prophase, metaphase and anaphase. However, during prophase homologous chromosomes attract each other and get paired forming bivalents. At the end of prophase genetic material gets exchanged between the chromatids of homologous chromosomes. Meiosis is also different from mitosis in that after anaphase, instead of nuclear membrane forming, there is a second division, which may be divided into metaphase II and anaphase II. The second division is terminated with the formation of four haploid nuclei, which develop into gametes.

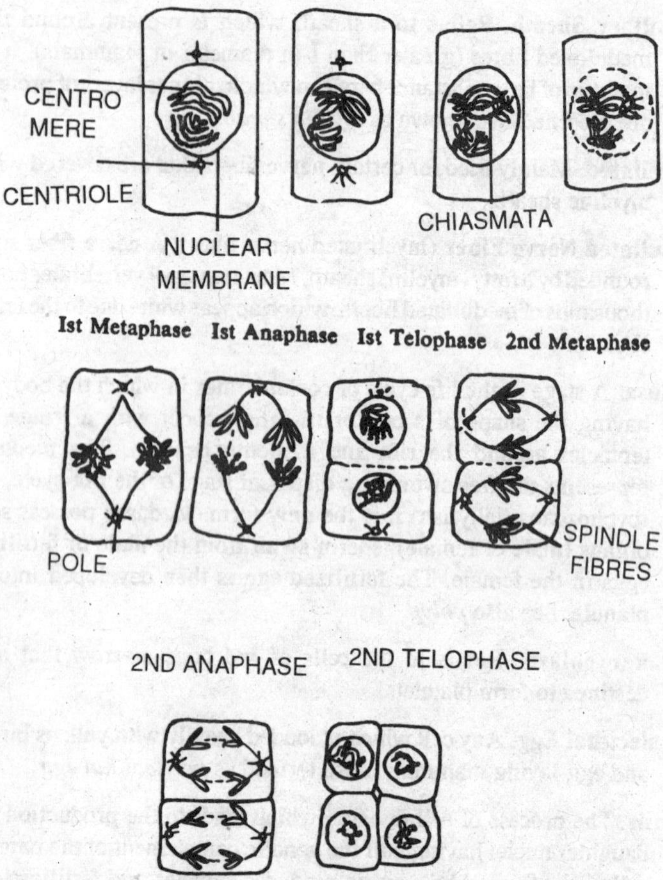

Fig. 1. The Stages of meiosis in a cell containing two pairs of homologous chromosomes

Meissner's Corpuscles. Encapsulated nerve endings that form corpuscles located in dermis. They respond to light touch.

Meissner's Plexus. Used for describing a nerve plexus between submucosa and circular layers of alimentary canal. Nerve fibers from plexus are able to control secretory activity of intestinal gland. It is concerned with sympathetic and parasympathetic systems.

Melanin. One of a group of pigments occurring in animals and plants. It is derived from the amino acid tyrosine. The colours range from black through brown to yellow, orange, or red. In animals melanin is found in *melanophores* (pigment cells) in the skin, usually below the epidermis. It is responsible for colour to the skin, hair, and eyes of animals and causes colour in various seedlings and roots of plants. The absence of the enzyme tyrosinase in animals results in a condition called *albinism,* in which no pigments develops in the eyes, skin, or hair.

Melanism. Refers to the dark appearance due to the presence of a dark brown pigment, melanin.

Melanoblasts. Branched cells in the skin of mammals. They control skin pigmentation. Branches twine around the epidermal cells in the neighbourhood and terminate in cup-shaped endings around an epidermal cell. They produce melanocytes.

Melanocytes. Cells in the skin of mammals that bear melanophores (produced from melanoblasts).

Melanocyte-Stimulating Hormone (Intermedin). A polypeptide hormone which is produced by the anterior pituitary gland in man. Its role is not known although its blood level is often disturbed in patients having endocrine disordered accompanied by pigmentation abnormalities. It is having marked action on pigmentation in the skin of amphibiants.

Melanosomes. Refers to irregular vesicle-like structures in melanocyte cells having tyrosinase enzyme.

Melatonin. A hormone, which is produced by the pineal gland. It is no known action in man but produces market lightening of the skin in embryonic fish and larval amphibians.

Membranes. Refer to the tissue which are surrounding structure, organ, cells etc. They are mostly semiprmeable permitting selecting substances to pass through them. Several types.

Unit membrane. Plasma membrane that is having sandwich-like structure, with lipid layer which is enclosed by protein layer.

Subunit membrane. Lipid-protein globular aggregates.

Mucous membrane. Made of layer of simple or starified epithelium. It is anchored to basal connective tissue e.g.,lumen of alimentary canal.

Serous membrane. Mesothelial membrane which is supported by thin connective tissue. It is present over alimentary canal.

Membrane Bone (dermal bone). A bone formed by ossification of connective tissue instead of cartilage.

Membranous Labyrinth. Membranous part of ear which has semicircular canals, utriculus and sacculus. It is associated with balancing of the body.

Menarche. The form used for the beginning of menstruation in a human female. It starts at the age of 14-16 years.

Mendelism (Classical genetics). Refers to the study of inheritance by controlled breeding experiments which were first carried out by Mendel in the 1860s. The characteristics studied are usually controlled by one gene and exhibit a simple dominant/recessive relationship between alleles.

Mendel's Laws. Two laws which formulated by Mendel to explain the pattern of inheritance he observed in plant crosses. The first law is known as the Law of Segregation which states that any character exists as two factors, both of which occur in the somatic cells but only one of which passes on to any one gamete. The second law is known as the Law of Independent Assortment which states that the distribution of such factors to the gametes is random; if a number of pairs of factors is considered each pair segregates independently. Today Mendel's characters' are known as genes and their different forms (factors) are known as alleles.

Meninges. The protective membranes surrounding the brain and spinal cord in vertebrates. in man and other mammals those are three: the stiff outer dura mater, the arachnoid membrane, and the soft inner pia mater. The latter two have been separated by the subarachnoid space, which is filled with cerebrospinal fluid.

Meningitis. Refers to the inflmmation of meanings which is of two types: primary (epidemic) and secondary (tubercular and purulent). Its common symptoms are headache, nausea, vomiting, occipetal rigidity, the blood shows neutrophilic leucocytosis (15,000-20,000) and enhanced erythrocyte sedimentation rate (ESR). Cerebrospinal Fluid (CSF) exhibits elevated pressure, increased proteins and large number of cells. Disturbances in consciousness are seen in severe cases.

Menix Primitiva. Refers to the single menix of crclostomes and fishes that fit closely over brain and spinal cord. Space between menix primitiva and endorachis is known as perimenigeal space and filled with lymph-like CSF.

Menopause. Refers to the time period when menstrual cycle in a woman terminates and reproduction is no more possible. It takes place at the age of 45-50 years in most cases.

Menorrhagia. Refers to abnornally profuse bleeding which takes place during menstrual cycles.

Menstrual Cycle. A cycle occurring in female of mammals belonging to primate group (apes and men). Involved four phases and completed in about 28-30 days in majority of cases. Its various phases are as follows :

Menstrual phase. It lasts for 4-5 days and there occurs discharge of mucous and from utrine wall.

Proliferative phase. It lasts between 5-13 days during which uterine wall gets repaired and ovum present in Graffian follicle matures.

Ovulatory phase. It lasts between 14-16 days during which ovum gets released in oviduct and fertilization may take place.

Premenstrual phase. During this phase the fertilization leads uterine wall to prepare for pregnancy. However, if it fails the uterine wall gets again subjected to changes and a fresh beginning is made up for starting the next menstrual cycles.

Menstruation. Refers to the discharge of blood and uterine tissue in female human beings and apes at the end of menstrual cycle in which fertilization does not take place. Also known as *menses.*

Mental Glands. They are below chin in some male salamanders. They are tubular glands and develop as a secondary sexual character. They are volatile exudations from these attract females.

Mentomeckelian. A bone which replaces cartilage in the lower jaw or frog.

Merckel's Discs. Used for describing the localised little disc-line ares of epidermis where many nerve endings terminate. Mechanoreceptors occurred in the mammals.

Meroblastic Cleavage. The egg cleavage in which division of egg is not complete i.e., only takes place in cytoplasm and not in yolk. It is common in egg having large quantity of yolk like hen's egg.

Merocrine Glands. Refers to the glands in which secretary cells discharge their products without making much damage to the cells and after discharge the cells undergo a period of rest during which they undergo refilling e.g. mucous and poison glands of amphibia.

Meromyosin. Refers to the protein present in myosin molecule fibers. It is differentiated in H-meromyosin and L-meromyosin.

Merozoites. Refers to the mature trophozoites of sporozoa like, *Plasmodium* which are formed after mitotic division.

Mesaxonic Foot. Refers to foot in which 3rd digit is carrying the weight while the other digits are either left as vestiges or lost e.g. horse.

Mesencephalon. Used for describing embryonic portion/division of neural tube which is forming optic/mid brain in adult vertebrate.

Mesenchyme. A loose network of cells which are usually underlying epithelial layers, in animal embroys. In some cases, it is divided into *primary mesenchyme* the first cells to invade the blastocoel, and *secondary mesenchyme,* later contributions of diffuse cells from other sources.

Mesentery. Refers to a thin transparent membrane consisting of a double layer of peritoneum which is surrounding all parts of the alimentary canal and gets attached it to the dorsal wall of the abdomen.

The blood vessels, lymphatics, and nerves that supply the alimentary canal are situated between two layers of mesentery.

Mesoblasts. Balastula cells which are destined to form mesoderm.

Mesobronchus. Refers to the main bronchus in the birds after entering lungs and loosing its rings.

Mesocoel. Refers to the paired coelomic cavities present in the collar of *Balanoglossus.* It occurs in form of two sac-like cavities between body wall and alimentary canal and occupied by connective tissue and muscles which are derived from coelomic epithelium. Collar coeloms of each side are opening to anterior through a canal and a pore located on the side of mid-dorsal line and collar coeloms do not have any connections with proboscis coelom.

Mesodaeum. The mid gut of alimentary canal which is derived from endoderm of embryo and represented by archenteron in embryo.

Mesoderm. Refers to the germ layer from which muscles, connective tissues, andblood system are usually developed. At gastrulation the mesoderm lies between ectoderm on the outside and endoderm lining the gut.

Mesodermal Crescent. Refers to the granular cytoplasm in the egg of *Amphioxus* present between clear cytoplasm and yolky cytoplasm which is developed into mesoderm.

Mesodermal Sacs. During embryonic stage the mesoderm gives rise to the series of segmentally arranged hollow sac-like structures which are known as mesodermal sacs.

Mesodermal Somites. During embryonic development the mesodermal material after differentiation and covering undergoes division thereby forming segmentally arranged block-like structure which are mesodermal somites.

Mesogenesis. The formation of mesoblastic somites and differentiation of lateral plate mesoderm in the embryo.

Mesogles. Refers to layer of jelly-like material that separates. It is secreted by the ectoderm and endoderm in coelenterates.

Mesometrium. Mesenteries that are suspending the uterus in the body cavity.

Mesonephors. Refers to the type of vertebrate kidney that is present in the embryo of most vertebrates. It persists in adult cyclostomes, fishes and amphibians but replaced by metanephros in reptiles, birds and mammals.

Mesophilic. Used for describing micro-organisms having an optimum temperature for growth between 25-45^{0}C.

Mesophragm. A line which is crossing the H-zone of A-band in striped muscle fibers. Also called *M-line*.

Mesopeterygium. Refer to the middle of three baslia present in pectoral girdle of fishes.

Mesoptiles. Refers to the feathers in young bird following nestling downs. They gets replaced by teloptiles in adult.

Mesorchium. Part of peritoneum which is suspending the testis in the body cavity in animals like fishes, amphibia and reptiles etc.

Mesoternum. Refers to the part of sternum in frog which is situated between xiphisternum and epicoracoid portion of pectoral girdle.

Mesothorax. Middle portion of thorax.

Mesotubaeia. Large mesemtery-like membranes which are able to suspend oviducts and uteri in the body cavity.

Mesoverium. Part of peritoneum which is able to suspend ovaries in the abdominal cavity in animals like fishes, amphibia and reptiles.

Mesozoic. Describing the middle ear in the geological time scale, which is dated from about 230-70 million year ago. It is known as the 'Age of Reptiles'. It has been divided into three main periods : the Triassic, Jurassic and Cretaceous.

Messenger RNA (mRNA). The form of RNA that is able to transfer the information necessary for protein synthesis from the DNA in the nucleus to the ribosomes in the cytoplasm. One strand of the double helix of DNA is acting as a temple alone which complementary RAN nucleotides are aligned. These give rise to a polynucleotide which is identical to the other DNA strand, except that the thymine bases have been replaced by uracil. This process is known as *transcription*. The new mRNA molecule is thus having a copy of the genetic code, which directs the formation of proteins in the ribosomes.

Metarterioles. Smaller muscles walled vessels which arise from arterioles.

Metabolism. The chemical reactions taking place in cells. The molecules taking part in these reactions are known as *metabilities.* Some metabolities are synthesized within the organism, where as others have to be taken in as food. It is metabolic reactions, especially those that produce energy, that keeps cells alive.

Metabolite. Defined as a substance which is taking part in a metabolic reaction, either as reactant or product. Metabolites thus act as intermediates in metabolic pathways. Some are being synthesized within the organism itself, while others have to be taken in as food. See also *metabolism.*

Metacarpal Bones. Rod-shaped bones that are present in the lower forelimb or forefoot of tetrapods; they form the palm of the hand in man they undergo articulation with the carpal bones proximally and phalanges distally.

Metacarpus. Refers to the collection of metacarpal bones that form part of the forefoot or lower forelimb in tetrapods; the palm of the hand in man.

Metacentric. Chromosome having centromere in the middle producing a chromosome having two arms which are equal in length.

Matachromatic Stains. Colouring agents which stain tissues a colour different from that of the stain. An example is methyl violet which is violet blue but stains tissues purple or red.

Metachronal Rhythm. Refers to a pattern of movement which is shown by cilia, parapodia of certain polycheates. etc. In this each beats one after the other in regular succession and gives the appearance of wave motion. A wave which passes forwards may propel the surrounding medium backwards or the organism forwards.

Metachronous. Means a wave-like successive beating of cilia in a row (one advance behind).

Metacromian Process. The backward process which arises from acromial process of mammals. This process is developed specially in cases which possess reduced clavicle.

Metameric Segmentation (metamerism, segmentation). The repetition of body parts of an animal along the logitudinal axis of the body to form a series of similar units (called *segments* or *metameres*).

Metamorphosis. A phase in the life history of many animals during which there occurs a rapid transformation from the larval to the adult form. Metamorphosis is quite common among invertebrates, especially marine organisms and arthropods, and is typical of the amphibians.

Metanephridium. An excretory organ in many Annelida. It consists of a tubule that opens into the coelom by a ciliated funnel. It conducts waste fluids to the exterior.

Metanephros. Means the third type of vertebrate kidney which develops the mesonephors, to which it is posterior, forming the functional kidney of reptiles, birds and mammals.

Metaphase. Refers to the stage in mitosis and meiosis when the chromosomes are aligned along the equator of the nuclear spindle.

Mateplasia. The change from one tissue type to another as observed response to certain diseases or abnormal conditions. An example is that the epithelium of a respiratory tract may exhibit metaplasia in response to irritants, e.g., smoke.

Metapleural Folds. Refers to the two folds on the anterior ventral surface of *Amphioxus* that extend between mouth and antriopore.

Metapodium. The distal portion of autopodium which is forming palm in hand and sole in foot. It is supported by the five long metacarpals or metatarsals.

Metatarsal Bones. Rod-shaped bones present in the lower hindlimb or hindfoot of tetrapods; they produce the arch of the foot in man. They undergo articulation with the tarsal bones proximally and phalanges distally. In the typical pentadactyl limb there are five, although there are modifications to this basic plan; in fast-running mammals (e.g. the horse) they are greatly enlongated and raised off the ground. *Compare* metacarpal bones.

Metatarsus. The collection of metatarsal bones that form part of the hindfoot or lower hindlimb in tetrapods. An example is the arch of the foot in man.

Metatheria. The mammalian sub class that includes the marsupials (prouched mammals). Marsupials are more primitive than the placental mammals.

Metazoa. A subkingdom of multicellular animals whose bodies are composed of specialized cells grouped together to form tissues and that have a coordinating nervous system. This subkingdom includes all animals except the Protozoa and Parazoa (sponges).

Metaencephalon. The posterior portion of neural tube (rhombencephalon) which produces cerebellum.

Methystylic. A variation of autostylic jaw suspension. It is present in teleosts in which anterior part of upper jaw gets reduced and its place has been taken up by membrane bone like palatines and pterygoid. The position part has been replaced by metapterygoid and quadrate bone. Quardate suspends the lower jaw.

Microdissection (micromanipulation). A technique of dissecting which is carried out under a microscope by using fine mechanically manipulated instruments. Such techniques find use when dealing with organisms at the cellular level, and have been applied successfully in transferring nuclei between species of *Amoeba*.

Microfilament. A minute filament which is about 6 nm wide. It is found in eukaryotic cells and having roles in cell motion and shape. Microfilaments are composed of globular protein subunits almost identical to the action of muscle.

Micromeres. The smaller cells which are formed by unequal 3rd cleavage during the early development. These cells finally give rise to the formation of ectoderm of adult.

Micronucleus. The small nucleus in ciliates like *Paramecium*. It carries out both vegetation and reproductive functions.

Microsomes. Fragments of endoplasmic reticulum and Golgi apparatus in the form of vesicles that ate formed during homogenization of cells.

Microtome. An instrument which is used for cutting thin sections (slices few micrometers thick) of biological material for microscopic examination.

Microvilli. Elongated slender projections (~2μ m long and ~0.2μ m diameter) of the plasma membrane which occur especially in secretory and absorptive cells. The closely packed arrangement of microvilli on the free surface of epithelial cells forms a brush border. Microvilli increases surface area for the exchange of molecules.

Midbrain (mesencephalon). Refers to one of the three basic anatomical divisions of the brain. It connects the forebrain and hindbrain. It is traversed by the cerebral aqueduct.

Middle Ear. (tympanic cavity). An air-filled cavity in the skull, which is present between the outer and inner ear in most tetrapods. It is connected to the back of the throat by the Eustachian tube. In mammals, it is having three small bones, the ear, ossicles (malleus, incus, and stapes) which link the tympanum (eardrum) with the oval window (fenestra ovalis).

Midgut. The central part of the alimentary canal of arthropod and vertebrate animals which is responsible for digestion and absorption.

Mild Gut Diverticulum. In animals like *Amphioxus* it refers to the ventral projection from the mid gut serving as so called liver-like structure.

Migration. Refers to an instinctive regular two way movement of part of all or an animal population to and from a given area, generally along well-defined routes. It is closely linked to the cycle of the seasons and is triggered off by seasonal factors like increasing and decreasing day lengths in spring and autumn. Many birds, hoofed mammals, bats, whales, fish and insects migrate, often covering immense distances. An examples is that the Arctic breeds on the northern most coasts of Eurasia and América and winter around the Antarctic pack-ice 11000 miles to the south.

Milk. White slimy, proteineous, fatty secretion by mammary glands in mammals which is used to feed young ones.

Milk Teeth. In diphyodont dentition they represent the set of teeth which fall off at early stage (6-8 years) and replaced by another set. Such teeth constitute incisors, canines and premolars.

Million's Reagent. A mixture of mercury, nitric acid, and water that finds use in testing for the presence of proteins. The regent is heated with the test material and the production of a brick red precipitate reveals the presence of protein.

Mimicry. Refers to the resemblance of one animal to another by which the mimic is having advantage from its resemblance to the model. In *Batesian mimicry* there are certain edible insects that mimic the warning colouration of noxious insects and so get avoided by their predators.

Mineralocorticoid. A type of steroid hormone which is produced by the adrenal cortex. Mineralocorticoids (e.g. aldosterone and deoxycorticosterone) control salt and water balance by acting on the kidney.

Mineral Salts. Used for inorganic Cl^-, SO_4^{2-}, PO_4^{3-}, CO_3^{2-}, of calcium, magnesium sodium and potassium. These are taken in the body along with diet and from electrolyte balance in the body.

Miocene. Refers to a epoch of the Tertiary, 25-7 million years ago. The climate became drier and the grasses evolved and spread rapidly, this probably is explaining the replacement of early mammals by more modern forms. About half to three quarters of exiting mammalian families are represented in rocks of the Miocene.

Miltochondria. 'Powerhouse of cell'. It is an organella of cell which is made up of an envelop like outer membrane and inner membrane which gets raised in form of cristae having a number of oxysomes which acts seat of oxidative enzymes that convert pyruvic acid to carbon dioxide and water and release of energy that is used for regenerating ATP.

Mitosis (karyokinesis). Used for describing the ordered process by which the cell nucleus and cytoplasm get divided in two. The chromosomes undergo replica ion prior to mitosis and are then separated during mitosis in such a way that each daughter cell is having a genetic complement which is identical to that of the parent cell. Although mitosis is a continuous process, it has been divided into four phases; prophase, metaphase, anaphase and telophase. (Fig. on p. 226).

Mitral Valve (biscupid valve). Refers to a valve with consists of two membranous flaps or cusps situated between the atrium and ventricle of the left side of the heart in mammals and birds. When the ventricle contracts, blood is not allowed from returning to the atrium by closure of the valve.

Modality. Quantitative aspect of physical means e.g. modalities of sensation include taste and smell.

Modifier Gene. An individual having a pair of alternative gene forms (alleles) heterozygous at a single locus under consideration. It also refers to a cross in which a single pair of alleles in being followed.

Molars. Refers to the grinding teeth in heterodont dentition. They are multirooted with broad crown having ridges furrows, grooves, cusps etc., and are used for grinding food. Permanent teeth.

Mollusca. A phylum of bilaterally symmetrical unsegmented invertebrates, including the aquatic bivalves, mussels, octopuses, squids, etc., and the terrestrial slugs and snails.

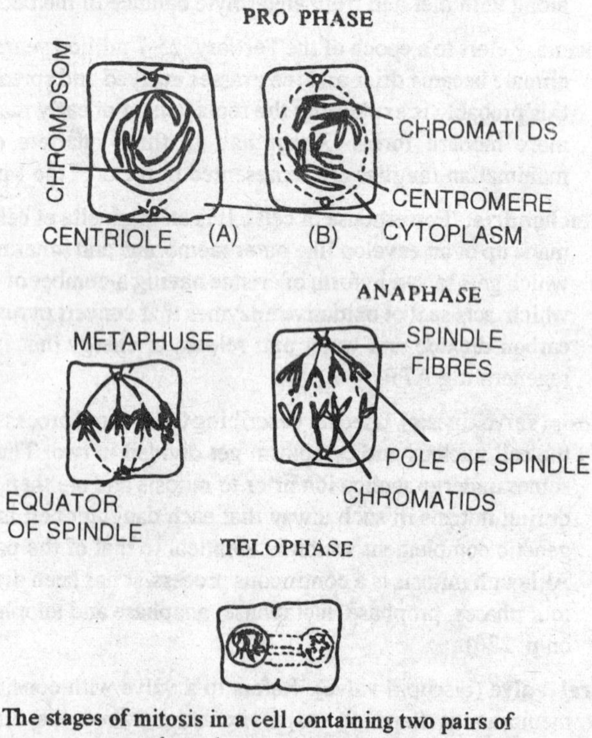

The stages of mitosis in a cell containing two pairs of homologous chromosomes

Fig. 2. Stages of mitosis

Molluscs. Non-segmented, schizocoelomate phylum which includes mussels, snails, slugs, octopus etc.

Moner. A major division of organisms that are characterised by absence of nuclei and chromosomes. It includes some bacteria and blue green algae.

Monimostylic Jaw. Refers to the autostylic jane suspension where quadratę in immovable. Found in lizards.

Monocular Vision. The vision when each eye is having independent visual functioning as in fishes, amphibia etc., except mammals.

Monocytes. Larged sized WBCs of mammals (10/15 μ in diameter) that have large extranuclear protoplasm and kidney shaped nucleus with kink or indentation one side 1-3% of WBCs and phagocytic. They help to engulf bacteria etc., to fight infection.

Monoecius. Used for describing a condition when both male and female reproductive organs are present in a single individual; (Also called *hermaphrodite*); opposite to *dioecious*.

Monohybrid. A hybrid with respect to only one pair of allelic genes e.g., Aa or Bb.

Monohybrid Cross. The genetic cross which involves only one pair of contrasted character e.g., cross between tali and dwarf plants.

Monophylectic. When evolved from a single known ancestral type.

Monophyodont Teeth. Refers to single set of teeth in life as molars of mammals. These teeth are not replaced.

Monosaccharide. A simple sugar which cannot be further hydrolysed e.g. glucose, pentose etc.

Monosomy. Refers to a chromosomes anamoly in which there is only one of the given kind of chromosomes instead of two.

Monotremata. Group of egg-laying mammals that are having a cloaca instead of separate anus and urinogenital openings.

Morphogenesis. Refers to the development of form and structure.

Morphology. Refers to the study of the form of organisms. It is possible to use this term synonymously with *anatomy* although generally the study of form is called *morphology* while the study of internal structure is called *anatomy*.

Morula. A loose aggregation of blastomeres that is formed from cleavage of the egg of mammals. It develops into the blasocyst.

Mosaic Egg. An egg whose cleavage products (blastomeres) are determined in their fate there occurs the second cleavage. If the blastomeres become separated, they give rise to the incomplete embryos.

Mosaicism. Occurrences of two or more genetically different cell types taking place within the body of an individual.

Motor Neurone (motoneuron). A nerve cell (neurone) that is able to transmit impulses from the brain or spinal cord to a musical or other effector.

Moulting. The casting off exoskeleron, to shed portions of skin, feathers or hair.

Mouth. Refers to the anterior opening of alimentary canal which is formed generally at the blastoporal end and is lined with skin (ectoderm). It leads to mouth cavty and involves in the ingestion of food.

Mouthparts. Joints appendages on the heads of arthropods which are modified in various ways for dealing with food. They comprise of the *labrum* (upper lip), which is a single plate; a pair of *mandibles* (upper jaws), which possess serrated edges; a pair of *maxillae* (lower jaws), which also have serrated regions;.and the *labium* (lower lip), which is a single plate.

Movement. Active spatial displacement in an animal.

Mucin. Refers to a glycoprotein (protein conjugated with glucose) which lubricates food. It is present in saliva, gastric secretion and large intestinal secretion.

Mucosa. A membrane secreting mucous. Usually used for innermost lining of certain organs like divisions of alimentary canal and uterus etc.

Mucous Glands. Glands in skin, mucosa of alimentary canal secreting mucopolysaccharide mucin which is a slimy structure.

Mucous Membrane. Refers to the tissue, in vertebrates that is lining many tracts that open to the exterior. It consists of surface epithelium having goblet cells, which secret mucus, and is underlaid by connective tissue.

Mucrone. In few sporozoa (e.g. *Monocystis*) is represents an anterior structure in some individuals which is modification of portion of ectoplasm.

Mucus. A slimy substance which is produced by goblet cells in mucous membranes of animal. It is viscous and insoluble, comprising mainly of glycoproteins. Its function is to protect and lubricate the surface on which it gets secreted.

Mullerian Duct. The oviduct of female jawed vertebrates developed in both sexes from embryonic mesoderm, in association with the Wolffian duct, but becomes vestigial in the male.

Multicellular. Consisting of many cells.

Multifactorial Inheritance. Refers to the control of one character by two or more genes. This produces continuous variation is such characters (e.g. height) and a continuous gradation is apparent from one extreme to the other.

Multiple Allelism. Refers to existence of the series of allele (three or more) for one gene. In man there are three alleles (A, B and O) which govern blood type. Only two alleles of the series can be present in a diploid cell. Dominance relationships within an allelic are usually complicated.

Multiple Factor Inheritance. Refers to the inheritance that involves many genes at different gene loci.

Multiple Fission. Type of asexual reproduction in which nucleus carries out several mitotic divisions before any cytokinesis.

Mumps. A pathogenic condition giving rise to inflamnation of parotid glands.

Muscone. A substance which is responsible for characteristic odour in musk. It is obtained from musk-deer (*Moschus moschiferus*) from a pouch under skin of abdomen of male.

Muscles. A type of connective tissue which articulates bones with each other. It gives rise to a definite shape to body and in locomotion and movements of body parts.

Muscular Dystrophy. Vitamin E deficiency may give rise to reduced and disturbed functioning of body muscles in most of animals. It is known as muscular dystrophy.

Muscular Twitch. The contraction which takes place by skeletal muscle with response to single stimulus. It is having latent period of 0.1 second, contraction period of 0.4 second and relaxation period of 0.5 seconds. Twitch may last as long as 60 milliseconds as in large cells and as small as 10 milliseconds as in muscles of eye-ball.

Muscularis Mucosae. Refers to the muscle in the middle layer of mucosa of alimentary canal which comprise of smooth muscles.

Mutation. Change in a gene that gives rise to a new hereditary variation (trait). It includes all kinds of hereditary variations that result from gross chromosome changes. Mutations may also take place in somatic cells, but these do not get inherited (unless individual reproduces asexually).

Mutation Pressure. Refers to the spontaneous rate of gene mutation which takes place continuously in a population.

Myelencephalon. Means the remaining hind brain of embryonic neural tube (rhombencephalon) after metaencephalon (from which arises cerebellum) which produces medulla oblongata (in adult).

Myeloblasts. Cells in red bone marrow which produce WBCs.

Myeloma. The malignant proliferation of clones of mature plasma white blood cells.

Myelon. Refers to the remaining part of neural after pro, mesen-and metaencephalon that produces spinal cord.

Myocardial Infraction. Used for describing the severe reduction in blood supply to a portion of myocardium due to arteriosclerosis, thrombus formation or other physiological disorder thereby death of myocardial cells and formation of necrotic area.

Myocardial Ischemia. Refer to the reduction in blood supply to a portion of myocardium which occurs die to formation of plaques, arterio-sclerosis or other physicological reasons.

Myocardium. Refers to the muscle layer of heart (derived from endocardial cells) as different form pericardium.

Myocommata. The connective tissue partitions that are present between two successive myotomes as in *Amphioxus* and fishes etc.

Myocyte. Contractile cells that surround pore and oscula sponges and serve to close them.

Myoepithelial. Smooth muscle-like cells lining the ducts of breasts.

Myofibril. Refers to one of the numerous longitudinal small fibrils present within protoplasm of muscle cell or fiber.

Myogenic Heart. The heart whose activity could not be directly controlled by nerves but controlled directly by muscles e.g., heart of all vertebrates.

Myoglobin. An oxygen binding pigment present in the red or slow muscles and in the respiratory enzyme cytochrome-C. Having haeme in the structure.

Myoneme. Type of contractile fibrils present in certain animals or cells.

Myopia. Defect of eye. Short sightedness when eyeball becomes relatively short resulting in the eye's point of focus lying behind instead of on retia.

Myosin. Thick protein of muscle that is involved in muscle contraction and forms A-band. Distinguished into light-and heavy meromyosins.

Myotones. Refers to the dorsal parts of embryonic mesodermal pouch which gives rise to muscles. Found in *Amphioxus*, fishes etc., where they appear like segmental muscle blocks which will be arranged on each side of body.

Myotonic Dystrophy. Refers of a disorder which is characterised by progressive weakness of muscles, development of cataract and also cardiac abnormality associated with a dominant autosomal gene.

Myriapoda. Collective term used for describing arthropods such as millipedes and pauropods.

Myxoedema. Thyroid deficiency disease which is characterised by swelling within and beneath skin.

N

Nacre. Mother of pearl.

Nacrolepsy. The disease in which a patient is having an eventually irresistible urge to sleep during day time activities.

NAD (Nicotinamide Adenine Dinucleotide). A derivative of nicotinic acid which acts as a coenzyme in electron-transfer reactions (e.g., the electron-transfer chain). Its role is to carry hydrogen atoms; the reduced form may be written as NADH.

NADH-(Dihydro Nicotinamide Adenine). The substance which is formed by accepting hydrogen by NAD (during electron transport).

NADP. A coenzyme which acts as hydrogen acceptor and is changed to NADPH (Dihydro-nicotinamide adenine dinucleotide phosphate) during electron transport system.

NADPH (Dihydro Nicotinamide Adenine Dinucleotide Phosphate). The substance which is formed by accepting hydrogen by NADP during electron transport system.

Naiad. Refers to an aquatic gill breathing nymph.

Nails. Refers to the characteristic structures which are present on the digits in primates like man; possess nail plate and subunguis very much reduced and represented by a narrow strips of less modified epidermis below the edge of the unguis. Dermis of nail bed which underlies the unguis is highly vascular which is evident from pinkish colour of nail.

Nasal Cavity. Refers to a paired cavity in the heads of vertebrates. It has thin bones covered with a mucous membrane and it is lined with epithelium, some regions of which have nerve endings which are sensitive to smell. In mammals the large area of mucous membrane warms and moistens air when it passes through on its way to the lungs. Hairs near the nostrils and mucus from the membrane filter dust and bacteria from the air.

Nasolacrymal Duct. Refers to the duct from lacrimal sinus. It terminants into sinuses inside certain bones of nasal passage.

Natural Pathenogenesis. Refers to the development of an egg without involving fertilization during normal life cycle of an animal as found in aphids (a type of insects).

Natural Selection. The process, which was called the 'struggle for survival', by Darwin. By this process organisms less adapted to their environment tend to perish, and better adapted organisms tend to survive. According to Darwinism, natural selection acting on a varied population gives rise to evolution.

Nature and Nurture. The interaction taking place between inherited and environmental factors (*nature* and *nurture* respectively) in determining the observed characteristics of an organism.

Nauplius. An early free-swimming larval stage of some crustaceans. It has only three pairs of appendages (antennules, antennae and mandibles) and later moults to form the cypris larva.

Neanderthal Man. A sub-species of man, *Homo sapiens neanderthalensis*, which was replaced by modern man about 40,000 years ago.

Necrosis. Means the death of living cell, tissue etc.

NEFA (Non-Essential Fatty Acids). Also termed *free fatty acids.* These are present in plasma and are bound to lipoproteinic complexes. These are not including linoleic acid, linoleinic acid and archdonic acid which are essential fatty acids.

Nekton. Animals of the pelagic zone of a sea or lake that are free swimming and independent of tides, currents, and waves, such as fish, whales, squid, crabs and shrimps.

Nelson's Syndrome. Refers to a pathological condition caused by ACTH-secreting pituitary tumours which are developed in some patients whose bilaterally hyperplastic adrenals have been removed. These tumours bring about hyperpigmentation of skin and neurologic signs due to pressure on structure in celler region.

Nematocysts. Refers to the stinging capsules that are found in *Hydra* like animals (coelentrates). Each is produced by a single cell called cnidoblast.

Nematomorphs. Used for describing a group of animal such as hair snakes having thread-like body, blunt anterior end rough cuticle, adults free living but larvae parasite in insects e.g. *Gordus*.

Neocortex. Refers to the part of cerebral hemisphere which are arranged in six layers numbered I-VI. These are:

Laser I. Molecular layer having numerous dendrites, axons and glial cells.

Layer II. External granular layer having small densely packed cells which are rounded and polygonal.

Layer III. External pyramidal layer having large pyramidal cells usually increases in size from without inwards.

Layer IV. Internal granular layer that resembles layer II.

Layer V. Internal pyramidal layer having typically cells which resemble those present in layer III.

Layer VI. Fusiform cell layer having long spindle cells which are arranged perpendicularly to the surface.

Its main connection is to receive information directly or indirectly from all receptors in the body and each muscle and gland cell gets influenced by impulses originating in it. Nearly every single act gets influenced by this part of brain.

Neo-Darwinism. Refers to the new theory which are obtained by modification of Darwin's theory of evolution through natural selection, by modern genetic studies arising from work of Mendel. Such studies have answered many questions which Darwin's theory raised, but failed to explains because of lack of knowledge at the time it was formulated.

Neo-Lamarckism. Refers to various modified versions of Lamarck's theory of evolution. It is sometimes incorporating ideas of natural selection, though generally still insisting that inherited changes could be linked directly to particular environmental stresses.

Neolithic. Refers to the recent Stone Age, from about 10000 years ago until the beginning of the Bronze Age. It is characterized by more advanced and often polished stone tools and the development of agriculture.

Neopallium (neocortex). Pertains to a type of nerve tissue that it constituting most of the cerebral cortex of mammalian brains.

Neoteny. Refers to the retention of larval or other juvenile features that are beyond the normal stage in the development of an animal. It may be either temporary due to the climatic or other factors, or permanent in which case the animal breeds in the larval stage.

Nephridiopore. The external opening of a nephridium.

Nephridium. An excretory organ which is present in many invertebrates (e.g. Platyhelminthes, Rotifera, Mollusca and Annelida) and *Amphioxus.*

Nepfrocoel. Refers to the space present inside nephrotome.

Nephromixium. Reproductive-excretory structure which is present in certain polychaetes.

Nephron. The excretory unit of the vertebrate kidney. It consists of a

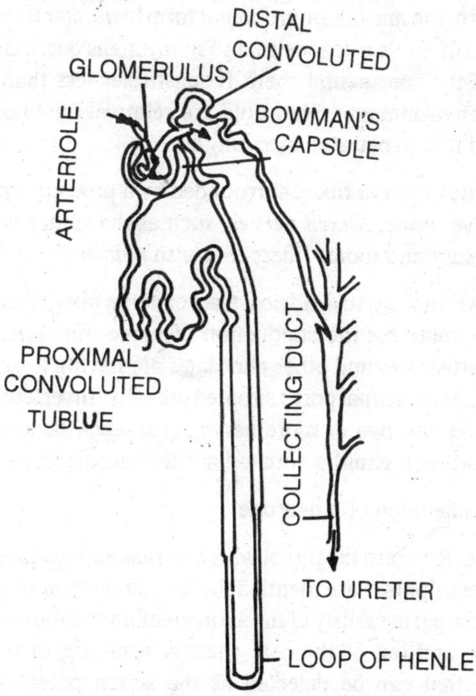

Fig. 1 Structure of a single nephron

Malpighian body (glomerulus and Bowman's capsule) and a uriniferous tubule. Water, salts, nitrogenous waste, etc., are filtered across the walls of the glomerulus and collected by the Bowman's capsule.

Nephors. A structure formed by several nephors in kidney and carrying out excretory function.

Nephrostome. Ciliated funnel-like opening present at the inner end of a nephridium through which excretory wastes enter the tubule.

Nephrotome. Mesoderm of vertebrate embryo that gives rise to nephron (kidney tubule)..

Nereis. A genus of polychaete worms, the ragworms that are found in the intertidal zone, where at low tide they form U-shaped burrows in the mud.

Neritic. Refers to the marine environment form low water level to a depth of about 200 m. It is a zone that in many areas corresponds to the extent of the continental shelf. It constitutes less than 1% of the marine environment. Nutrients are relatively abundant in this zone and it is penetrated by sunlight.

Nerve. A bundle of nerve fibers surrounded by a protective covering of connective tissue. *Mixed nerves*, such as the spinal nerves, have both sensory and motor fibers. See also *neurone*.

Nerve Cord. An enclosed cylindrical tract of nerve fibers that is forming a central route for the conduction of nerve impulses within the body. Vertebrates and other chordates are having a single hollow nerve cord (the spinal cord) situated dorsally. Invertebrates generally are having two or more nerve cords, each lacking a ventral cavity and with ganglia situated at intervals along its length.

Nerve Fibre. The axon of a neurone.

Nerve Impulse. Refers to the signal which is transmitted along neurones. All nerve impulses are identical in form and strength and involve changes in permeability of the axon membrane followed by flow of ions into and out of the cell. thereby resulting in the potential changes that can be detected as the action potential which is passing along the axon. The energy required to pass the impulse is obtained from the neurone itself, not from the stimulus.

Nerve Net. A net-like layer of interconnecting nerve cells. It is present in the body wall of certain groups of invertebrate animals; the most primitive type of nervous system. It is found in coelenterates, echinoderms, and hemichordates.

Nervous System. A ramifying system of cells. It is found in all Metazoa. It forms a communication system between receptors and effectors and permits varying degrees of co-ordination of information from different receptors and stored memory, giving rise to the integrated responses to stimuli. The system comprises of neurons, supportive glial cells, and various fibrous tissues, surrounds the softer matter. Impulses get transmitted through the neurones.

Neural Arch. An arch of bone of cartilage. It is arising dorsally from the centrum of a vertebra and enclosing a canal—the spinal or neural canal—through which the spinal cord is running. It may bear a number of projections, such as a neural spine, for attachment of muscles.

Neural Canal. A canal which is formed by neural arches. It extends through the vertebra by which spinal cord passes.

Neural Crest. Refers to a crest of specialized cells at the edge of the neural plate in vertebrate embryos that comes to lie above the neural tube when it closes over. The cells of the neural crest migrate separately thereby forming pigment cells, Schwann cells, and gill arches and aggregate to form dorsal root and sympathetic ganglia; neural crest also contributes most of the head mesenchyme.

Neural Gland Complex. In protochordates like *Herdmania* it represents a specialised complex that comprises of excretory system, nerve ganglion and dorsal tubercle.

Neural Plate (medullare plate). Refers to the area of the outer layer of vertebrate embryos that gives rise to the neural tube. The notochoridal tissue, which comes to the beneath this area, gives rise to substances that restrict the fates of the overlying cells to neural elements rather than epidermal.

Neural Spine. A spine-like structure/outgrowth of vertebra which is formed by union of a pair of rods called laminae that are present on dorsolateral side of it.

Neural Tube. Pertaining to the first formed element of the spinal ford and brain of vertebrate embryos. It is generally formed by the neural plate rolling up and sinking beneath the surface. The neural tube usually opens to the outside anteriorly by the *neuropore,* posteriorly it may communicate with the archenteron via the *neurenteric canal* (until the tail gets formed).

Neurentric Canal. Refers to the narrow connection which exists between neural tube and archenteron of neurulla in the posterior region.

Neurelemma. The outermost sheath of a nerve fiber.

Neurin. Refers to a filamentous which is attached to the inner surface of plasmalemma of nerve endings to the brain. It is similar to muscle action and may be involved in exocytosis.

Neuroblast. Used for describing the cells of animals embryos that become or produce nerve cells, frequently by under going unequal mitotic divisions. Although many of the cells of the neural tube of vertebrates will produce nerve cells, the term neuroblast is usually limited to those cells whose immediate progeny will differentiate as nerve cells.

Neurocoel. Refers to a cavity in chordate nerve cord.

Neurocranium. A structure consisting of cartilagenous brain box and three sense capsules developed in connection with sense organs. It gets developed in embryonic stage only.

Neuroectoderm. Pertains to the prospective area of embryo that is destined to form central nervous system.

Neuroendocrine Mechanism. A mechanism which involves regulation of some body function involving endocrine secretions which in turn get regulated by some part of the nervous system (according to homeostatic requirement of organism). When body blood sugar level has been regulated by hormones secreted by islets of Langerhans, the secretion of which is in turn controlled by PSH (pancreas stimulating hormone) from pituitary released under the influence of PSH-RF (pancreas stimulating hormone releasing factor) from hypothalamus.

Neuroendocrine Systems. The systems that are involving both nervous and endocrine factors. They control functions of the body. Many examples are known, especially involving the pituitary gland, from which hormones are secreted under direct nervous stimulation.

Neurogentic Pacemaker. A pacemaker which is specialised nerve cell.

Neuroglia (Glia). A specialised tissue which is present in the central nervous system of vertebrates. It supports and protects the nerve cells. It comprises of various types of cells, including *astrocytes,* which have many fine processes, smaller *oligodentrocytes,* and *ependymal cell,* which may be ciliated and the line the brain cavities and spinal canal.

Neurohormone. A hormone produced by specialized nervous tissue. Examples are noradrenaline, serotonin, vasopressin, and oxytocin. See Neuroendocrine systems.

Neurohumor. Refers to the synaptic transmitters and neuro-secretory hormones.

Neurohypophysis. Used for describing a neurally derived reservoir for hormones with antidiuretic and oxytocic actions. It comprises of neural stalk which is connected to and pass neurosecretions from hypothalamus and neural lobe which constitute the bulk of neurohypophysis.

Neuromast. Refers to one of numerous groups of sensory cells that are present in pits or canals, scattered or arranged in rows over the head and along the body (lateral line) of fish and aquatic amphibians. The sensory cells are bearing hair like processes, which detect vibrations of a frequency too low to be perceived by the ear.

Neuromuscular Junction. Refers to the specialized region in which a nerve ending is able to make close contact with a muscle. Impulses arriving at the nerve ending bring about a release of a chemical transmitter, which diffuse across the intervening gap and stimulates the muscle to contract.

Neurons. Refers to the constituent cells of nervous tissue. Each possesses cell body/cyton or perikaryon having nucleus and Nissl's granules etc., dendrites —which constitute one or more processes arising from cyton and axon which is a long process terminating at affector or synaptic junction. Length of neuron may be from few mm to several cm.

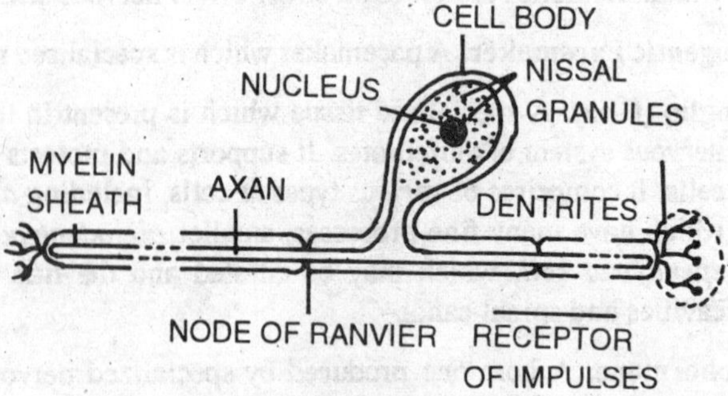

Sensory Neurone

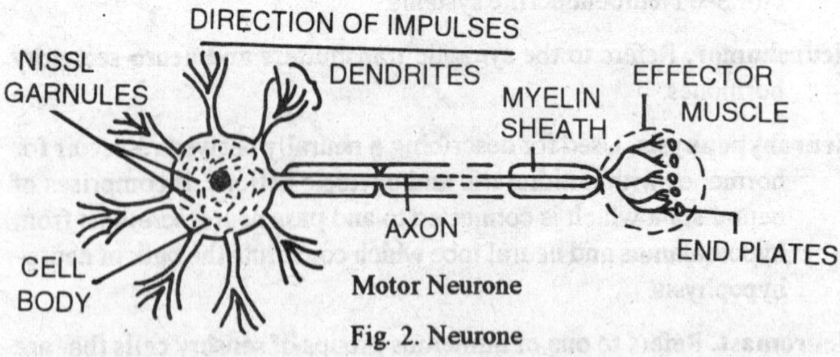

Motor Neurone

Fig. 2. Neurone

Neuropophysins. Proteins which are associated with neurohypophysial hormones stored in granulesin neurosecretory terminals. They are cleaved from the hormones before secretion.

Neuropore. Closure of neural tube in neurula does not take place uniformly at all places. It starts from middle which then precedes both anteriorly and posteriorly. Thus, for sametime the neural tube remains open in front of the prospective brain box by the opening which is called neuropore.

Neurosecretion. Neurons having garnules that stain with Gomori stain are generally known as neurosecretory neurons and the secretion of chemical substance at their endings is known as neurosecretion.

Neurostennin. A complex of two proteins present in brain nerve cell endings and more possibly associated with exocytosis.

Neurotransmission. Refers to passage of impulse which takes place from nerve to muscles. It is chemical at synapse, electrical in pre-synaptic region and it is both chemical and electrical at few of the conjoint synapses.

Neurotransmitter. A chemical released from neurone endings to bring about either excitation or inhibition of an adjacent neurone or muscle cell. It is stored in minute vesicles near the synapse and release when a nerve impulse arrives. The most common neuro-transmitter is acetylcholine.

Neurula. Refers to that stage of vertebrate embryos, following gastrulation, when the neural tube is formed (the process is known as *neurulation*).

Neuter. Describing individuals that do not have sex organs.

Neutrophies. Type of lecucoytes which constitute 65-75% of total WBCs with granular cytoplasm and many lobed nucleus. They are phagocytic in function.

Neutrophil. A white blood cell (leucocyte) having granules that do not stain with either acid or basic dyes. Neutrophils are having a many-lobed nucleus and are therefore known as *polymorphonucleur leucocytes* or *polymorphs*.

Nexal Junction. A small gap between cells.

Nexus. Refers to a gap junction between cells characterised by 2nm space between opposing membranes filled with densely packed particles through each of which there is a channel that connects each cell. It allows direct transfer of ions and others substances (small molecules) between cells that are traversing the extracellular space.

Niacine. It is a member of vitamin-B complex group. It is a constituent of NAD and NADP. It occurs in yeast, lean,meat, liver etc. Its deficiency causes pellagra. It could be synthesised in body from tryptophan.

Niche. Constellation of environmental factors into which a species could fit or which is needed by a species.

Nicotinamide. Part of NAD molecule which is involved in various biological reactions.

Nictiatating Membrane. Transparent, thin membrane which many vertebrates are having. Each opens and closes laterally across the cornea.

Nidamental. A gland in female cephalopod which is able to secrete a protective egg capsule.

Nidicolous. Young ones in birds that are helpless for some time and cannot walk and look after themselves soon after their birth e.g., chicken in sparrow.

Nidifugous. Also termed as precocious i.e., young ones in birds that are able to walk and look after themselves soon after their birth. Chicken in *Gallus* (fowl) is an example of this category.

Night Blindness. Means nyctopia.

Nipple. Refers to the structure which is formed by raising of skin at places and at which the ducts of mammary glands open. It is made up of erectile tissue. Their number and position are different in different mammals.

Nissl's Granules. Refer to the numerous bodies in the cytoplasm of cyton of a nerve cell. These have ribosomes and RNA and are produced under the influence of nucleus. They play active role in metabolism of nerve cell. They are not present in dendrites especially in the region of their origin.

Nitrogen Narcosis. If a diver breathes compressed air, the increased partial pressure of nitrogen can bring about symptoms resembling alcohol intoxication at greater pressures in which manual dexterity is maintained but intellectual functions become impaired.

Nociceptive Stimuli. It is possible to produce flexor responses by innocerous stimulation of skin or by stretch of muscles but strong flexor response with withdrawal get initiated only by stimuli that are noxious or at least potentially harmful to animal, these are therefore known as nociceptive stimuli.

Nocturia. A renal disease in which urine volume gets increased which is resulting in walking up at night to void.

Nocturnal . Active at night. It is *opposite* to diurnal.

Nocturnal Enuresis. One of the sleep disorders which results in bed wetting.

Nodal Point. The optical centre of eye. It also refers to the point through which light rays from object may pass without refraction.

Node of Ranvier. Axon of nerve cell gets constricted at places to form this structure. The portion of axon between two such modes is known as internode.

Non-Assortative Mating. That type of mating in which mate selection is random.

Non-Chordate. An animal that does not have notochord in entire life cycle, e.g., earthworm cockroach etc. *Compare with* chordata.

Non-Competitive Inhibition. Enzyme inhibition which occurs due to destruction of active site.

Non-Disjunction. Refers to the failure of chromosomes, chromatids to become separate at the mitotic or meiotic anaphase. Non-disjunction is responsible for several human chromosome anamolies.

Non-Secretor. A person having A & B antigens on the RBCs but not in any body secretion.

Non-Sense Mutation. A mutation resulting in change from a codon specific for an amino acid to a terminating chain — terminating codon which will bring about translocation of a premature halt.

Nor-Adrenaline. A catecholamine that is secreted by adrenal medulla and also liberated at nerve eadings. It increases blood pressure by vasoconstrictions free fatty acid level, CNS stimulation, heat production and peripheral resistance and decrease cardiac out put.

Nor-Ethindrone. An orally active synthetic drug acting as a progestaional agent; gestagen or progestin. It blocks the midcycle burst of LH secretion that causes ovulation and is thus used as contraceptive agent.

Normoblast. When a pre-erythroblast becomes rounded and possesses intense synthetic activity then it is known as normoblast.

Notochord. A cylindrical chord/string-like structure which is along mid-dorsal line in all the chordates. It is also known as *chorda* *dorsalis.* It consists of a mass of vacuolated cells which are surrounded by three layered sheath (made up of external elastic membrane, middle fibrous layer and internal membrane).

Nrem Sleep (Non-Rapid Eye Movement Sleep). Also known as slow wave sleep which is characteristic of deep sleep having a pattern of synchronised slow waves.

Nuclear Bag Fibres. The intrafusal fibres in mammals in which many nuclei are present in a dialated central area.

Nuclear Chain Fibres. Refer to the type of intrefusal fibres in mammals which are thinner and shorter and lack a definite bag. Ends of this connect to the sides of nuclear bag fibres.

Nuclear Membrane. A thin double layered protein-lipid membrane that is enveloping nucleus. It is perforated at intervals having pores of 500 A^0 in diameter.

Nuclease. Eneyme acting on nucleic acids yielding formation of nucleotides.

Nucleolus. A spherical well-defined body present inside the nucleus of cells in most cases.

Nucleoplasm. Refers to the viscous ground substance which is filled in the vicinity of nucleus.

Nucleoprotein. Molecular complex which is formed by nucleic acids and proteins.

Nucleosidase. An enzyme which on acting on nucleosides in able to break union between purine or pyrimidine combined with pentose sugar (ribose or deoxyribose).

Nucleoside. Refers to the combination of purine or pyrimidine with pentose sugar (ribose or deoxyribose).

Nucleotidase. An enzyme which on acting on nucleotides (purine or pyrimidine combined with ribose or deoxyribose sugar and phosphoric acid is able to release phosphoric acid and resulted in the formation of nucleosides.

Nucleotide. A molecule that consists of phosphate, pentose sugar and purine or pyrimidine. (Also see *DNA*).

Nucleus. Cell organelle present in eukaryotic cells. It consists of genetic material in the cell which is separated from cytoplasm by a nuclear membrane. It is able to control all the activies of cell including reproduction. It divides amitotically, mitotically or meiotically. During cell division the nucleus divides and one of the products (daughter nuclei) goes to each new cell.

Nucleus of Pander. In the center of oocyte a mass of white yolk is present that continues to grow vertically as neck-like structure and the neck expands into a disc-like area lying just beneath the blastodisc in the egg of a bird. This is known as nucleus of pander.

Nucleus Pulposus. In mammals the mesenchyme cells present between vertebral centra produce discs which are obliterating the aotochord. However, it persists as nucleus pulposus or pulpy nucleus of intervertebral discs.

Nutrition. Defined as the sum of the processes that are connected with growth, maintenance and repair of living body as a whole or its constituent parts.

Nyctalopia. A vitaminosis of vitamin A causes disturbance in synthesis of rhodopsin and iodopsin resulting in night blindness (a visual abnormality).

Nystagmus. The term used for the jerky movements of eye which take place at the start and end of a period of rotation. It is actually a reflex which is able to maintain visual fixation on stationary points where as the body rotates.

Nyetatine. Rod-shaped antibiotic molecule which is able to create channels through membranes that permits passage of molecules of diameter less than 4 A^0.

O

Obelia. A genus of marine coelenterates that are attached to seaweeds in small branching colonies. Each colony possesses a stem bearing feeding polyps and protected by a skeletal covering of chitin. The polyps repreoduce asexually, giving either more polyps or free-swimming umbrella-shaped medusae. The medusae produce polyps by sexual reproduction, thereby dispersing the species.

Obesity. A pathological condition in human beings characterised by excessive deposition of fat in the subdermis and much gain in body weight and dimensions. It is due to disordered fat metabolism and usually starts with hyperhagia. This may ultimately lead to disorders of cardiovascular and respiratory systems as well as diabetes etc.

Oblique Septa. Refers to a pair of tough elastic septa that are able to separate thoracic cavities from abdominal cavity in case of birds.

Observation. Used for describing perception either quantitative or qualitative, of a set of phenomenon (the set may be having only one member).

Obtect. Describing pupae (of Lepidoptera) in which the wings and legs get fused to the body Compare coarctate, exarate.

Obturator Foramen. Refers to the foramen between ischium and pubis in the pelvic girdle.

Occipetal Arches. Pertaining to a pair of cartilagenous structure that arise from parachordals behind the otic capsule and get attached to hind part of otic capsules during development of skull.

Occipetal Condyle. Used for describing a rounded prominence on the posterior portion of the tetrapod skull. In mammals there are two occipital condyle; they articulate with the first (atlas) vertebra to allow noddling of the head. There are also two in amphibians, but only one in reptiles and birds.

Occipetal Lobes. Refers to the posteriormost region of cerebral hemispheres.

Occipetal Vertebra. Mean 2-4 cartilages which develop behind parachordals during development of skull and later get incorporated into skull.

Occiput (1). Refers to the posterior portion of the vertebrate skull where it gets joined the vertebral column.

(2). Refers to a skeletal plate at the back of an insects head.

Oceanic. Means the marine environment beyond the continental shelf, which is generally deeper than 200 m. It makes up about 99% of the total marine invironment.

Oceanic Islands. Refers to islands of volcanic origin. Faunal complements of which are depending on motality of animal e.g. few mammals but frequently large bird's species.

Oceanic Region. Means deep water beyond neritic zone in the sea.

Ocellus. Used for describing simply eye of some invertebrates (e.g. arthropods). It consists of a group of light-sensitive cell (photoreceptors). Ocellia are involved in perceiving the direction and intensity of light and do not form an image.

Occlusion. Used for describing decrease in expected response which takes place due to presynaptic fibres sharing postsynaptic neurons.

Occular Muscles. Those muscles which are associated with eye-ball. They help in its movement.

Occulocardiac Reflex. Acetylcholine liberated at vagal endings have been found to depress conduction in atrial musculature and AV-node thereby stimulating reflex vagal discharge by pressing on eye-ball. It is known as *occulocardiac reflex*.

Occulomotor Nerve (Cranial nerve III). One of a pair of nerves that are arising from the ventral midbrain in vertebrates for supplying the muscles of the eye balls. It consist of chiefly motor nerve fibres.

Odonata. The order of insects including the dragonflies, brightly coloured carnivorous insects having a long thin abdomen and two pairs of equal-sized elongated wings. The head bears compound eyes, reduced antennae, and biting mouthparts.

Odontoblast. A cell secreting the dentine of a tooth. Odontoblasts are situated around the outside of the pulp cavity and their long fine processes are extending into fine canals (canaliculi) in the dentine.

Odontognathae. Used for describing a group of extinct toothed birds that had been more advanced than *Archaeopteryx* but flightless. Incorporates *Hesperornis* and *Ichthyornis* (both aquatic), former was a diver.

Odontoid Process. A tooth-like, bony, roughly rectangular structure which is present on the anterior phase of axis vertebra of most vertebrates.

Oesophagus. The tube connecting the buccal cavity or pharynx to the stomach. The lining of mucous membrane is folded to allow expansion during the passage of food. In vertebrates, layers of circular and longitudinal muscle surround the tube; their rhythmic contraction brings about peristaltic waves to move the food down to the stomach. In birds and insects the oesophagus may include the crop.

Oestradiol. The most active oestrogen which is produced by the body. It is metabolized to oestrone and then oestriol.

OeStriol. A metabolite of oestradiol and oestrone which occurs in the urine in high concentrations. It has only weak oestrogenic activity.

Oestrogen. A female sex hormone (a steroid) which is involved in the development and maintenance of accessory sex organs and secondary sex characteristics (e.g. growth of the breast). During the menstrual cycle oestrogen acts on the female sex organs to produce an environment which is suitable for fertilization of the egg cell and implanation and growth of the embryo. It finds use therapeutically to correct oestrogen deficiency, for example at the menopause, and to treat some forms of breast cancer. In oral contraceptives, synthetic oestrogens and progesterone prevent ovulation and the release of gonadotrophins e.g., follicle-stimulating hormone.

Oestrone. A metabolite of oestradiol, which occurs in the urine in high concentrations. It is having only weak oestrogenic activity. See *oestrogen*.

Oestrous Cycle. A rhythmic cycle that is varying from 5 to 60 days and occurring in sexually mature females of most mammal species in the absence of pregnancy. It can take place either throughout the year or only during a breeding season. The central event, oestrus, is a brief period during which the females is both receptive as well as attractive to males and during which both ovulation and copulation normally take place.

The oestrous cycle can be separated into phases.

(1) *Follicular Phase*. Graafian follicles grow in the ovary, and secret the hormone oestradiol, which makes the uterine lining to proliferate and thickens.

(2) *Oestrus*. It refers to the period of 'heat' in which ovulation occurs, stimulated by a rise in the secretion of pituitary luteinizing hormone, and copulation also occurs. In some mammals, such as the cat and the rabbit, ovulation is delayed until copulation has taken place.

(3) *Luteal Phase* (*metoestrus*). A corpus luteum is formed from the follicle and secretes progesterone. Oestradiol secretion decreases. If fertilization follows oestrus, the cycle is suspended in this phase for the duration of the pregnancy.

(4) *Regression Phase* (*dioestrus*). In the absence of fertilization, the corpus luteum and endometrium diminish, ovarian hormone levels fall, and a new follicle starts growing. In some animals this phase is lengthened if pregnancy has not taken place, giving rise to symptoms of pregnancy (known as *pseudopregnancy*).

Off Response. Increased ganglion cell discharge which is produced when light is shined on the area surrounding the retinal spot.

Olecranon Process. A projection on the upper end of the ulna of tetrapods; it forms the point of the elbow in man. It is an attachment site for the triceps muscle—the main extensor muscle of the forelimb.

Olfaction. Refers to the process of smelling objects.

Olfactory. A sense organ which shows capability of smelling objects.

Olfactory Capsules. Refers to the cartilagenous structures in the skull of cartilagenous fishes like *scoliodon enclosing* olfactory lobes and open on outer and inner lower sides and present in antilateral part just anterior to orbits.

Olfactory Epithelium. Sensory epithelium which forms olfactory organ. It is also known as schniderian membrane. The olfactory receptor cells are specialised neurosecretary cells (in fact bipolar neurons). These cells are spindle shaped and their cytoplasm has Nissl's granules. Outer end of cell is long, thin olfactory rod bearing bunch of 5-12 delicate olfactory hair at tip and from other and arises axonic fiber and all such fibers are said to constitute olfactory nerve.

Olfactory Gland. Olfactory epithelium has many mucous producing structures that are called olfactory glands. They are also known as Bowmann's glands.

Olfactory Lobes. Anteroventral parts of fore brain which have power of smell (constitute rhinencephalon).

Olfactory Nerve (cranial nerve I). The nerve connecting the olfactory cells of the nose with the olfactory bulb at the front of the brain in vertebrates. The cell bodies of its sensory fibers are not laying in a ganglion but in the olfactory membrane.

Olfactory Tract. Refers to the posterior stout portion of olfactory lobe that leads to a slightly rounded elevation called tuberculum olfactorium.

Olfactory Ventricles. Extension of paired cavities of cerebral hemispheres into olfactory lobes. They are also known as *rhinocoel*.

Oligocene. Means the epoch of the Tertiary 38-25 million years ago. It is represented in Britain by deposits around Hampstead in London. It is characterized by the gradual disappearance of earlier mammal groups that include primitive insectivores and primates, and their replacement by more modern forms.

Oligochaeta. A class of annelid worms that include the terrestrial earthworms (e.g. *Lumbricus*) and many freshwater species. Earthworms are adapted for burrowing and lack parapodia or head appendages and only a few setae. They are hermaphrodite but crossfertilization is usual. The gonads are restricted to a few fixed segments. The eggs develop in cocoons produced by the clitellum, a glandular structure near the reproductive segments. Development is direct, with no larva.

Oligodendrocytes. Myelin having neurons within the central nervous system.

Oligodendroglia. Myelin secreting cells of glia or neuroglia.

Oligolecithal Egg. The egg present in protochordates and mammals. It is small (0.1-0.2 mm) with low concentration of yolk. The yolk is also dispersed.

Obligomenorrhea. An abnormality of menstruation in which there is scantly menstruation.

Oligospermia. A sperm count below 20 million/ml in man.

Oligotrophic. Describing lakes that deficient in nutrients and consequently low productivity. Compare eutrophic.

Oligouria. A fall in urine volume which takes place due to destruction of most nephrons.

Olivocerebellar Tract. An afferent tract in cerebellum that is able to transmit proprioreceptor inputs from whole body via relay in inferior olive. It distributes all cerebeller cortex and deep nuclei.

Olivocochilear Bundle. Refers to the prominent bundle of efferent fibres in each auditory nerve which is arising in clivary nucleus of opposite side and ending around the bases of the hair cells of organs of corti.

Omasum. Refers to the third region of the specialized stomach of ruminants (e.g. the cow). Here surplus water is absorbed. This part of the stomach is very muscular and possesses a lining of tough cornified stratified squamous epithelium.

Omentum. The structure (remnant of ventral mesentery) by which the alimentary canal in some regions gets jointed to an organ.

Ommatidium. One of the units that make up the compound eye of arthropod animals. Each ommatidium is having a cornea and lens, which focus light onto a sensitive rod (rhabdom). A chemical change takes place in the rhabdom and this stimulates the adjacent retinal cells to send nerve impulses along fibres to the brain.

Omnivore. An animal that is able to eat both animal and plant material. An example is man.

Omosternum. Inverted 'Y' shaped bony structure which gets attached to clavicles tipped by cartilage called episternum.

Onchosphere. The protective layer which is present around hexacanth embryo of tape-worm *Taenia solium*.

Onchogenic Virus. Any virus inducing cell proliferation and tumour formation.

Onchology. Refers to the study of cancers.

One Gene-One Enzyme Hypothesis. According to this theory, each gene is able to control the synthesis of one enzyme. Thus by regulating the production of enzymes, genes are able to control the biosynthetic reactions which are catalysed by enzymes and ultimately the character of the organism. Genes also code for proteins or polypeptides that form proteins, other than enzymes, so the idea could be perhaps more accurately expressed as the one gene-one polypeptide chain hypothesis.

Ontogeny. The course of development of an organism from fertilized egg to adult. Occasionally ontogeny is restricted to describe the development of an individual structure.

Oocyte. Means a reproductive cell in the ovary of an animal that produces an ovum. The primary oocyte is developed from an oogonium, which has undergone a period of multiplication and growth. It undergoes division by meiosis and the first meiotic (or reduction) division gives rise to a secondary oocyte, having half the number of chromosomes, and a small polar body.

Oogamy. Sexual reproduction which involves the fusion of two dissimilar gametes. The male gamete is generally motile and smaller than the female gamete, which is usually non-motile, has a food store, and a retained by the parent.

Oogenesis. Refers to the formation of ova within the ovary of female animals. Precursor cells in the germinal epithelium multiply by mitosis to form oogonia, even before the animal is born. In humans, about 150-500 thousand oogonia are present in the ovaries at birth, each surrounded by a small cluster of cells forming a primary follice.

Oogonium. A cell in the ovary of an animal that undergoes a period of multiplication and growth to give rise to an oocyte.

Ootheca. Covering over the egg of insects such as cockroach formed in oothecal chamber.

Ookinete. Movable zygote of *Plasmodium* (fertilized egg) which is formed of a gametocyte within a mosquite or suitable host.

Oozoid. Tailed larva which is developed from ovum of a sexual zooid in some protochordates such as *Doliolum*.

Open System. A system which is not self-sufficient. Most biological systems are open because they are energy requiring (obtains energy from environment).

Operant Coditioning. A learning under which a response is reinforced thereby resulting in its most frequent occurrence.

Operator Gene. A gene regulating synthetic activity of closely linked structure genes via its association with a regulator gene.

Operculum. (1). The bony plate that covers the gill chambers of a bony fish.

(2). The horny disc closing the aperture of the shell of some gastropod molluscs.

(3). The fold of skin growing over to enclose the gills during the development of a frog tadpole.

Opercular Chamber. Refers to the spaceous cavity in bony fishes e.g., Labeo, which lies between gill slits and operculum communicating to exterior through a single aperture.

Opercular Plates. The four plate-like structures which support opercular flap. These are opecular, preopercular, subopercular and interopercular plates.

Operon. A group of clasely liked genes that act together and code for the various enzymes of a particular biochemical pathway. At one end there is an *operator*, which may under certain conditions be repressed by another gene outside the operon, called the *regulator gene*. The regulator gene produces a substance that binds with operator, makes it inoperative, and so prevents enzyme production. The presence of a suitable substance does not allow this binding, and so enzyme production can start. Another site in the operon, the *promoter*, which initiates the formation of the messenger RNA that carries the code for the synthesis of the enzymes which are determined by all the *structural genes* of the operon.

Ophidia. A group of animals to which snakes belong.

Ophthalmic. Anything referring to eye such as optahlmic artery, ophthalmic nerve.

Ophthalmic Nerve. Refers to the sensory nerve and one of the branches of 5th cranial nerve (trigeminal) supplying skin of upper eye lid, skin of front of head and inside as well as outside of nose.

Opisthonephros. Kidney in some fishes like *Scoliodon* is basically mesonephros but the tubules in adult shift towards the posteriorly side of body giving kidney a shaped that very much looks similar to that of metanephros. In such a kidney anterior portion is non-renal and generally reduced while the posterior portion is functional and contains renal tubules.

Ophiuroidea. The largest class of Echinodermata, including the brittle stars (e.g. *Ophiothrix*). The body is covered by articulating skeletal plates and has a small central disc with long fragile arms, which are used in locomotion. Feeding is carried out by tube feet, which convey food to the ventral mouth.

Opsin. The protein component of the retinal pigment rhodopsin, which gets localized in the rod cells of retina Opsin gets released from rhodopsin when light strikes the retina.

Opsonin. A type of antibody molecule adsorbed on the surface of bacteria, combining with their antigens in such a way that the bacteria get more easily engulfed by neutrophils.

Opsonization. The process by which a host coats the surface of an invading cell with antibodies to make it susceptible to phagcytosis.

Optic Axis. An imaginary straight line which passes through the center of curvature of simple lens.

Optic Capsules. Refer to paired cartilagenous structures which are formed on each side of cranium during development of skull. Unlike olfactory capsules these do not undergo fusion with brain box and remain cartilagenous in many vertebrates. Free optic capsule is forming the sclerotic coat of the eye-ball.

Optic Chiasma. Used for describing the site in the floor of the vertebrate forebrain at which fibers entering via the optic nerves cross to the opposite side of the brain, thereby forming an X-shaped pattern. In man and other mammals having good stereoscopic vision, only half the fibres from each side may cross over. In lower vertebrates, nearly all the fibers may cross over.

Optic Lobes. Refers to the thickenings on the dorsal surface of brain. they are usually two (forming corpora bigemina) but in mammals they are four in number (forming corpora quadrigemina).

Optic Nerve (cranial nerve II). The tract of sensory nerve fibres that is running from the retina of each eye. It enters the floor of the forebrain via the optic chiasma.

Optic Stalk. Narrow passage which is formed by construction of proximal portion of optic vesicle. It serves to attach the eye-ball.

Optocoel. Pertaining to cavity/ventricle/space in the optic lobes or mid brain.

Oral. Refers to mouth.

Oral Cavity. Mouth cavity.

Oral Cirri. From the edge of oral hood, II pairs of stiff, tentacle-like structures get attached which are known as oral cirri and they are sensory structures (chemoreceptors).

Oral Hood. In animals like *Amphioxus* the ventral side of the rostrum gets expanded into a hood-like structure called oral hood.

Oral Groove. Means groove heaving ciliated ridge which gets started at anterior one-third of ventral part of body and runs upto almost middle in some animals such as *Paramecium*.

Oral Valves. Refer to a pair of folds in mouth of certain bony fishes like

Labeo. They arise from maxillary and mandibular region and allow water entry in buccal cavity but do not allow back-flow.

Ora Serrata. In the retina of eye light sensitive part gets demarcated from the non-sensory part by an irregular and serrated area which is called ora serrata terminalis.

Orbit (eye socket). One of two cavities or depressions in the vertebrate skull that have the eye balls and their associated muscles, blood vessels, nerves, etc.

Order. A collection of similar families.. Plant orders generally end in *ales* (e.g. Liliales) but animal orders do not have any particular ending. Orders may be divided into suborders. Similar orders constitute a class.

Organ. A part of an organism that is made up a number of a number of different tissue specialised for carrying out a particular function. Examples are the lung, stomach, wings and leaf.

Organ Culture. Refers to the maintenance or growth of living organs, which are usually embryonic, in vitro. In animals the organ must be small to enable diffusion of nutrients and excretory product to and from the inner cells and is thus usually embryonic.

Organelle. A discrete subcellular structure that has a particular function. The largest organelle is the nucleus; other examples of organelles are chloroplasts, mitochondria, Golgi apparatus, vacuoles and ribosomes. Organelles bring about division of labour within the cell. Prokaryotic cells have very few organelles in comparison with eukaryotic cells.

Organizer. A part of an embryo whose presence makes neighbouring tissue to develop in a particular way. Examples of organisers are the eye-cup of vertebrates which causes lens, and later cornea, to be produced; the gut of snail embryos, which organizes shell gland and mantle.

Organism. Any living individual such as plant or animal.

Organ of Corti. Sensory organ which lies within the cochlea of mammalian ear.

Organum Vasculosm of Lamina Terminalis. When an acidic dye is injected into an animal, a small area among five area among five areas in brain stains like tissues outside the brain. It is also known as supra optic crest and is forming part of circumventricular organ.

Ornithine Cycle. Also called *urea cycle.* A cyelic succession or reactions in which there occurs elimination of ammonia and urea produced in the liver of ureotlic animals like frog and eutherian mammals such as man.

Orthogenesis. Refers to evolution that is taking in a definite direction and along a predetermined route, irrespective of natural selection. As this theory involves that evolution continues independently of natural laws, it is impossible to test experimentally. Orthogenesis therefore conflicts with conventional evolutionary theory such as neo-Darwinism.

Os-Cloacae. Hypoischium in *Sphenodon* is bony structure and also known as oscloacae or cloacal bones.

Os-Cornua. Used for describing the two bony buttons present in newly born calf. These can be felt under skin just opposite to frontal bone.

Osculum. Refers to the wide opening at the distal end of a sponge, by means of which water leaves the sponge after it has passed through the chambers and canals inside.

Os-Innominate. Refers to each half of pelvic girdle in animals such as *Varanus* or rabbit.

Osmoconformer. As organism exhibiting little or no osmoregulation so that the osmolarity of its body fluids is following change in osmolarity of environment.

Osmolar Concentration. The number of osmols present per litre of solution.

Osmolarity. Defined as the number of osmols (molecular weight of substance in grams divided by number of freely moving particles each molecule liberate in solution) present per kg of solvent.

Osmoregulation. Refers to the maintenance of internal osmolarity with respect to environment by an animal. Balance or water and salts taking place within the body of an animal.

Osmorgulator. An organism which is able to control its internal osmolarity in case of changes taking place in the environmental osmolarity.

Osmosis. Refers to the movement of solvent molecules which takes place across a semipermeable membrane when two solutions of different concentrations have been separated by such a membrane or a solution has been separated from solvent by semipermeable membrane.

Osmotic Potential. Defined as the depression in chemical potential of a solvent due to presence of solute molecules in it. It is always negative value.

Osmotic Pressure. The hydrosatic pressure which is sufficient to stop the movement of solvent molecules across a semipermeable membrane in a system where osmosis is taking place. It is having a positive value.

Os-Penis. A small bone which is present in the septum between corpora cavernosa

Osphradium. Sensory area which is present in incurrent siphon in some molluscs, epithelial and frequently yellow.

Ossein. The protein which is found in the matrix of bone.

Ossification. Refers to the process of bone material formation.

Osteichthyes. Refers to a group of animals including bony fishes (like *Labeo, Anabas*).

Osteoblast. Refers to any of the cells that are forming layers of bone in the early stages of ossification.

Ostoclast. Refers to any of the cells that attack and erodw the calcified cartilage or membrane formed in the early stages of ossification of bone.

Osteocyte. Refers to any of the cells that secrete the hard matrix of bone. They are present in small spaces (lacuna) between the concentric lamellae of bone that are forming the Haversian systems.

Osteology. Branch of science dealing with the study of bones.

Osteomalacia. Refers to disorders like rickets in which amount of mineral secretion in matrix of bone is deficient thereby resulting in the deformations in bones.

Osteon. In mammals large bones possess a system formed by a single Haversian canal and surrounding lamellae having osteocytes.

Osteoporosis. The decrease in bone mass by preserving the normal ratio of minerals to the matrix.

Osteoscelerosis. Refers to the presence of increased amount of calcified bone taking place in matastatic tumous, thereby leading poisoning and hypoparathyroidism.

Osteoscutes. Refers to large dermal scales present in turtles and tortoises where they form a box-like shell. Usually they are of small size and unfused and present over the throat of crocodilians. Lizards and snakes are having these under epidermal scales.

Optium. Used for describing a mothlike opening; for example, any of the lateral openings present in the heart of an arthropod or any of the openings through which water enters the body of a sponge.

Otic Capsules. Cartilagenous structures that have been developed around the internal ear during the development of skull in the vertebrates.

Otoconia. Small crystals of calcium carbonate and protein which get deposited in outer part of otolith membrane in the macula of mammalian ear.

Otocyst. A balancing sense organ which is present in certain organisms such as trochophore of annelids.

Otolith. Refers to one of several granules of calcium carbonate that are present in a gelatinous mass and attached to hair- like processes of sensory cells within the utriculus and sacculus of the vertebrate inner ear. They respond to changes which are taking place in the position of the head and so stimulate the sensory cells.

Otolithic Organ. In macula of mammalia near a gelatinous substance found among hair cells, in cuter portion of which otoconia are found.

Outbreeding. Breeding taking place between individuals that are not closely related. In animals behavioural mechanisms often promote Outbreeding. The most extreme form crossing between species—

usually produce sterile off spring and various mechanisms are known to discourage it. Outbreeding increases heterozygosity, producing more adaptable and more vigorous population.

Outer Ear. The region of the vertebrate ear which is external to the eardrum (tympanum). It is found in birds, mammals, and some reptiles and has a passage, the external auditory meatus, and the pinna (in mammals only).

Outer Segment. Refers to the portion of a vertebrate visual receptor which has pigmented receptor membranes. It gets attached to the inner segment by a thin cilium-like connection.

Oval Window (fenestra ovalis). A membrane separating the middle and inner ear in tetrapods. It transfers vibrations from the stapes—an ear ossicle of the middle ear—to the liquid (perilymph) of the inner ear.

Ovarian Follicle. Refers to a group of cells from germinal epithelium that grow into connective tissue (stroma) of ovary. One of the cells from this group gives rise to an oocyte.

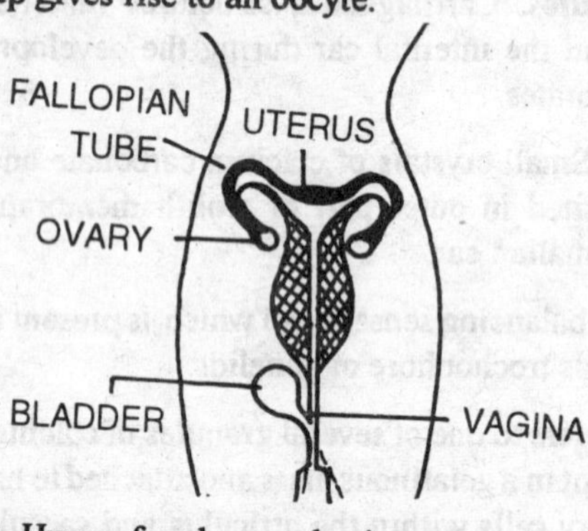

Fig. 1. Human female reproduction system

Ovary. Term used for describing the female reproductive organ in animals, which produces egg cells (ova). In vertebrates there is usually a pair of ovaries (in birds, only the left is functional); they also produce sex hormones. In humans, the ovaries are cream-coloured oval structures which are about 4 cm long, and are attached to the posterior wall of the abdominal cavity, below the kidneys. There are about 150-500 thousand follicles present at birth but only 300-400 ever undergo maturity to become ova.

Oviduct. A duct in animals which is able to convey ova from the ovaries to the exterior. It may or may not connect directly with the ovary.

Oviparity. Means the production of undeveloped eggs, which are laid or spawned by the female. Fertilization may take place before their release, as in birds and some reptiles, or after, as in most invertebrates, fish and amphibians.

Ovipositor. An egg-laying structure which is found at the hind and of the abdomen of female insects. It is formed from modified paired appendages. It is generally long and needle-like to make the piercing of animal and plant tissues to lay eggs. In wasps, bees, and ants it gets modified into a sting.

Ovotesis. The reproductive organ of certain hermaphrodite animals (e.g. the snail) which is able to function as both an ovary and a testis.

Ovoviviparity. Term used for describing the condition in invertebrates, fish and reptiles, in which eggs are produced and retained within the body of the female during embryonic development. The embryo obtains nourishment from the yolk store and so only depends on the mother for physical protection. Ovoviviparity is considered to be an evolutionary stage leading to viviparity.

Ovulation. Refers to the release of an egg (ovum) from a Graafian follice at the surface of a vertebrate ovary. In humans it first takes place at the onset of sexual maturity and a single ovum gets released about every 28 days from alternate ovaries until menopause, at the age of about 45.

Ovum (egg cell). The term used for describing the immotile female reproductive cell (gamete) that is produced in the ovary of an animal. It has a central haploid nucleus surrounded by cytoplasm, having a variable amount of yolk and a vitelline membrane. Size varies between species; in humans, it is about 0.15 mm in diameter and in chickens it is about 30 mm in diameter. A single ovum is released from the ovary at regular intervals; in humans, about once every 28 days. If it is fertilized by a spermatozoon it is developed into a new individual of the same species.

Oxidative Phosphorylation. The formation of ATP from phosphate and ADP in aerobic respiration, Oxidative phosphorylation takes place in mitochondria, the energy being provided by steps in the electron-transport chain.

Oxygen Debt. Defines as physiological state that takes place when a normally aerobic animal respires anaerobically during a temporary shortage of oxygen (anoxia), e.g. due to violent muscular exertion. Pyruvate, a product of the first stage of internal respiration, is converted anaerobically to lactic acid, which is toxic and needs oxygen for its breakdown, thereby building up an oxygen debt. The debt is repaid when oxygen is available and brings about the oxidation of the lactic acid in the liver.

Oxygen Dissociation Curves. The curves describing the relationship between the extent of combination of oxygen with the respiratory pigment and partial pressure of oxygen in gas phase.

Oxygen Quotient (QO_2). Refers to the rate of oxygen consumption of an organism or tissue. It is generally expressed in micro-litres of oxygen per milligram of dry weight per hour. Small organisms possess higher oxygen quotients than large ones.

Oxyhaemoglobin. An unstable chemical compound which is formed in blood by combination of oxygen with heamoglobin. When it goes to tissue, the oxygen gets released due to dissociation of the compound.

Oxyntic Cells. Large sized cells which are present in epithelium that secrete HCL.

Oxytocin. A peptide hormone which is secreted by the hypothalamus and posterior pituitary gland. It acts on smooth muscle. At the end of pregnancy the uterus becomes very sensitive to Oxytocine, which promotes labour and also the release of milk from the mammary gland. It also induces labour artificially.

P

PABA (Para Amino Benzoic Acid). A component of folic acid which is necessary for its formation. Also considered as antigraying factor in rats. It is found in many plants' and animals' tissues.

Pacemaker (Sinoatrial node, SAN). A small area of specialized cardiac muscles fibres in the wall of the right atruim of the heart in mammals. It spontaneous rhythmical electrical activity is able to initiate and maintain the contractions of the heart (the heart beat). However, the rate of heart-beat becomes under nervous control. An *artificial pacemaker* is an electronic device is used in certain cases of human heart disease for assuming the functions of a natural pacemaker.

Pachytene. In meiosis it refers to the stage in mid prophase I which is characterized by the contraction of paired homologous chromosomes. At this point each chromosome has a pair of chromatids and the two associated chromosomes are called a tetrad.

Pacinian Corpuscles. Heavy touch receptors.

Paedogenesis. Reproduction in the larval or other juvenile stage by animals that are exhibiting permanent neoteny. For example, the axolotl—the larva of a salamander— is able to produce offspring similar to itself.

Pain. A sense of discomfort/non-euphoric state which occurs due to internal/external stimulation perceived through nacked nerve endings. Internal stimulation may take place due to some pathological/ physiological disorder but external due to environmental changes. Pain is localised to an organ/part of body or is vague in localisation.

Pa-interval. A unit in electrocardiogram which indicates depolarisation and conduction through AV node (auriculo ventricular node).

Normal time (duration) is usually 0.18 second.

PAH (Para Amino Hippuri Acid). A substance which is infused to measure plasma flow by determining the urine and plasma concentration of substance.

Palaeocene. The oldest epoch of the Tertiary, 65-55 million years ago which is represented in Britain by marine sand deposits at Thanet. It is characterized by the absence of dinosaurs and the presence of various primitive mammals, now extinct. The first insectivores, rodents, and primates are also found.

Palaeoecology. Refers to the investigation of prehistoric ecology as revealed by studying fossils and their artifacts, pollen (palynology) samples and the mineral deposits in which such structures are present.

Palaeolithic. The older Stone Age, when stone tools could first be recognised as such, extending from about 2 million years ago to about 10000 years ago.

Palaeontology. Deals with the study of extinct organisms, including their fossil remains, and impressions left by them. Sometimes the subject has been divided in *palaeobotany*, the study of fossil plants, and *palaeozoology*, the study of fossil animals.

Palaeozoic. The term used for describing the first and oldest era in which life became abundant, about 590-230 million years ago. It has been divided into six main periods: the Cambrian, Ordovician, Silurian, Devonian, Carboniferous and Permian. Beginning with aquatic invertebrates and algae, the era had ended with the invasion of land by tree ferns and reptiles.

Palate. Refers to the roof of the mouth; also refers to a partition that separates the nasal passage from the buccal cavity in mammals. The anterior portion is supported by bone and is known as the *hard palate.* Behind there is *soft palate* which ends in the uvula.

Palatine Bone. Bone serving as an anterior brace of upper jaw in the skull of frog and other vertebrates.

Palatine Rugae. The series of transverse ridges present in the anterior part of palate.

Palaptopterygoquadrate Bar. Mandibular arch is gnathostomes is divided into a dorsal structure called palaptopterygoquadrate bar.

Palaeontology. Science dealing with the study of ancient life of earth as revealed by fossils, impressions and other remains present in earth's surface layers.

Palp. Projecting part of process which is sensory in function and often near the mouth.

Palpus. Palp i.e., process of an appendage as in insects.

Pampas. It is also known as Savanna which is North American grassland.

Pancreas. A gland which lies between the spleen and the duodenum. It has duct that enters the duodenum. It secretes pancreatic juice which have enzymes of three types: trypsin for breaking down proteins to amino acids, amylase for converting starch to maltose, and lipase for changing emulsified oils to glycerin and fatty acids. The gland is also having endocrine tissues and produces insulin.

Pancreatic Juice. Secretion of pancreas which is having digestive enzymes like trypsin, amylase and lipase etc.

Paneth Cells. Among cells of Krypts of Lieberkuhn these are some argentaffin cells which probably are able to produce enzymes.

Pangenesis. The theory, no longer accepted, introduced by Darwin to explain the inheritance of variation. According to him the body fluids carry particles from all over the body to the reproductive cells where they affect the hereditary material and thus the characters car inherited by the next generation. Pangenesis was used to explain the erroneous theory of the inheritance of acquired characteristics.

Panniculus Adiposus. A layer of adipose tissue present in the subcutaneous layer of mammalian skin.

Panting. Increased respiration state in which breathing is carried out at a higher rate so that the loss of water vapours gets increased taking away excess of heat along with it.

Pantothenic Acid. (vitamin B_5). One of the water-soluble B-group of vitamin. Sources of the vitamin include egg yolk, kidney, liver, and yeast. As a constituent of coenzyme A, pantothenic acid is essential for several fundamental reactions in metabolism. Its deficiency results in symptoms that affect a wide range of tissue; the overall effects are fatigue, poor motor coordination, and muscle cramps.

Papilla. A small nipple-like projection from a surface of an animal tissue or organ.

Parabiosis. The experimental or natural union of two similar animals (each a *parabiont*) so that their blood circulations become continuous. Experimental parabiosis is often carried out in insects to study what effects chemical in one individual have been passed to the other. Natural parabiosis takes place in Siamese twins.

Parabronchus. Refers to the branches of ecto-and endo-bronchi in the respiratory tree of birds which enclose air capillaries and where the exchange of gases occurs.

Paracentric Inversion. An inversion which involves single chromosome arms, both breaks being to one side of centromere.

Parchordals. Refer to a pair of flattened and curved cartilages that are differentiated from mesenchyma on each side of notochord and beneath the brain during development of skull.

Paracoels. The paired cavities of cerebral hemispheres. Also known as *lateral ventricles*.

Pardoxical Sleep. Sleep state in which rapid eye movements take place (REM-sleep); Electroence Phalogram tracings look as if organisms awake.

Parafollicular Cells. The ultimobranchial bodies (which form calcitonin in non-mammals) get incorporated into tiyroid gland and the tissue as distributed around follicles as Parafollicular cells.

Paraflocculus. Part of cerebellar area which cordinates movement.

Paraganglia. Small group of cells which resemble those in adrenal medulla and occur near the thoracic and abdominal sympathetic ganglia.

Parahaemophilia. Congenital disease which occurs die to deficiency of blood clotting factor V (proaccelerin, labile factor, accelerator globulin).

Parallel Evolution (parallelism). The development of similar features in closely related organisms due to strong selection in the same direction. There are few examples of this phenomenon.

Paralysis. Partial incapability of individual to move or feel due to loss or impairment of nerve action.

Paramecium. A genus of Protozoa common universally in fresh water having decaying vegetable matter. *Paramecium* is slipper-shaped and covered with cilia, whose beats are responsible for rapid locomotion. It reproduces asexually by binary fission and sexually by conjugation.

Paramylum. Starch-like substance which is produced by photosynthetic activity in *Euglena*. It serves to act as a reserve food material.

Parapodium. A lobed lateral appendage, pairs of which are present on all segments of a polychaete worm (e.g. *Nerelis*) except the first two and the last one. Parapodia serve to act as paddles for swimming and provide a large respiratory surface. The surface of the parapodium is well supplied with blood capillaries for increasing the rate of exchange of respiratory gases.

Parapsid. Reptiles whose skulls possess a single superior temporal fossa on each side above the post-orbital and squamosal bones. This group includes the Ichthyosauria, lizards and snakes.

Parasitism. An association taking place between two organisms in which one, the *parasite*, is benefited at the expense of the other, the host. The tolerance of the host varies form being almost unaffected to serious illness and often death.

Parasympathetic Nervous System (craniosacral nervous system). Refers to one of the two divisions of the autonomic nervous system, which supplies motor nerves to the smooth muscles of the internal organs and to cardiac muscles.

Parathormone (Parathymine). A hormone which is produced by the parathyroid glands of vertebrates. It helps to regulate the amount of calcium and phosphorus in bones.

Parathyroid Glands. Four small oval shaped structures which are embedded in the thyroid gland. They produce a hormone that is able to control the blood calcium level. The glands are composed of columns of cells having vascular channels between the columns.

Parathyroid Hormone. Linear hormone which consists of 84 amino acid residues. It increases Ca^{2+} of plasma and decreases phosphate level and besides increases phosphate excretion in urine. It also increases renal tubular reabsorption of calcium and formation of a substance called 1, 2, 5-dihydroxycholcalciferol (physiologically active metabolite of vitamin-D).

Paraventricular Nucleus. Refers to the group of neurosecretory neurons in supraoptic hypothalamus that send their axons into neurohypophysis.

Parenchyma. Loose connective tissue formed of large cells. It packs the space between the organs in some simple acoelomate animals such as the Platyhelminthes. The name is also used fro similar tissue in plants.

Paresis. Leisions limited to pyramidal tracts producing weakness rather paralysis which is known as paresis.

Parietal Eye. A third eye developed from an anterior outgrowth of the forebrain of early vertebrates.

Parotid Gland. A set of salivary glands present in the region of ear just beneath external autitory maetus. Its duct is known as Stenson's duct in man.

Paraxonic Foot. Foot in which third and fourth digits are dominant and are able to carry most of body weight while 2nd and 5th digits are reduced or vestigial and displaced posteriorly and the first digit gets disappeared e.g., pig.

Pars Basilaris. An outgrowth/outpocketing of lagena in membranous labyrinth of certain frogs in which basilar papilla (macula) may be present.

Pars Ciliaris. A part of the retina which extends forwards over the ciliary processes in the eye of a vertebrate.

Pars Distils. The anterior lobe of the pituitary organ which is secreting a number of hormones including *gonadotrophins*, *adrenotrophins*, *thyrotrophins* and growth hormones.

Pars Intermedia. The central part of the pituitary organ that is prominent in lower vertebrates but absent in birds and mammals. It secretes a hormone *intermedin* which brings about darkening of the skin by expansion of melanophores.

Pars Iridica. A part of the retina which is extending forward over the inner surface of the iris in the eye of a vertebrate.

Pars Nervosa. Refers to the posterior lobe of the pituitary gland in direct nervous communication with the brain.

Pars Optica. Refers to the light-sensitive part of the retina in the eye of a vertebrate.

Pars Tuberalis. A part of the pituitary organ which is attached to, and in some higher vertebrates surrounding, the infundibulum to form the stalk.

Parthenogenesis. Refers to the development of an unfertilized into a new individual that is genetically identical with the parent. Such development is common among insects and in many cases may be brought about artificially by various chemical or physical stimuli.

Parthenogonidia. Zooids or units which carryout a sexual reproduction in certain protozoa.

Parturition. Refers to the process of giving to the foetus at the termination of pregnancy in viviparous animals.

Patagium. (1) A membrane of skin which is stretched between the limbs to form the wing of a bat or the 'parachute' of such animals as the flying squirrel.

(2). A short wing-like outgrowth present on eachside of the prothorax in certain insects.

Patella. The keen-cap; a flat bone which is embedded in the tendon of the keen in birds and mammals.

Patheticus (Trochlear) Nerve. Refers to the fourth cranial nerve of a vertebrate; a motor nerve which innervates the superior oblique muscle of the eye.

Pecten. Any comb-like structure or organ.

(1) A vascular comb-like process from the choroid in the eye of a bird.

(2) A comb-like stridulating organ on some insects.

(3) A row of spines or hairs on a palp or chelicera of various arthropods or on the respiratory tube of a mosquito larva etc.

Pectoral Girdle (shoulder girdle). A bony or cartilaginous skeletal structure present in the anterior region of the vertebrate body, to which the forelimbs or fins get attached. In mammals, it usually has two ventral clavicles and two dorsal scapula. The clavicles are generally joined mid-ventrally to the sternum and the scapula get attached to the vertebral column by muscles, which give mobility to the shoulders. Each scapula bears an articular surface, the glenoid cavity, for the forlimbs. In birds and reptiles the clavicles have been functionally replaced by the more prominent coracoids. Lower vertebrates may be having both clavicles and coracoids.

Pedal. Means foot.

Pedicellaria. Small pincer or scissor-like process which are present on the surface of certain echinoderms like sea-stars seaurchins etc.

Pedicle. Narrow waist present between cephalothorax and abdomen in spiders.

Pedipalps. A pair of appendages present on the fourth segment of the head of arachnids. They function in the place of antennae (absent in this group) in spiders and are used for grasping prey scorpions. The segment nearest to the body has a hard plate which is used for chewing food.

Pedogenesis. Refers to the reproduction by larvae.

Pelagic. Inhabiting the open upper waters rather than the bed of a sea or ocean. Pelagic animals and plants are divided into the plankton and nekton.

Pellagra. Disease which is caused by deficiency of niacin in body. This disease is characterised by dermatitis, diarrhoea and dementia (3 D-disease).

Pellicle. Refers to thin flexible transparent outer protective covering of many unicellular organisms, especially flagellated and ciliated protozoans,-e.g. *Euglena* and *Paramecium*. It is made of protein, and is able to maintain the shape of the body. It may thus be considered to be an exoskeleton.

Pelvic Girdle (hip girdle). Refers to a rigid bony or cartilaginous skeletal structure which is present in the posterior region of the vertebrate body, to which the hindlimbs or fins get attached. In tetrapods, it is having two sides, each consisting of, ventrally, an anterior pubic bone and a posterior ischium and, dorsally, an ilium. They meet at the acetabulum— an articular surface for the hind limb. Each ilium gets fused to the sacral vertebrae to permit the transmission of thrust from the hindlimbs to the body to produce movement, and also to allow the weight of the body to be supported by the girdle.

Pelvis (1). The pelvic girdle.

(2). The basin-like cavity which is produced by the bones of the pelvic girdle, especially in man and other mammals.

(3). (*renal pelvis*). The central cavity of the mammalian kidney into which the urine drains. It forms the expanded upper end of the ureter.

Penetrance. Refers to the expression of dominant homozygous recessive genes in the phenotype. When some of the organisms known to have such genes in the genotype do not exhibit their characteristics in the phenotype penetrance is said to be incomplete. Penetrance, which is assumed to be 100% in Mendelian genetics, is measured by scoring the number of individuals exhibiting a character against the number known to have the gene controlling that character. Incomplete penetrance may occur due to environmental factor or the effects of other genes.

Penis. The male copulatory organ which is used by many animals with internal fertilization (mammals, some reptiles and invertebrates, etc.) to introduce sperm into the female reproductive tract. In mammals, both urine and semen (the fluid containing spermato-

zoa) pass, via the urethra through the penis. The penis has a layer of skin and connective tissue surrounding three cylindrical bodies of spongy tissue having numerous blood spaces—two corpora cavernosa dorsally and a single corpus spongiosum ventrally which surrounds the urethra. Prior to copulation the spaces get filled with blood, making the penis more rigid (i.e. erect).

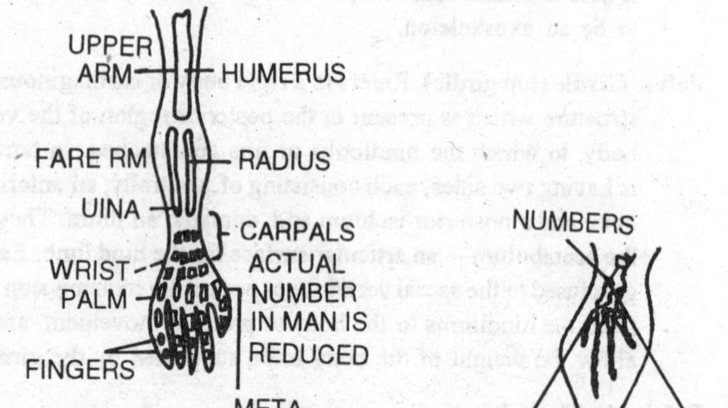

A basic pentadactyl forelimb, as exemplified by the human arm

Flipper of whale

Foreleg of pig

Wing of bat

Fig. 1 : The modified pentadactyl forelimbs of various vertebrates

Pennae. Fathers which are visible over the body of a bird.

Pentadactyl Limb. A limb having five digits; its bone components have a basic arrangement characteristic of all tetrapod vertebrates.

Pentose. A sugar having five carbon atoms in its molecules.

Pentose Phosphate Pathway (hexose monophosphate shunt). A pathway of glucose breakdown in which pentoses are produced in addition to reducing power for many synthetic reactions. It is an alternative to glycolysis.

Pepsin. An enzyme catalysing the par tial hydrolysis of proteins to polypeptides. It is secreted by the gastric glands in an inactive form, pepsinogen, and is activated by hydrogen ions. At pH values of 4.6 and less pepsin activates pepsinogen, i.e. it is autocatalytic. Pepsin is able to initiate the digestion of proteins, splitting them into smaller fragments. The extent of this action has been proportional to the length of time the protein is in contact with the enzyme.

Peptidase. An enzyme that is catalysing the hydrolysis of certain peptide bonds. The peptidases help break down peptides into amino acids.

Peptide. A type of compound which is formed of a number of amino acid molecules linked together. Peptides are regarded as formed by a reaction in which the carbonyl group of one amino acid reacts with the amino group of another amino acid with the elimination of water. This link between amino acids is known as a *peptide bond.*

Peptide Bond. Refers to the bond—CO—NH—which is formed by condensation of amino acids into peptides.

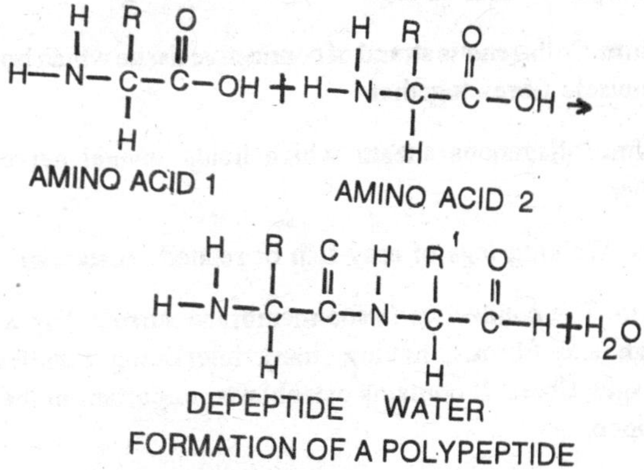

Fig. 2. Formation of peptide bond

Perching. A tendency of griping objects by toes in birds, so that they do not fall from branch even at rest/sleep.

Perfusion. Means passage of fluid over or through an organ, tissue or cell.

Periblast. Layer of cells which are present under blastocoel in fishes.

Pericardio-Peritonial Canal. The perforation in the transverse septum separating abdominal cavity with pericardium in fishes.

Pericardium. Closed membrane surrounding the heart. The part in contact with the heart is called visceral pericardium; the other the parietal pericardium.

Pericentric Inversion. An inversion which involves both arms of a chromosome, the breaks being on the opposite side of centromere as a result centromere's position can be shifted.

Perichondrin. Protein (mucopolysaccharide) present in the matrix of cartilage.

Perichondrium. Fibrous connective tissue layer which is covering the surface of cartilage.

Periderm. Cornified layer which is present around the feather papilla during the development of feather.

Perilymph. A fluid in between membranous labyrinth and periosteum of bony labyrinth of ear.

Perimycium. Collagenous strand of connective tissue which holds bundles of muscle fibres together.

Perinerium. Collagenous sheath which holds several nerve fibres together.

Periopods. Walking legs of cray-fish or related crustacean.

Periosteum. The connective tissue membrane surrounding a bone. It is tough and fibrous, having many interlacing bundles of white collagen fibres. It contains osteoblasts, important in the formation of bone.

Periostracum. Distal chitinous layer in mollusc shells; they are protective in function.

Peripheral Nervous System. The system of nerves and their ganglia that run from the central nervous system to the organs and peripheral regions of the body. It constitutes all parts of the nervous system which are not included in the central nervous system. In vertebrates it consists of the cranial and spinal nerves with their many branches. These are able to convey impulses from sense organs for processing by the central nervous system and transmit the consequent motor impulses to muscles, glands etc.

Parisodactyla. Odd-toed ungulates that include horse, rhinoceroses etc.

Peristalsis. Waves of musclular contraction that pass along tubular organs of the body, primarily the alimentary canal. It occurs by the sequential contraction of circular muscles in the wall and serves to force the food contents along.

Perkinje Fibres. Myocardial conducting tissue which forms net work in the ventricular walls.

Pernicious Anaemia. The deficiency of cyanacobalamine (vitamin-B_{12}) due to inadequate absorption leads to this disease which is characterised by a fall in the haemoglobin contents of blood. It could be cured by injecting vitamin B_{12} in the body.

Peroxidase. The enzyme which is capable of breaking peroxide (super-oxide radical O_2^-) present commonly in WBCs like neutrophils. The granules in neutrophils produce H_2O_3 which is able to kill most of bacteria and later the compound gets decomposed by peroxi-dases that break it into water and nascent oxygen. The nascent oxygen is used to form water during electron transport.

Peroxisomes. Microbodies having amino acid oxidases, urate oxidases and catalases. They catalyse the decomposition of hydrogen per-oxide to water and nascent oxygen.

Pessulus. A cartilage-like structure present in the voice box of birds (syrinx) over which a semilunar membrane is mounted.

Petrohyals. Refers to the set of five bands of muscles which arise from outer surface of otic capsule and pass in fan-like manner around floor of pharynx and esophagus and gets inserted into the side of hyoid. They are able to raise floor of mouth cavity during pulmonary respiration in frog.

Peyer's Patches. Numerous nodules of lymphoid tissue which are found in form of large oval areas in the region or intestine. These help in fighting infections.

Phagocytes. Type of WBCs that are able to engulf, digest bacteria and other foreign materials.

Phagocytosis. Engulfing of food particles due to membrane invagination proximal to particle and subsequent membrane closure over the particle.

Phalanges. Mean structural elements of acropodium of forelimb. They form fingers in hand and toes on foot and are in form of series of small bones.

Pharyngohyal. One of the segments of hyoid arch which forms suspensorium and gets articulated with otic region of skull on one side and the mickel's cartilage on other side. It is also known as hyomandibular arch.

Pharynx. Refers to the anterior portion of the digestive tract between mouth cavity and oesophagus.

Phenetic. The observable similarities and differences between organisms that may be used for indicating natural rather than evolutionary relationships between groups.

Phenocopy. Refers to a change in the appearance of an organism which is caused by the environment, but which is similar in effect to a change caused by gene mutation. Such changes which are not inherited, are generally caused by environmental factors, (e.g. radiation) affecting the organism at an early stage of development.

Phenotype. The appearance of an organism, which could be determined by the interaction of the genotype with the environment. Many genes present in the geno-type do not exhibit their effects in the phenotype because they get masked by dominant alleles. Genotypically identical organisms may be having very different phenotypes in different environments, an effect particularly noticeable in plants grown in various habitats.

Phenotype Ratio. Ratio of phenotypes which results from a particular cross.

Phenylalanine. An essential aromatic amino acid which is converted to tyrosine, a precursor for dopamines.

Phenylketoneuria. A recessively inherited condition which is characterised by the inability of organism for converting phenylalanine to tyrosine resulting in accumulation of toxic products which may result in mental retardation unless dietary precautions are taken at proper time.

Pheromone. A substance which is excreted by an animal and cause a response in other animals of the same species (e.g. sexual attraction development).

Phlorizin. A glycoside inhibiting active transport of glucose.

Phosphoprotein. A conjugated protein which is formed by the combination of protein with phosphate groups. Casein is an example.

Photic Zone. The surface layer of an ocean or lake that is penetrated by sunlight and in which the phytoplankton flourish. The diatoms, which are the main components of phytoplankton, may be found down to 80m. Beyond 200m the water is perpetually dark.

Photophores. Modified mucous glands of fishes found in deep sea. They produce light or become luminous organs to attract the prey and also warn the enemies.

Photoreceptors. A light sensitive cell or organ-like eye which is present in most animals and eyespot in case of *Euglena*.

Phosphatases. Refer to the hydrolytic enzyme. They act on acid and alkaline phosphatases and release phosphates from such compounds. They take active pact in metabolism, secretion and in excertory activities.

Phosphagens. Compounds that store light energy phosphates like phosphocreatine.

Phosphodiesterase. Hydrolytic enzyme which degrades c-AMP to AMP.

Phospholipids. Refer to the type of lipids conjugated with phosphoric acid replacing one of the fatty acid of normal lipid. They occur in cells, tissues and in the nervous system.

Phophoproteins. Refer to proteins which are conjugated with phosphoric acid. They occur mainly in the nervous system.

Phosphorylase Kinase. An enzyme that when phosphorylated by protein kinase is able to convert phosphorylse-b to more active form phosphorylase-a.

Photoreceptor. Any light-sensitive organ or organelle. The eyes of vertebrates and the ocelli and compound eyes of insects are photoreceptors, as are the organelles of such protozoans as *Euglena.*

Phylogeny. The evolutionary history of a group of organisms.

Phylum. One of the major groups into which the animal kingdom is classified. The phylum is divided into subphyla, e.g. the phylum Chordata is split into the subphyla Acrania and Craniata.

Physiological Saline. A solution of sodium chloride and various other salts in which animal tissues are bathed *in vitro* so as to keep them alive during experiments. It must be isotonic with, and of the same pH as, body fluids.

Physiology. The way in which organisms or parts of organisms function.

Physostomous. Fishes with duct connecting air bladder with oesophagus.

Pia-Arachnoid Membrane. In amphibians, reptiles and birds it refers to the inner vascular membrane (derived from cells of neural crest) below endorachis.

Pia Mater. The soft delicate innermost membrane that surrounds and protects the brain and spinal cord in vertebrates.

Pigeon's Milk. Crop glands (in both the sexes) of pigeons secret a fatty, slimy, proteinous substance that is regurgitated by parents into their beaks and then poured into the mouth of nestlings so as to nourish them.

Pinacytosis. Means cellular drinking or intake of fluid by the cell.

Pineal Apparatus. Behind the anterior choroid plexus the roof of diencephalon are giving off two median projections, anterior parietal body and posterior pineal body. The two collectively form pineal apparatus or epiphyseal apparatus.

Pineal Body. A structure which is located on the roof of the brain of vertebrates. It is phylogenetically associated with median eye.

Pineal Eye. Refers to a structure which is derived from the pineal gland and thought to have existed in fossil vertebrates. The only known living example is the tuatara lizard of New Zealand in which eye is situated on the top of the head and seems to be light sensitive. It is having a lense and retina and nerve fibres connected to the brain.

Pineal Gland. A gland arising from an outgrowth on the dorsal surface of the forebrain of the vertebrate brain. Its function is not known, although is has secretory cells. It is thought to have given rise to the third eye in fossil vertebrates.

Pinna. The external part of the outer ear in mammals which is often generally referred to as the 'ear'. It has a flap of skin supported by cartilage and surrounding the external opening of the ear, into which it helps deflect sound waves. It is movable in some mammals, such as dog and rabbit.

Pituitary Gland. An endocrine gland in the vertebrate brain which is situated beneath the thalamencephalon behind the optic chiasma. It is thought to be as the 'master' endocrine gland because many of its hormones are able to control the secretions of other endocrine glands. The more important pituitary hormones are as follows:

(1) Growth hormone (somatotrophin), which affects protein meta. Excess production causes gigantism and deficiency results in dwarfism.

(2) Vasopressin (antidiuretic hormone), which is able to stimulate reabsorption of water from the kidneys.

(3) Adrenocorticotrophic hormone (ACTH), which is able to stimulate the secretions of the adrenal gland.

(4) Gomadotrophic hormones (e.g. follicle-stimulating hormone. luterinizing hormone), which are able to stimulate gonad development.

(5) Oxytocin, which is able to stimulate the uterine walls to contract during birth.

(6) Prolactin, which is able to stimulate milk production by the mammary glands.

(7) Thyrotrophin (thyroid-stimulating hormone) which is able to stimulate the secretion of the thyroid glands.

Placenta. A disc-shaped organ that get developed within the womb (uterus) of a pregnant mammal and establishes a close association between embryonic and maternal tissues for the exchange of material.

Planaria. A genus of small free-living freshwater platyhelminths (flat-worms) of the class Turbellaria. Their body is covered with locomotory cilia and the mouth is present in the centre of the ventral surface at the end of a tubular pharynx, which can be thrust out to seize food and leads into a branched gut. *Planaria* has remarkable powers of regeneration, almost any piece of the body can grow into a complete individual, and shows indifference to starvation-various body organs are absorbed when food is in short supply.

Plankton. A variety collection of aquatic organisms that drift freely, not being attached to any substrate and not having any organs for locomotion.

Plantigrade. Refers to the mode of progression is some mammals (e.g. bears, rabbits, and man) in which the entire sole of the foot, i.e. digits and metatarsals (or metacarpals), remains in contact with the ground.

Planula. Refers to a small ciliated larva of a coelentrate. After swimming to a suitable site, it settles and is developed into a polyp.

Plaque. A film which covers the teeth. It is made up of mucus from saliva with dissolved sugar and other food, and it provides a breeding ground for bacteria. If left on the teeth for any length of time, the bacteria produce acids from the food and these eat into the tooth enamel. This allows bacteria to reach the dentine and brings about tooth decay.

Plasma Cells. White blood cells (leucocytes) which are derived from lymphocytes in the lymphoid tissue. They have clear cytoplasm and a large nucleus and they form antibodies important in defence against disease.

Plasmagene. A gene present in self-replicating cytoplasmic particle. Inheritance of the characters controlled by such genes is not Mendelian because appreciable amounts of cytoplasm are passed on only with the female gametes. Mitochondria, plastids, and centrioles all have plasmagenes.

Plasma Membrane (cell membrane, plasmalemma). The membrane that surrounds all living cells.

Plasmasol. Relatively fluid cytoplasm as found in the protozoa like *Amoeba.*

Plasmin. Active compound of fibrinolytic system which is able to lyse fibrin and fibrinogen to degradation products and limit clotting.

Plastron. The ventral shell of tortoise or a turtle.

Platelet (thrombocyte). A tiny particle which is found in blood plasma. Platelets are 2-3 μ in diameter and there are about 250 000 per cubic milimetre of blood. They are made in red bone marrow.

Platythelminthes. A phylum of primitive worm like invertebrates the flatworms including the classes Turbellaria (aquatic free-living planarians), and the parasitic Trematoda (flukes) and Cestoda (tapeworms).

Pleistocene. Refers to the first epoch of the Quaternary, from about two million years ago until the last glaciation ended about 10 000 years ago. The four Ice Ages drove many organisms towards the equator while others, (e.g. mammoth) became extinct. Many present-day mammals of South America and Africa resemble pre-Ice Age mammals of Europe. Madern man (*Homo sapines*) was evolved during this period.

Pleura. A double membrane succounding the lungs and lining the walls of the thorax in mammals. The narrow space between the two membranes— the pleural cavity—is filled with air and help cushion the lungs against damage. Cells in the membrane secrete pleural fluid, which is able to lubricate the pleural where they touch, so decreasing friction during breathing movements.

Pleuron. The plate, stiffened with chitin, that is able to form the protective lateral covering on either side of a body segment of an insect.

Plexus. An intricate system of interconnections taking place between nerves, blood vessels, or lymph vessels; for example, the *brachia plexus* of interconnecting spinal nerve branches supplying the forelimbs of vertebrates.

Pliocene. Refers to the epoch of the Tertiary, about 7-2 million years ago, which followed the Micene. In the Pliocene the hominids such as *Australopithecus* and *Homo,* which eventually resulted in man, became clearly distinguishable from the apes.

Pluteus. A form of dipleurula larva which is characteristic of brittle stars and sea urchins, in which the ciliated band is continuous, with a small pre-oral lobes supported by calcareous ribs.

Podsol. The type of soil present under heathland and coniferous forests in temperate climates. It is strongly acid and often deficient in nutrient due to leaching.

Poikilothermy. Refers to the condition of having a body temperature that varies approximately with that of the environment. Most animals other than birds and mammals are poikilothermic ('cold blooded').

Polar Body (polocyte). A minute cell which is produced during formation of an ovum when the oocyte undergoes two meiotic division.

Pollex. A first digit of the forelimbs of tetrapods; it forms the thumb in man. In the typical pentadactyl limb it has two phalanges.

Polychaeta. A class of marine annelid worms, the bristle worms. Many are carnivorous and active crawlers. Some, e.g., Nereis (ragworm), are able to burrow in sand or mud while others are able to build tubes of sand or mucus, which they rarely leave.

Polygene. A gene having an individually small effect on the phenotype that undergoes interaction with other polygenes controlling the same character to yield the continuous quantitative variation typical on such traits as height, weight, and skin colour.

Polymorph (Polymorphonuclear leucocyte). A white blood cell (leucocyte) with a lobed nucleus and granules in the cytoplasm. The term could be used for any granulocyte but is usually restricted to the neutrophils. See *granulocyte, neutrophil.*

Polyp. Refers to a stage in the life cycle of coelenterates in which the body is tubular, having a mouth surrounded by tentacles at one end; the other end gets attached to a fixed surface.

Polypeptide. A compound having many amino acids which are linked together peptide bonds.

Polyphyodont. A type of dentition in which the teeth could be replaced throughout the animal's lifetime if damaged or broken. it is present in frogs and lizards.

Polysaccharide. High molecular polymers of the monosaccharides or sugars. They have many repeated units in their molecular structures. They are able to be broken down to smaller polysaccharides, disaccharides, and monosaccharides by hydrolysis or by the appropriate enzyme. Important examples of polysaccharides are inulin (hydrolysed to fructose), starch (hydrolysed to glucose), glycogen (also known as animal starch), and cellulose (hydrolysed to glucose but unlike stretch all the linkages are 'β', - not metabolized by man).

Polytene. Describing the chromosome condition which is caused by chromatids not separating after duplication. It results in the formation of *giant chromosomes* consisting of numerous identical chromatids lying parallel to each other. Giant chromosomes have characteristic bands, Polytene chromosomes are common in the salivary gland cells of dipterous insects, e.g. *Drosophila.*

Pons (Pons cerebelli, pons Varolii). A thick band of nerve fibre present in the mammalian brain that passes across the medulla oblongata to link the two hemispheres of the cerebellum.

Population. A group of organisms of the same species (or other groups within the which individuals may exchange genetic information) occupying a particular space.

Porifera. A phylum of primitive multicellular animals, the sponges, that probably evolved a multicellular structure independently of the other multicellular animals (Metazoa). All are sessile and almost all are marine.

Porphyrins. Cyclic organic structures having the important characteristic properly of forming complexes with metal ions. Examples of such *metalloporphyrins* are the iron porphyrins (e.g. haem in haemoglobin) and the magnesium porphyrin chlorophyll, the photosynthetic pigment in plants. In nature, the majority of metalloporphyrins get conjugated to proteins to form a number of very important molecules, e.g., haemoglobin, and the cytochromes.

Portal Vein. A vein connecting capillary networks of two particular regions which allows blood from one to be regulated by the other.

Portal System. Refers to a double capillary circulation. A vein draining some organ is not opening into heart directly but it is opening in an intermediate organ where it distributes its blood capillary formation. From this organ another set of capillaries are able to collect blood and pass it to heart by another vein. There are two such systems which are common amongst like the vertebrates—hepatic portal system and renal portal system. Of these the hepatic portal system is present in all chordates.

Post Caval. The large vein which drains the posterior body and opening into right auricle. It is also called inferior venacava.

Posterior. Designating the hind end of an animal; in bilaterally symmetrical animals this is the end directed backwards during locomotion. However, in bipedal animals, such as man the posterior side corresponds to the dorsal side of other animals.

Postzygapophysis. Refers to articular processes near the posterior phase of vertebrae and with articular surface directed downwards.

Precavals. Refers to symmetrical large veins draining head, neck and forelimbs are pouring into right article separately. (Also known as *superior vena cava*).

Pre-Chordals. A pair of elongated cartilages which are differentiated from mesenchyme in front of prechordals during the development of skull in vertebrates.

Precipitin. An antibody which combines with and precipitates soluble antigen. It is used in the *precipitin reaction* for identifying antigens.

Predaceous. Means capturing of living animals by an individual for food.

Predator. An animal that preys upon other animals for its food.

Preen Glands. The branched tubular glandular structures in birds which are placed at the end of tail. They are also *called* pyginial glands. The oily secretion of these glands is used to coat over the feathers by birds (by rubbing the secretion with the help of its beak) to make them water proof.

Preformation. According to this the embryonic development of animals and plants involves merely growth or extension of a preformed germ or programme. Early adherents of the theory assumed the presence of tiny human figures (*homunculi*) in the heads of the sperms, while modern biochemical preformationists assume that the complexity of organisms is only the complexity of nucleic acid extended.

Pregnancy. Defined as the series of processes right from fertilization to the maintenance of embryo upto the birth in human beings.

Pregnenalon. One of the intermediates produced during the biosynthesis of testesrone from chlesterol.

Prehensile. A structure that is adapted for holding/grasping some object like prehensile tail in many animals such as *Chamelion. Hippocampus*, langoor etc.

Premaxilla. A paired membrane bone which forms the anterior region of the upper jaw in most vertebrates. In mammals, it bears the incisor teeth.

Premetamorphosis. Refers to the developmental stage which is just preceeding amphibian metamorphosis during which iodine binding and hormone synthesis take place in the thyroid gland.

Premolar. A mammalian tooth which is situated between the canine teeth or incisors in front and the molars behind. Premolars are multiple-rooted and ridged and are used for grinding food.

Prepubis. Small tubercle-like structure present at the pubis in case of reptiles.

Presbyopia. The term used for describing the condition in human beings which leads to difficulty in reading and close work by the age of 40-45 years due to receeding of near point of eye. It can be corrected by using convex glasses.

Presumptive. Describing embryonic tissues that are presumed to develop in a certain way. For example, presumptive neural plate of amphibians lies towards the animal pole of the blastula, while notochord will induce it from cells of the outer surface.

Presynaptic. Located proximal to synaptic cleft.

Prezygapophysis. The articular surface near the anterior phase vertebrae and directed upwards.

Primary Ammenorrhea. The term used for a abnormality of ovarian function in certain women during which bleeding never take place in life. Such women are having small breasts and other signs of failure to mature sexually.

Primary Groove. As the migration of cells from lower lateral borders of primitive srreak is grater than convergenee of cells over it from upper layer there occurs the formation of a groove along its entire length in the blastoderm of chick which termed as primary groove or primitive groove.

Primordium. Refers to the earliest stage of formation of body organ or part.

Proboscis. Refers to the tubular extension of nose or lip or phyarynx. Also refers to the extended beak-like mouth part in certain insects.

Procencephalon. One of the divisions of neural tube that gives rise to olfactory lobe, cerebral hemispheres and diencephalon.

Procoelous. Vertebrae having concave centrum at anterior end as found in case of frog and reptiles.

Proctodaeum. Refers to the terminal part of digestive tract near anus. It is lined with ectoderm.

Progesterone. Vertebrate female sex hormone which is secreted by corpus luteum and placenta (during pregnancy). It is responsible for maintenance of uterine cycle, maintenance of pregnancy and change which are associated with reproductive cycle in female.

Progestins. Refers to the substance mimicing the action of progesterone in female. They are also known as progestational agents, gestgens etc. When administered in the body orally in appropriate dose, they tend to block mid cycle burst of LH secretion and act as contraceptives.

Prolactin. Hormone from pituitary (also secreted by corpus luteum of pregnancy). It stimulates secretion of milk from mammary glands. The hormone secretion is influenced by PRH from hypothalamus.

Prolactin Inhibiting Hormone. Hormone from hypothalamus which blocks prolactin secretion from pituitary gland.

Prometamorphosis. Refers to the first stage of amphibian metamorphosis during which there occurs increased development of thyroid gland.

Prometaphase. Stage during which there occurs disappearance of nuclear membrane and completion of spindle formation during mitosis/meiosis.

Pronephric Duct. The duct connecting the pronephric tubules. This function as ureter in such cases.

Pronephros. Refers to the first kidney embryologically formed in the vertebrates by tubunephrotomes of anterior part of embryo. It is present in primitive vertebrates.

Prong Horn. Horn in antelope. It resembles the hollow horn.

Pronotum. Refers to the dorsal surface of an insect's prothorax including sclerites.

Pronucleus. One of the nuclear bodies of fertilized egg (male Pronucleus and female pronucleus) fusion of which gives rise to cleavage nucleus.

Prophase. Refers to any of the first stages of mitosis/meiosis during which the chromosomes become distinctly visible.

Proprioreceptors. The sense organs that respond to internal stimuli.

Prorenin. Inactive precursor of renin which occurs in the kidneys. Its molecular weight is 60000 and is converted into renin (40,000) which is active state.

Prosopyle. One of the surface pores through which a sponge's incurrent canals are connected to its radial canals.

Prostate Glands. Glands which are associated with urinogenital system of mammals. These are present at the junction of fundus and urethra of urinary bladder. They secrete fluid portion of semen.

Prostaglandins. Unsaturated fatty acids' derivatives which are in high concentration in the semen and also in little concentration in other tissue. They are divided into three main groups —PGA, PGE, PGF. The biologically active prostaglandins are PGA_1 PGA_2, PGE_1, PGE_2 PGF_{12} and PGF_{22}. They have effects on various body functions and physiological activities by modulating reactions which are mediated by c-AMP and c-GMP. They are able to produce contraction in uterine muscles, lower the blood pressure and modify) and modify actions of certain homones.

Prosthetic Group. An organic compound which is essential to the function of an enzyme in the organisms. It is different from coenzyme in the fact that it is more firmly attached to enzyme protein.

Prostomium. Anterior portion of the segment of earthworm which is hanging over the mouth.

Protandrous. The term used for describing a condition in which there occurs early maturation of male gonads and formation of sperms earlier than ova.

Pratagonist. Co-ordinated activity of muscle which is responsible for the movement. It is also called antagonist.

Protein. Organic compound having C, H, O, N and other elements. It is made up of amino acids. It forms the essential componant of protoplasm.

Protein Kinase. Any enzyme which catalyses transfer of PO_4 group from ATP to a protein forming a phosphoprotein.

Proteolysis. Refers to the splitting of proteins into peptide by hydrolysis.

Proterozoic. Geological are that dated back some 925 million years ago. It was Characterised by many protozoa.

Prothoracic Glands. Ecdyosone secreting tissues that are situated in the anterior thorax of insects.

Prothorax. Refers to the anterior portion of thorax in an arthropod.

Prothrombin. Refers to the precursor of thrombin in the blood. In blood coagulation it can be converted to thrombin by activated factor-X, factor-V, and Ca^{2+} ions etc., during blood coagulation.

Prothrombinase. The enzyme converting prothrombin into thrombin in presence of Ca^{+2} ions during blood coagulation process.

Protochordates. A collective term used for including lower chordates and cephalochordates, urochordates and hemichordates.

Protogynous. A condition in which there occurs early maturation of ova than sperms. (*Opposite* to protandrous).

Prototheria. A group of egg-laying mammals e.g., *Echidna, Tachyglossus*.

Protozoa. Group of first known animals. Acellular having simple body structure e.g., *Amoeba, Euglena, Paramecium* etc.

Prototriches. The sense organs which are found in fishes either scattered or a large surface or localised at certain places.

Proximal. Refers to the nearest point of an organ with respect to origin e.g. , shoulder is proximal end of the arm.

Proximal Tubule. Refers to the proximal portion of proximal convoluted tubule which terminates in the thin segment of descending limb of loop of Henle.

Proventriculus. Refers to the anterior portion of avian stomach; gizzard in stomach of insects; crop preceeding the gizzard in earthworm.

Pseudobranch. An opercular gill which is present in lung-fishes and many teleosts on the posterior side of hyoid arch. It receives oxygenated blood.

Pseudocaudal Tail. In cod (*Gadus*) tail fin gets merged with dorsal and ventral median fins and such a tail is known as pseudhcaudal tail.

Pseudocel. A body cavity present in nematodes. It is not lined with a membrane derived from mesoderm.

Pseudopodia. Used for the blunt temporary protoplasmic projections that are pushed out from an amoeba-like structure during feeding or locomotion.

Pterodactyl. Extinct reptiles. They were fish-eating type. They flied over water and diving for capturing small fishes. Their body was bat-like and they hung suspended from trees over hanging rocks during their time of rest.

Pterygiophores. Cartilagenous rod-like elements that are able to support the dorsal and vent ral fins of fishes.

Pterylae. Areas in avian skin where feathers grow.

Ptyalin. An enzyme of saliva present in some vertebrates. It acts on starch.

Puberty. Means the changes in male and female children occurring due to hormones culminating in their maturation to adult stage where reproduction becomes possible. The age for this is 16 years in males and 14 years in females.

Pubis. Constituent bone in the inniminate of pelvic girdle.

Pulmocutaneous. Blood vessel (artery/vein) in frog which supplies to lungs; it carries deoxygenated blood.

Pulmonary Artery. An artery which supplies blood to lungs from heart. It carries deoxygenated blood.

Pulmonary Vein. A vein carrying pure blood from lungs to heart.

Pulp Cavity. Refers to the space inside the dentine of tooth and placoid scale filled with pulp.

Pulse. The blood forced into aorta during systole produces pressure wave in arterila walls as it travels. The pressure wave is expanding the arterial walls as it travels and this expansion is known as pulse. In normal man it is usually 72/minute.

Pulvilus. Fleshy pad or lobe-like structure present between claw pair on the terminal segment of tarsus in some insects.

Pupa. Refers to the development stage of holometabolous insects usually encapsulated or in coccoon.

Pupil. Hole in the eye through which light travels. It appears black due to lack of illumination from within the eye.

Pupillary Reflexes. Contractions of pupil of eye when light is directed into one eye.

Purkinjie Cells. One of the layers in cerebellar cortex. It is single layered and these cell are among biggest neurons in the body. They have very extensive dendritic arbores which extend throughout the molecular layer of cerebellar cortex.

Purkinjie System. Specified and specialised conducting tissue that is able to transmit impulses from SA-node to ventricle. It exists through interventicular septum and terminates in the wall of ventricle.

Pus. The term used for describing the mass of dead leucocytes and some of damaged removed from the body.

Pygostyle. A plough-shaped structure in birds which is formed by fusion of last 3-4 caudal vertebrae around which tail feathers are arranged in a fan-like manner.

Plyangium. Refers to the proximal most muscular part of conus arteriosus in frog.

Pyloric Caecum. The digestive glands which are found in animals like insects.

Pyloric Sphincter. The constriciton in pyloric part of stomach which acts like a valve between stomach and intestine.

Pyramidal Tract. A bundle of nerve fibres which originate in the motor cortex and descend down the brain stem to the medulla and spinal cord. It is responsible for mediating control of voluntary muscle movements.

Pyramid of Kidney. The medulla of mammalian kidney has been divided into pyramid like areas by renal columns of Bertini.

Pyridoxine. It is also known as vitamin-B_6. It forms prosthetic group of certain decarboxylases and transaminases, converted into puridoxal phosphate within the body. Its deficiency is responsible for convulsions and hyperirritability. It is present in yeast, wheat, corn and liver.

Pyriformis. A muscle which arises from urostyle in the frog.

Pyrimidine. Heterocyclic compound ($C_4H_4N_2$) derivatives of which (thymine and cytosine and uracil) are found in nucleotides.

Q

Quadrate. A bone of suspensorium which is able to suspend lower jaw in authostylic jaw suspension. (It is present in amphibia, reptiles and birds).

Quadratojugal. The posterior bone of upper jaw in some vertebrates.

Qudruped. The four legged animals like cow, sheep, goat, elephant etc.

Quartenary Malaria. A type of malaria in which the recurrence of fever takes place after every 72 hours (every fourth day). This is due to infection of *Plasmodium malariae.*

Quotidian Malaria. The malaria with daily recurrence of fever and occurs due to mixed infection of *Plasmodium species.*

Queen. Reproductive female of social insects e.g., honey-bees and ants.

Quill. Proximal hollow shaft of feather of bird. It is also known as calamus.

R

RAS (Reticular Activating System). Refers to the system in brain which is responsible for conscious, alert state. It makes perception and location of individual sensation possible. It is present on the midventral portion of medulla and mid—brain and is made up of myriads of small neurons which are arranged in complex interwinning nets. In this system there are centers which regulate respiration, blood pressure, heart rate and other vegetative functions .

RBCs (Red Blood Corpuscles). They are known as erythrocytes; they are specialised blood cells which are able to carry oxygen due to the presence of haemoglobin in them. They are formed in the red bone marrow cells (haemocytoblasts) in case of mammals but in non-mammals they are produced within the liver as well as spleen and liver. They are rounded, oval and biconcave in mammals but biconvex in non-mammals. They are usually nucleated except in mammals (excluding camel where they are nucleated). Life of these cells is 100 days in non-mammals and 120 days in mammals. Their number being 54 million/ml in main and in 4.8 million/ml in woman. Mammalian RBC's are 7.5 μ m. in diameter and 2 μ m. thick. They have 29 μg of haemoglobin in mammals. About $3x10^{13}$ RBCs/900 gm of haemoglobin are present in circulating blood of man.

RNA (Ribose Nucleic Acid). One of the nucleic acids which exists in single strand. It consists of pentose sugar ribose, phosphoric acid and four heterocyclic bases same as in DNA except uracil in place of thymine. Its three forms are known with are as follows:

m-RNA. It is present in the nucleus and is able to copy the code given by DNA strand.

t-RNA. It is present in the cytoplasm and is able to bring the required amino acids to the sites of protein synthesis.

r- RNA. It is present in ribosomes within the cytoplasm and acts as the site for protein synthesis.

Race. A population of individuals in which the frequency of certain genes in gene pool is different from that of other population.

Rachis. Refers to the distal tapering portion of shaft of feather which is elongated and is bearing barbs, barbules etc., on it.

Radials. Refer to the rest of pterygiophores other than basalia (at the base) to which fin rays get attached.

Radial Canal. Means canal which radiates from disc of sea stars and related animals. This canal runs through arm and is supplying tube feet.

Radial Cleavage. Means cleavage in which successive tiers of equal sized blastomeres are situated directly above and below one another.

Radial Links. Refer to extensions from peripheral doublets to central sheath in cilia and flagella.

Radial Symmetary. Means condition in which similar parts have been arranged about common centre like wheel's spokes.

Radical Mastectomy. Refers to an operation of breast for cancer in which *axillary* lymph nodes have been removed.

Radiogenetics. Science which deals with study of effects of radiation upon heredity.

Radioulna. Fused radius and ulna bones present in frog and toads.

Radius. The bone of lower arm which is located on thumb side of vertebrates.

Radula. Rasping organ which is present in the alimentary canal of certain molluscs like *Pila*.

Ramus Communicans. Branch of trunk of spinal nerve, which later undergoes joining ganglion of autonomic nervous system, having sensory and visceral motor fibres and grey ramus of non-medullated fibres.

Ramus Dorsalis. The dorsal branch of trunk of spinal nerve which goes to skin and the dorsal (epiaxial) muscle of body and has medullated somatic sensory fibres.

Ramus Ventralis. Refers to the ventral brnach of trunk of spinal nerve which goes to ventral (hypoaxia) and lateral voluntary muscles of body, having somatic motor fibres.

Rathike's Pouch. Refers to the dorsal evagination of buccal diverticulum (cctoderm) in form of vesicle which forms non-nervous part of pituitary or adenohypophysis. It breaks and gets united with neurohypophysis to form pituitary gland complex of adult.

Ratitae. A group of flightless birds e.g. *Struthio* (ostrich).

Rat-Snake. Common name for *Zamenis mucosus* (dhamin). It is non-poisonous snake of plains that lives in the vicinity of human dwellings.

Recapitulation Theory. According to this theory, the individual during its development passes through ancestral history of species. In other words, it means 'Ontogeny repeats Phylogeny'.

Receptacula Seminalis. The seminal receptacle which is attached to female reproductive organs and used for storing sperms..

Receptacle. A blind structure which is attached with some part as seminal receptacle in case of tape-worms. It is used for storage of sperms.

Receptive Field. Refers to that area of a sensory field which on stimulation is able to influence the activity of a given neuron.

Receptor. A sensory end organ. It is device which is used for registering changes in internal/external environment of the organism.

Receptor Molecule. Molecule/molecules that is/are lived on the outer surface of cell membrane. They interact especially with messenger molecules like hormones or transmitters.

Receptor Potential. A membrane potential change that is elicited in receptor cells by the flow of receptor current.

Recessive Trait. Refers to that trait which appears only when both members of an allelic pair of genes are alike i.e., homozygous condition. The term is also applicable to genes.

Reciprocal Crosses. Two crosses similar to each other. In these some traits are being followed but the sexes of parents get interchanged.

Recombination. Means the formation of new combinations of genes which follow meiotic events.

Rectal Gland. In fishes like *Scoliodon* it refers to a gland of like structure which is attached to rectum. It is probably having some obscure function.

Rectum. Refers to the terminal portion of large intestine in vertebrates; also refers to the posterior region of intestine in some higher invertebrates.

Recurrent Bronchi. Mean the branches of respiratory tree through which inspired air from inspiratory air sacs of birds gets expelled into expiratory air sacs during expiration.

Red Bodies. One or more retia mirabilia which are fused into patches. They are usually present near the opening of pneumatic duct and covered with internal epithelium of bladder.

Red Gland. In cods and perches the epithelium of red body becomes glandular and undergoes folding to form composite structure called red gland.

Red Marrow. Refers to bone marrow. Active cellular marrow is termed as red marrow. It is opposite to inactive slow marrow which is called yellow marrow.

Red Muscle. Muscle which solely consists of B-fibres (intermediate size and ATPase contents). It is also known as slow muscle of back.

Red Nucleus. The structure which is associated with basal ganglion.

Red Pulp. Refers to the mass of reticular tissue of spleen which has free erythrocytes apart from macrophages.

Red Reaction. When skin is touched with a pointed instrument there occurs reddening that appears at the site in about 10 seconds. This is known as red reaction.

Redia. One of the several types of larvae occurring in the life-cycle of most trematodes like *Fasciola*.

Reduced Penetrance. Refers to the failure of a genotype, when present, to express itself in any way in an individual.

Reduction Division. Refers to the maturation division in which there occurs the separation of homologous chromosomes from daughter cells; in this division the parental chromosomes of a given homologous pair get separated so that they go to different daughter cells. It is also called *meiosis*.

Reflex Action. Relatively simple autonomic response taking place to a stimulus. It is independent of higher nerve centres in the brain.

Reflex Arch. A neural pathway which is used in reflex action. It consists of afferent nerve input to a nerve centre that gives rise to activity in efferent nerve to effector organ.

Reflexes. Actions which are reflected involuntary motor responses. They are mediated by neural arch in response to sensory inputs.

Reflexes Grasp. Reflexes related to grasping of objects which are brought about in contact with it.

Reflexes Monosynaptic. Reflex arches in which there is only a single synapse between afferent neurons and reflexes occurring in them are monosynaptic. e.g., stretch reflex.

Reflex Smyenteric. Refers to the response of intestine to stretch with respect to peristalsis permitting them to pass at a rate which varies from 2-25 cm/second.

Reflexes Postsynaptic. Refers to reflexes in which one or more interneurons get interposed between afferent neurons and reflexes taking place in them e.g., withdrawal reflex.

Refractory Period. Means the period of increased membrane threshold which immediately follows an action potential.

Regeneration. Means the period of increased membrane threshold which immediately follows an action potential.

Regeneration. Means the replacement of a body part, by growth, which has been lost. The power can be seen in the tongue of man, body of sponges, *Hydra* etc.

Regeneration Blastema. Bulge in which proliferating cells get accumulated during regeneration.

Regulator. A gene which is able to produce some substance such as a repressor that could affect transcription of an operon by interacting with an operator.

Regurgitation. Means taking back something from alimentary canal into mouth e.g., birds are able to regurgitate milk from crop into mouth and the feed it to young ones through their beak.

Reissner's Membrane. Refers to the thin walled sloping of scala media of mammalian cochlea which separates it from scala vestibuli.

Relaxin. Ovary secretion which takes place during pregnancy. It is a polypeptide that is able to loosen ligament of pulic symphysis (often softens cervix) and facilitates the delivery.

Remiges. Contour feathers having equal vane. They are present on wings or patagium.

Rem-Sleep. Refers to the sleep state in which eyes move rapidly (rapid eye movement sleep).

Renal. Any structure like artery/vein which pertains to kidneys.

Renal Blood Flow. The blood to renal tuble has been supplied by afferent renal arteriole from renal artery which are branched into Bowman's capsule to form capillary network called glomerulus from which efferent renal arteriole arises. The efferent renal arteriole undergoes division into many capillaries which supply the tubule before draining into renal vein. The branches of efferent renal arteriole on joining with venule opens into renal vein.

Renal Columns of Bertini. Connective tissue partitions which separate medulla of kidney of mammals into a number of pyramidal areas.

Renal Erythrpoitic Factor. Refers to the substance which is secreted by kidney. It forms [illegible]thropoitin (which regulates RBC's level in the blood) by act[illegible] with plasma globulin. Its production gets increased by hypoxia, cobalt salts and androgens.

Renal Portal System. The portal system which involves veins from posterior limbs and posterior part of the body opening into the kidney. These veins are able to distribute blood into kidney tissue from where it gets recollected from kidney that gets joined to form renal vein/veins that open into posterior vena cava.

Renal Threshold. Refers to the minimum concentration of substance which is necessary to produce a detectable response in kidney.

Renal Uremia. A disorder of renal function where there occurs accumulation of breakdown products of protein metabolism. It has been characterized by anorexia nausea and vomiting mental deterioration and confusion, muscle twitching, convulsions and coma.

Renette. In some nematodes it refers to an excretory cell with short tube attached.

Renin. A proteolytic enzyme which is secreted by kidneys into blood stream (molecular weight approximately 40,000). It attacks a glycoprotein in a 2-globulin fraction of proteins in the circulating plasma and liberates decapeptid angiotensin-I which gets converted to angiotensin-II dipeptidyl-carboxypeptidase and reninangiotensin system is able to regulate aldosterone secretion and is able to regulate renal blood pressure etc. With reference to digestive enzymes it is considered to be milk coagulation enzyme which is responsible for milk curdling.

Repair Enzymes. Refer to those enzymes which remove and replace segments of damaged DNA.

Replacing Bones. Refer to the bones which are developed from pre existing cartilage. They are also known as cartilage bones or endochondral bones e.g., bones of cranium.

Replication. Means the formation of a large molecule e.g., DNA under guidance of a template gives rise to identical molecules, the copy and template.

Representational Hemispheres. The cerebral hemispheres which are specialized for spatio-temporal relations.

Repressor. Means substance which is produced by a regulatory gene and can combine with an operator and regulator activity of an operon.

Reproduction. Defined as the capability of an organism to produce similar offspring.

Reproductive Cycle. Includes structural functional and behavioural cyclic changes in the body resulting in the conditions that are able to bring about reproduction.

Reproductive Migration. Migration to area where it is expected to have copulatory activity.

Reptiles. Crawling vertebrates leading terrestrial life, having incompletely divided four chambered heart, pentadactyl limbs and development of extraembryonic membranes during development e.g., lizards, alligators, snakes etc.

Respiration. Means use of oxygen by cells and usually called cellular internal respiration. *Compare* with breathing.

Respiratory Dead Space. The space occupied by gas which does not exchange with blood in pulmonary vessels.

Respiratory Organ. Organ/some part of body which is set aside for exchange of gases with blood surrounding. It is also called breathing organ/respiratory surface.

Respiratory Pigment. Means organic pigment which is able to transport oxygen in the body e.g., haemoglobin, haemocyanin etc. In nonchordates it is dissolved in plasma but in chordates it is present in RBCs.

Respiratory Surface. Refers to that part of body where exchange of gases (oxygen and carbon dioxide) gets facilitated during breathing. It is present in form of lung in most of terrestrial vertebrates and gill in aquatic vertebrates.

Response. Reaction to stimulus (either external or internal).

Restiform Body. The inferior peduncle which connects cerebellum with brain stem.

Rete Mirabile. Refers to an extensive counter current arrangement of arterial and venous capillaries.

Rete Testes. System of tubes in testes of mammals to which are attached proximal ends of vas efferentia.

Reticular Tissue. Fibres in this tissue are composed of compounds of a protein called reticulin; they are branched but do not form fascia and exist as sheets forming basement membranes below epithelial tissue. They are present in sarcolemma around muscle fibres, neurilemma around nerve fibres and frame-work of many lyphoid organs.

Reticulum. A division of stomach in ruminants like cow, sheep, goat etc. which is present between rumen and omasum.

Retina. Light sensitive layer of the eye.

Retinal. A part of rhodopsin, a visual pigment. It is aldehyde of vitamin A_1. It combines with protein opsin.

Retinula. Group of pigmented elongated cells in arthropods which are present in basal portion of the compound eye.

Retinens. Aldehydes of vitamin-A_1. They are also known as etinals. Retinene-1 occurs in all the eyes and retinene-2 occurs in some eyes only.

Receptor Muscle. A muscle that is also to pull back some structure opposite to protractor that is able to pull forward.

Retrogressive Metamorphosis. There are certain animals like *Herdmania* in which the larva has more advanced chordates' characters like notochord, tubular nerve cord and pharyngel slits etc. They are lost during metamorphosis and adult shows an organisation which is totally different and that of primitive type. Such a metamorphosis is known as retrogressive metamorphosis.

Reversal Potential. Means the membrane potential at which current through an activated synapse or receptor membrane becomes null.

Reverse Mutation. means the mutation from a mutant gene from back to the form which is considered standard or wild. The rate of reverse mutation has been found to be generally lower than mutation from wild to mutant.

Rh-Factor. Refers to the kind of antigen present in some RBCs. It was first found in RBCs rhesus monkeys therefore, it was named Rhesus factor (Rh-factor). About 97% of Indian population is Rh-positive and 3% is Rh-negative. The antigen plays vital role in blood transfusion and marriages of individuals. Rh^+ is a dominant trait and marriage of such individual with a Rh^{--} female would result in senstisation of the body and would bring about agglutination of foetal cells in the pregnancies after first or maximum second one. These complications lead to abortions, miscarriages etc.

Rh-Incompatibility. Rh^+ is a dominant trait and Rh-incompatibility occurs when marriage between Rh^+ father and Rh^- mother takes place as the child is Rh^+ the foetal blood is able to produce sensitisation of mother's blood. It may not having any bad effect in first pregnancy but in later pregnancies the sensitised mother's blood bring about haemolytic disease in new-born (which is called *Erythroblastosis foetalis*). The disease may result in the abortions, miscarriages, natal jaundice etc. Blood transfusion of Rh^+ to Rh^- may cause sensitisation of blood of recipient, in one or two transfusion no significant complication may occur but later such transfusions may be responsible for the death of the patient.

Rhabdom. Rod-like structure which is formed of inner surfaces of adjacent sensory cells in the commatidium of a compound eye (as found in arthropods).

Rheoreceptor. A receptor which responds to water current as lateral line organ in fishes.

Rhinal Process. Small cartilage which closes nostrils of frog during breathing.

Rhinarium. Small moist hairless area of skin as present in many mammals like rabbit.

Rhinencephalon. Refers to the anterior ventral part of the anterior end of fore brain that is differentiated into olfactory lobes.

Rhodopsin. Retinal pigment is rod which is sensitive to low levels of illumination.

Rhythm. Refers to periodically occurring events, act. Also refers to α & β rhythms of electroenoephalogram.

Riboflavin. It is also called vitamin B_2 which is a constituent of flavoproteins. It occurs in liver, milk. Its deficiency causes diseases like glossitis and cheilosis.

Ribonuclease. Refers to the pancreatic enzyme which acts on RNA nucleotides.

Ribonucleic Acid. A complex organic compound (a nucleic acid) in living cells that is concerned with protein synthesis. A RNA molecule consists of a long chain of nucleotides in which sugar is ribose and the bases are adenine cytonine, guanine and uracil. It is of the following types, massenger RNA, ribosonal RNA and transfer RNA.

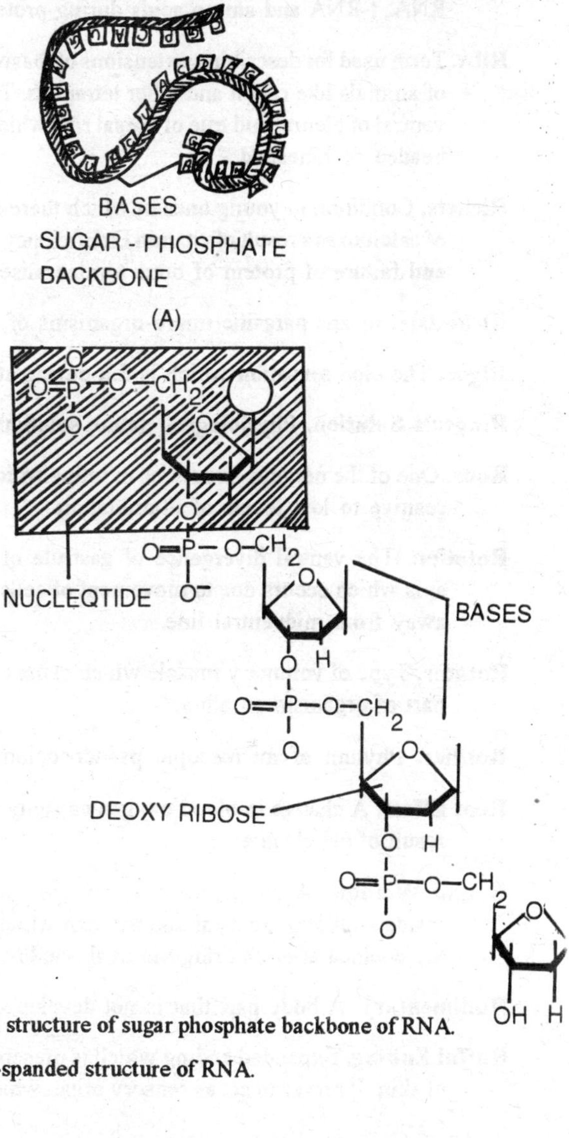

Fig. 1. (A) Detail structure of sugar phosphate backbone of RNA.

(B) Single-spanded structure of RNA.

Ribose. Pentose monosaccharide. It is constituent of RNA. Its chemical formula is $HOCH_2 (CHOH)_3 CHO$.

Ribosomes. Ribonucleoprotein particles which are present in the cytoplasm. They are sites of protein synthesis (sites of interaction of m-RNA, t-RNA and amino acids during protein synthesis).

Ribs. Term used for describing extensions of basiventralia in the sternum of animals like rabbit and other tetrapods. They are of two types—ventral of pleural and true of dorsal ribs which may further be single headed or biheaded.

Rickets. Condition in young ones in which there occurs poor absorption of calcium as a result of vitamin D deficiency leading to hypocalemia and failure of protein of bone tomineralise.

Rickettsia. means parasitic micro-organisms of insects and ticks.

Rigor. The clod sensation which is felt during attack of material fever.

Ringer's Solution. Physiological saline solution.

Rods. One of the nervous structures in retina which has rhodopsin and is sesitive to low levels of illumination.

Rotation. The ventral divergence of gastrula of frog about horizontal axis which occurs due to movement of cells outwards and dorsally away from midventral line.

Rotator. Type of voluntary muscle which causes rotatory motion of one part of organism on other.

Rotifier. Phylum of microscopic pseudocoelumates.

Root Effect. A change in blood oxygen capacity which takes place as a result of pH change.

Round Window. A membrane-covered opening. It is situated in the inside wall of middle ear and through which pressure waves leave the cochlea after entering via oval window.

Rudimentary. A body part that is not developed completely.

Ruffni Ending. Expanded ending which is present in deep fibrous tissue of skin. It serves to act as sensory organ which is able to respond to one or more type of stimuli.

Rugae. Ridges or folds as in the lining of vertebrate stomach.

Rumen. Sac-like anterion part of stomach in ruminants. It is lined with stratified epithelium and in which cud, ingested in hurry has been stored for sometime.

Ruminant. Group of mammals having hooves, horns and ruminating stomach. Also known as cud chewing mammals as cow, sheep, goat etc.

S

Saccades. Sudden jerky movements which take place when the gaze shifts from one subject to another.

Saccadic Movements. Rapid ballistic movements. Unlike slow ramp movements they are not subject to feed back control.

Saccharide. Any of series of sugars e.g., disaccharides (maltose, sucrose), trisaccharides, polysaccharides (starch, glycogen) etc.

Sacculus. Ventral part of internal ear. It is connected with utriculus through a sacculo-utricular canal.

Sacculus Rotundus. Refers to the rounded sac present at the junction between ileum and colon an a caecum. It has definite valves which are able to direct food either through caecum or into colon.

Sacculus Vasculosus. Refers to the thin walked dorsal expansion of diencephalon. It acts as a pressure receptor.

Sacral Ribs. Transvere processes which are present in the sacral vertebra in frog.

Sacral Vertebrae. 3-4 vertebrae that follow lumber vertebrae. They are often fused to form synsacrum.

Sacro-Iliac Join. A joint which is formed by articulating the sacral ribs with ilium bone of pelvic girdle.

Sacrum. Region in vertebral column which follows lumber region. It consists of 3-4 vertebrae. It is followed by caudal region.

Sagittal. In a bilaterally symmetrical animal it refers to the median antero-posterior plane,

Saliva. The watery, colourless secretion of salivary glands. It is having water, mucous, amylase and mucopolysaccharidase. The substance is able to lubricate food. It also helps in digestion of starch and kills harmful bacteria.

Salivary. The saliva secreting glands of mouth.

Salivary Glands. Glands within mouth cavity. They secrete saliva.

Salt Gland. There are many birds like herrings, panguines and albatroses living at sea and very far from land that have a pair of glands in the head called salt glands. They secrete salts to maintain osmoregulation.

Sand-Lizard. Common name for *Uromastix herdwickii.*

Sprotroph. An organism living on dead organic matter.

Saprozoic Nutrition. Type of nutrition by which an organism feeds upon decaying organic matter or by absorption of food through general body surface as in parasites.

Sarcocyte Zone. Ectoplasm in certain protozoa e.g.; *Monocystis* which gets differentiated in an outer zone called sarcotyte zone.

Sarcolemma. An extremely thin membrane (1μ thick) present on the periphery of muscle fiber.

Sarcomere. Area in striped muscle fibre present between adjacent Z-lines.

Sarcoplasm. Refers to the cytoplasm within the muscle fibre. It exhibits longitudinal striations due to myofibrils or sarcostyles.

Sarcoplasmic Reticulum. Means a net—work of tubules present within the striped muscle fibres. An uptake of calcium ions by this brings about muscle contraction.

Sarcostyles. Longitudinal striations or myofibrils present in the sacroplasm of muscle fibre.

Sarcotubular System. System of membranes within the muscle fibre. Under electron microscope, it appears as vesicles and tubules. It consists of sarcoplasmic reticulum and 'T' tubules.

Satiety Centre. The hunger controlling centre present in hypothalamus in vertebrates.

Saw-Fish. Common name for a cartilagenous fish *Pristis*. It is having a saw-like structure which is formed by elongation of ethmoid region of skull.

Scala Tympani. Lower tubular portion of cochlea in mammals which is filled with perilymph.

Scala Vestibuli. Upper part of cochlea in mammals. It is separated from scala media by Reissner's membrane.

Scales. Skin derivatives which are mainly found in fishes and reptiles and form exoskeleton.

Scaly Ant-Eater. Common Name for *Manis* (a mammal).

Scaphopoda. Tooth-shelled molluscan class of non-chordates.

Scapula. Refers to shoulder-blade in a vertebrate. Refers to the bone in the half of pectoral girdle.

Scavanger. Animal feeding on dead organisms which it has not killed.

Scent Glands. Skin glands. They occur in mammals and secrete scent which is characteristic of every species and sex. It helps in recognising members of same species, attracting the other sex or repelling the foes e.g., musk glands of musk-deer.

Scizocoel. Coelom which is developed from originally solid mesoderm and differentiated on the sides of notochord (also called *lateral plate mesoderm*) as in chick.

Schizogony. Multiple fission in asexual phase, e.g., Plasmodium.

Schwann Cell. A neuroglial cell wrapping its membrane around axons during development to form a myelinated insulating sheath between nodes of Ranvier.

Scientific Name. Binomial or trinomial designation of an animal.

Sclerites. Hardened body wall plate which is bounded by sutures as in arthropods.

Scleroblast. Amoebocyte like cell in sponges which is secreting spicules.

Sclerotic Coali. Means the white of an eye or outside layer.

Sclerotic Ossicle. Overlapping ossicle—like bodies that are present in sclerotic coat of the eye of reptiles.

Scotomas. Refer to blind spots in eyes due to diseases.

Scotopic Vision. Means night vision. Its main cause is that rods are extremely sensitive to light and are receptors for this vision. However, this visual apparatus is not able to resolve the details and boundaries of objects or determining the colour.

Scotopsin. Photosensitive pigment in rods of retina of eye is known as rhodopsin or visual purple. It opsin is known as scotopsin.

Scotopism. The night vision.

Scroll Valve. Watch spring like fold of mucous membrane present in the intestine of sharks like *Scoliodon*. It increases area of absorption for intestine.

Scrotal Sacs. Refer to the two leathery thick walled hairy sac—like structures outside the body that lodge the testes in mammals.

Scurvy. A disease which is caused due to the deficiency of vitamin-C in the body and is characterised by bleeding from gums, muscles and joints.

Scutes. Horny plates which are arranged symmetrically as in reptiles like alligators, crocodiles etc. However, they do not correspond to plastron and carapace.

Seal. Aquatic mammals that belong to group sirenia.

Sea-Anemone. Common name for anthozoan coelentirate *Metridium*.

Sea-Cow. Common name for manatees. An aquatic mammal that belongs to group sirenia.

Sea-Cucumber. Common name for echinoderm *Holothuria*.

Sea-Gooseberry. Common name for comb-jellies having spherical bodies. It belongs to phylum ctenophora.

Sea-Hare. Common name for mollusc *Aplysia*.

Sea-Horse. Common name for fish *Hippocampus*.

Sea-Lion. Common name for aquatic mammals that belongs to order pinnipedia of eutherian mammals.

Sea-Mouse. Common name for annelid *Aphrodite.*

Sea-Mussel. Common name for mollusc *Mytilus.*

Sea-Pansies. Common name for members of pinnatulacea, an anthozoan cnidarian.

Sea-Snakes. common name for group of snakes that belong to group hydrophidae. They are having compressed tail and venom. They are dangerous to human beings.

Sea-Spider. common name for arachnid *Nymphon* that belongs to class pychgonida.

Sea-Squirts. Common name for Ascidians.

Sea-Star. Common name for asteroid echinoderm star-fish.

Sea-Urchin. Common name for echinoderm *Echinus*. The animal has globular test and with long prominent spines.

Sea-Whip. Common name for gorgonian anthozoan coelentrate.

Sebaceous Glands. Glands which consist of few small sacs or alveoli. They secrete an oily substance called sebum (formed by esters of cholesterol) and are associated with hair follicle.

Sebum. Refers to the fatty sebaceous glands.

Second Messengers. A term applicable to c-AMP, c-GMP, Ca^{2+} or other intracellular regulatory agent which is itself under the control of an extracellular 'first messenger' such as hormone.

Secondary Sexual Characters. The characters which get developed in male and female young one due to secretions of formers so as to make them distinct from each other e.g., beard hard voice, skin hair etc., of male and breast, broad hips, well—developed shoulders etc. in female.

Secondary Spermatocytes. In meiosis (during spermatogenesis), they refer to the cells which are formed after cytoplasmic division during diakinesis.

Secretin. Duodenal hormone which makes pancreas to pour pacreatic juice through pancreatic duct.

Secretogouge. A substance stimulating or secrefing secretin.

Secretor. An individual having a water soluable form of antigen-A and or B. Its presence can be detected in certain body secretions such as saliva and amniotic fluid.

Secretory Granules. Membrane-bound cytoplasmic granules having secretory products of a cell.

Sedentary. Means staying in one place.

Selachii. Group of cartilagenous fishes to which sharks and rays are belonging.

Selection Pressure. Means the operation of natural selection on frequency of genes in a population. Selection pressure undergoes interaction with mutation pressure and is different from one environment to the next. The net result is that certain genes increase in frequency in some population but are kept at lower frequency in others.

Self-Fertilization. Means the fertilization of an egg by a sperm from same individual. It is *opposite* to cross fertilization.

Sella Turcica. A depression present on upper surface of basisphenoid which receives pituitary body.

Semen. Fluid that is having sperms by tests and related structures in most animals.

Semicircular Canals. Three canals that are perpendicular to each other so that they get oriented in planes of space. They are associated with internal ear of vertebrates.

Semi Conservative Replication. Refers to the synthesis of two new DNA molecules by making use of each of two single DNA strands composing the original molecule as a template for the assembly of complementary strands.

Seminal. Structure which pertains to spermatozoa. It may be in form of seminal receptacle which is a sac-like organ. It receives and stores sperms after their release in most animals and is regarded as a part of female genital system. It may also be in form of seminal vesicle which is also a sac-like organ in several animal in which seminal fluid is stored before it is discharged and regarded as a part of male genital system.

Seminal Vesicles. Sac-like structures which store sperms in most animals such as mammals.

Seminiferous Ampullae. Small seminiferous tubules present in lower vertebrates.

Semineferous Tubules. A tube conducting seminal fluid.

Sense Organ. An organ which is sensitive to a particular type of stimulus.

Sensilium. Chitinous hollow hair of like projections of arthropod skeleton which serve as auxillary structures for sensory neurons.

Sensory Epithelium. A specialised epithelium to receive stimuli, to conduct impulses etc. Such cells are present in receptors derived from surface e.g., retina of eye.

Sensory Fibres. The afferent fibers in nervous system of animals. They carry impulses from sense organs to the brain or spinal cord.

Sensory Filter Network. Neural cicuit that is able to transmit some selected features of sensory input and neglects other features.

Sensor Neurons. Neurons by which impulses are travelling from a receptor system e.g., skin to brain or spinal cord. They are called *afferent neurons*.

Sensory Papillae. Small projections which are present around body or some organs carrying tactile functions e.g., sensory paillae around lips of *Ascaris* or fleshy margins of *Petromyzon*.

Sensory Tentacle. Small projection like structures which are present externally or internally and bear sensory cells. They are useful for tactile function.

Septomaxillary. Small bones that are covering each nasal capsule in many animals as in frog.

Septum. A partition separating two cavities or structures.

Septum Mediale. Median septum in frog's heart which separates aperture leading to left systemic arch.

Septumpellucidum. Refers to thin membranous septum below corpus callosum, between it and fornix. It separates two cerebral hemispheres in mammals.

Sere. Set of communities formed during ecological succession in an area that ultimately results in the extreme conditions . Sere has one pioneer, many transitional and one climax.

Serial Homology. Refers to occurrence of homologous structures in different segments of the same individual.

Serology. Deals with the study of sera and their action.

Serosal. The side of an epithelial tissue which faces the blood. It is *opposite* to mucosal side which faces the exterior or luminal space.

Serotonin. Chemically the substance is 5-hydroxytrytamine. It is a mediator in the central nervous system which is present in high concentration in blood platelets and gastrointestinal tract where it occurs in the enterochromaffin cells and myenteric plexus. It is found in lesser amounts in brain particularly hypothalamus and retina.

Serpentes. Common name for a group of reptiles that are limbless and popular as snakes and may be poisonous or non-poisonous.

Sertoli Cells. In vertebrates they refer to the non-gametic cells in the seminiferous tubules which are able to provide nourishment for developing sperms and secrete inhibin.

Serum. Blood plasma separating from clot. It is free from cells or fibrin.

Setae. Bristles which are embedded in body wall of earth worm . They are used mainly as locomotory organs.

Sex Cells. The gametes which are produced by gonads (testes/ovaries). They included sperms and ova. Their formation is known as gametogenesis in general and termed spermatogenesis/ovogenesis for formation of sperm/ova respectively.

Sex Chromatin. Refers to chromatin in Barr body and F-body.

Sex Chromosomes. The XY chromosomes which are specially concerned with sex determination.

Sex Determination. Means the genetical basis of determination of sex of individual in a dioecous animals such as human beings. The male individuals are having XY chromosomes and female individuals are having XX chromosomes. This sex determination is of XY type . The other type of sex determination may be XO which is found in other animals.

Sex Hormones. The hormones which are secreted in the body that are able to influence sexual functions and secondary sexual characters in the individuals. The hormones secreted by interstitial cells of gonads (i.e., Leydig cells in males and corpus luteum in females in case of human beings) are known as *primary* sex hormones. The hormones secreted from *other* places like placenta and adrenals etc., are known as secondary sex hormones. They affect the development of embryo in the desired direction (according to genetic constitution XX/XY), bring about the development of child into boy or girl accordingly , cause puberty and ensure reproductive function, and are responsible for physical/physiological/ behavioural changes in males/females.

Sex-Limited Character. Characters that belong to only one sex but do not find their expression in opposite sex because the expression depends on a particular and specific hormone. For example baldness is a character that is able to find its expression only when testosterone hormone is present and due to this reason it is only present in males and not in females.

Sex-Linked Characters. Refer to the characters with their linked genes which are located in sex chromosome. Many such characters like haemophilia, night blindness, colour blindness etc., are linked to X-chromosomes where as characters like hypertirichosis are linked to Y-chromosome. These characters generally find their expression in males due to heterozygotus genetic constitution but rarely appear in females as females require double dose of such X-chromosomes which is a very rare incidence.

Sexual Cycle. Refers to the formation of offspring from parental organisms by processes which involves sexual reproduction.

Sexual Dimorphism. Means differentiation of two sexes by the appearance of the external characteristics in case of dioecious individuals. For example male frog is having copulatory pads, vocal cords but female frog does not have these characters.

Sexual Reprcduction. The process in which a vegetative body is formed from another vegetative body which involves fusion of sex cells or gametes.

Sexual Union. Means the temporary union of male and female individuals which involves transfer of sperm from male's body to female reproductive tract.

Shaft. Refers to the slender, middle portion of a long bone as femur. Also refers to the middle keratinised portion of avian feather and the portion of hair projecting outside the body in mammalian hair.

Shank. The zygopodium of hind limb which is supported by tibia and fibula bones. It is articulated anteriorly with thigh and with foot posteriorly.

Shark. Cartilagenous fishes having lateral gill slits e.g. Scoliodon.

Sheath of Henle. Thicker collagen fibre layer which is present outside the neurilemma and is continuous with endoneurium.

Sheath of Schwann. The tough, inelastic scleroprotein (reticulin) membrane which consists of fine circular fibres present outside Schwann cell in neuron. It is also known as neurilemma.

Shell. Outermost protective covering which is usually inorganic material e.g., calcium carbonates silica. often chitinous or pesudochitinous.

Shell Glands. Glands which are present mainly within reproductive system and secrete shell over the egg.

Shell Membranes. Paired membranes in birds' eggs which are able to secrete shell over the egg.

Siamese Twins. Conjugated twins that remain connected by some portion of their bodies due to the failure of complete separation of early embryo into two unattached embryos.

Siblings. Brother and sisters in same family.

Sigmoid Curve. 'S' shaped curve.

Silent Mutation. A point mutation in which there occurs a change from codon specifying an amino acid to other codon which is specific for same amino acid.

Silicosis. Lung disease in which silicon fibres are formed in the lung.

Silver-Fish. Common name for apterygote insect *Lepisma.* It is found in bookshelves and objects kept in dark for sometime.

Sinuatrial Node. In the heart it refers to the principal pacemaker near the tricuspid valve. It is able to initiate a wave of contraction across the atria when it receives stimulation from vagus nerve.

Sinuauricular Aperture. Refers to the opening of sinus venosus into auricle of heart. It is guarded by a valve.

Sinus. A thin walled cavity such as olfactory sinus, nasal sinus etc.

Sinus Glands. In some arthropods, there are the glands at the base of eye stalk which are able to control chromatophore dispersion, under influence of X-gland of brain.

Sinusoids. Large irregular toruous blood spaces. They could be compared to capillaries in that their walls form endothelial tube, but they are different from capillaries in the fact that their lumina are having greater size.

Sinus Venosus. Refers to the large chamber which receives blood from veins and passes it to auricle (usually right auricle where the auricles are divided) through sinuatrial opening which is guarded by a valve. The structure is lacking in avian and mammalian heart (present in fishes, amphibia, reptiles etc.).

Siphonoglyph. Ciliated groove in sea-anemone. It serves to create a water current into gastrovascular cavity.

Siphonozooid. Specialised polymorphic form which is able to produce water current to help in breathing in certain coelenterate colonies.

Siphuncle. Gas filled canal passing through coiled and chambered shell of *Nautilus* (a cephalopod mollusc).

Skeleton. External/internal hardened frame work of an animal body. It gives support and protection to the soft parts of body. The exoskeleton is inducing usually epidermal derivatives like scales, feathers, hair, nails, claws, horns, etc., whereas endoskeleton comprises of an axial skeleton that is including skull and vertebral column and appendicular skeleton that includes skeleton of girdles and limbs. Endo-skeleton is dermal and may be cartilagenous or bony or made up of both.

Skin. Refers to the outermost investment over the body of an organism. It is made up of epidermis, derived from mesoderm. Epidermis may be stratified but dermis is always a composite layer. It may or may not be having glands and other derivatives.

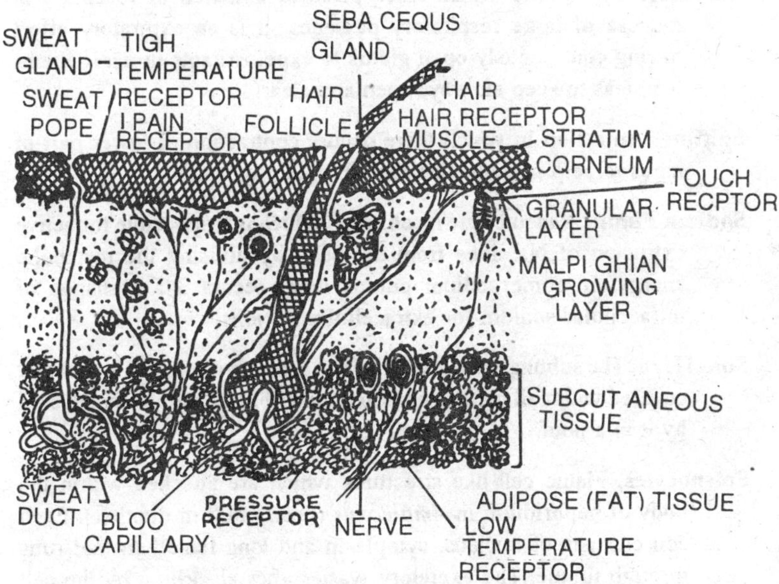

Fig. 2. Structure of a mamalian skin

Skull. Anterior portion of axil skeleton that enclose brain, sense organs and support mouth nostrils etc. (joined with vertebral column).

Sliding Filament Theory. According to this theory there occurs shortening of muscle sarcomeres by active sliding of thin (actin) filaments towards mid region of myosin (thick) filament. The theory was given by Huxley.

Slime Glands. Mucous producing glands in the skin, e.g., mucous glands in skin of fishes and amphibia. The secretion of these glands is able to protect skin from effect of water. It keeps skin free of bacterial action and makes body slippery to escape from grip of an enemy.

Smooth Muscles. Muscles that lack sarcomeres and thus do not carry striations. Myofilaments do not distribute uniformly within small uninucleated spindle-shaped cell. They are also known as involuntary or unstriped muscle e.g., muscles in the urinary bladder.

Snakes. Reptils lacking fore and hind limbs. Also known as *Serpentes* e.g., cobra.

Sneezing. A response which takes place to irritation of receptors in mucosa of large respiratory passages. It is an expiratory effort having continuously open glottis to expel irritants in nasal chamber so as to keep air ways open and clear.

Sniffing. An action in which there occurs contraction of lower part of nares to help in deflecting the air stream upwards.

Sodium Pump. A membrane mechanism which is responsible for active extrusion of Na^+ ions from the cells by utilising the metabolic energy. In some sodium pumps there occurs 3:2 exchange of intracellular sodium for extracellular K^+ ions.

Sole-Horn. The subunguis below nail plate at the tip of digit. It is partially covered with nail plate in nails and claws but is completely covered by it in a hoof.

Solenocytes. Flame cell-like structures which are attached within the body of nephridium in *Amphioxus* (nearly 500 in number). Each cell consists of nucleus, cytoplasm and long flagellum that runs through tubule. The excretory wastes after absorbing by the cell body are passed to tubule in form of aqueous solution by involving the movement of flagellum from where they pass out of nephridium through its pore.

Somatic. The term used for the entire body exclusive of reproductive cells.

Somatic Motor. Efferent nerve fibres which run between brain/spinal cord and the effector organs like muscles. They are able to respond and displace body organs in response to external stimuli.

Somatic Sensory. Afferent fibres which run between cutaneous sense organs and brain/spinal cord. They are able to receive external stimuli.

Somatodin. Factor/growth hormone which acts on cartilage's linear growth and other effects. There are three polypeptides with somatodin activity such as somatodin-A, somatodin-B and somatodin-C. Somatodin is having insulin like properties including inhibition of lipolysis.

Somatosensory Cortex. The region of cerebral cortex which is concerned with sensory input from body surface.

Somatostatin. Hypothalamic inhibiting hormone. It exerts its action on anterior pituitary secretion and inhibits extra growth. It is also known as growth inhibiting factor or growth inhibiting hormone.

Somnambulism. Sleep disorder in which person is having sleepwalking. It is more common is children than adult. The patient walks with eyes open so as to avoid obstacles but cannot recall episode when he is awakened. It may last for several minutes.

S-Phase. Phase of cell cycle during which chromosomes undergo replication.

Sparrow. Common name for *Passer domesticus* (common house sparrow).

Spasm. The convulsive, involuntary muscular contraction.

Spawn. The whole egg mass which forms jelly—like irregularly shaped flotilla as frog.

Speciation. Evolutionary changes that take place in species characteristics usually highly correlated with environmental pressures). When these are complete, this process evolves the formation of a new species.

Sperm. The male sex cell which is produced by testes. Generally it consists of an anterior head containing male pronucleus and a long posterior tail. The formation of sperms is known as spermatogenesis.

Spermatheca. Seminal receptacle in some female insects.

Spermatogenesis. Defined as the process of sperm formation within the testes. During this process the primary sperm mother cell undergoes division meiotically to form four spermatids and each of them metamorphoses into a sperm by involving the process of spermateleosis.

Spermateleosis. Defined as the process of metamorphosis of spermatids into spermatozoa. It is also known as spermiogenesis.

Spermatocytes. The cells which are formed by division of spermatogonia. The primary spermatocytes are formed by mitotic division, and secondary spermatocytes are formed by meiotic division of primary spermatocytes.

Spermatophores. Packets of sperms which are transferred by male to female as in insects.

Spermatozoa. The mature sperms which are formed due to spermatogenesis (mature male gametes).

Sperm Sacs. A pair of blind club-shaped sac like structure which are attached with the ventral wall of urinogenital sac in male *Scoliodon* and have probably sea water. The water discharges from sacs thereby resulting the flush out of the sperms from cloaca of male.

Sphenethmoid. Bone of cranium which is formed by fusion of sphenoid bone with ethmoid. It is replacing bone which is developed around and front part of brain.

Sphincter. Ring of smooth muscle which surrounds a tube or opening by which its contraction is able to close the lumen.

Sphincter of Oddi. Sphincter which is present at the entrance of bile duct into duodenum.

Sphingolipids. Lipids which are formed by fatty acid attached to nitrogen atom of sphingosine, a long chain oily amino alchohol ($C_{18}H_{37}O_2N$). They are primarily present in brain and nerve cells.

Spicule. One of the many solid structures that compose the skeleton of a sponge.

Spicule Penial. In male Ascaris, it refers to a chitinous rod (of which there are two) which is extending from cloacal opening from cloacal opening. It takes art in copulation.

Spinal Column. A series of vertebrae in the vertebrates enclosing the spinal cord. It is also known as vertebral column.

Spinal Cord. Part of nervous system which extends along the length of vertebral column in the vertebrates and continues anteriorly into brain.

Spinal Nerves. Nerves which are given out from spinal cord in the different body segments forming reflex arc from which ramus dorsalis, ramus ventralis and ramus communicans are arising.

Spindle. The structure which is formed within the cytoplasm during mitosis. It is made of fibres (asters) which are arranged in form of an elliptical structure. With this the chromosomes get attached in its equatorial part.

Spinnerets. Small structures which are present at the abdomen of spiders in form of finger like organs having tiny tubes at their bases. They secrete fluid-like secretion that get hundred when come in contract with air to form silk threads (forming spider web).

Spinule. Refers to the small projections in the cuticle of parasitic flat-worms like *Fasciola.* They help in anchoring the body within the body of host and do not allow the parasite being washed away.

Spiny Ant-Eater. Common name for egg laying mammal called *Echidna.*

Spiracle. In insects this term is used for an external opening of tracheal or respiratory system; in cartilagenous fishes this term is used for the opening of Ist gill cleft and in tadpole the atriopore.

Spiral Valve. In some amphibians like frog it refers to the twisted structure present in the truncus arteriosus while in cartilagenous it refers to the spiral partition in the intestine.

Spirochete. The term used for the long ciliated microscopic spiral-shaped organisms which may be pathogenic, e.g., syphilis producing spirochete *Treponema pallidum.*

Splanchnic. Refers to any structure which pertains to viscera as splanchnic mesoderm.

Splanchnocoel. Refers to the space in the hypomere (ventral portion of enteric mesodermal pouch).

Splanchnocranium. Mesenchyme surrounding gut results in a series of paired arches called visceral arches. These arches collectively are able to form splanchnocranium. It suspends the jaw with the uanium.

Spleen. A large organ which is characteristic of most vertebrates. This organ is situated near the stomach and known to destroy RBCs in mammals.

Splenial Bone. Small flattened bone which is present on the inner face of dentary of frog, reptiles and birds.

Splint Bone. Reduced 2nd and 5th metatarsals in horse which are associated with proximal part of canon bone (3rd and 4th metatarsals undergo union to form a long bone).

Sponges. Group of diploblastic animals having porous body, central cavity (spongocoel) and choanocyte cells and regulate current of water through the body. They have high power of regeneration.

Spongin. Chemical which is present in fibres of poriferan skeleton and is secreted by spongioblasts.

Spongioblasts. Cells in sponges (belonging to group demospongiae) secreting spongin fibres.

Spongy Tissue. A tissue which has spaces. It is not made up of compact material e.g., lungs of vertebrates.

Spores. Refers to the special reproductive bodies of lower organisms. They are produced generally in unfavourable conditions and provided with protective coverings and develop into a new individual under favourable conditions.

Sporocyst. Refers to the larval stage in the life cycle of *Fasciola* and other flukes.

Sporogony. The multiple fission in zygote which results in the formation of spores. It is also known as sporulation.

Sporozoite. Refers to the young infective stage of the sporozoan (as that or *Plasmodium*).

Sporulation. Refers to the process of spore formation; it means reproduction by asexual multiplication.

Spur. Horny structure which is attached with tarsometatarsus of male fowl. It is used for fighting and sexual selection.

Squalene. A hydrocarbon which is present in shark liver oil in 80-90%.

Squamata. Group of reptiles to which lizards and snakes belong.

Squamosal. 'T' shaped bone of skull of vertebrates which braces posterior end of jaws.

Squamous Epithelium. An epithelium which is made up of plate-like/ scale-like and very closely set (like tiles on floor) cells. It occurs in Bowman's capsule, walls of membranous labyrinth etc. It takes part in processes like ultrafiltration.

Staircase Phenomenon. If a muscle gets stimulated repeatedly after regular intervals, the initial few stimulation exhibit increasing amplitude and thus curves on kymograph resemble a staircase.

Stapedial Plate. Columella is communicating with fenestra ovalis by a cartilagenous nodule which is known as stapedial plate (a detached part of otic capsule).

Stapes. Third ear bone which is attached to incus and oval window of cochlea,.

Starch. A polysaccharide of plant origin whose chemical formula is $(C_6H_{10}O_5)_n$. When it is digested by amylase, it forms di-and monosaccharides which are used as food in the body. It is homocyclic polysaccharide.

Star-fish. Common name for echinoderm *Pentaceros* or *Asterias*.

Statocyst. Organ of equilibrium in animals such as cray-fish.

Statodont. Mean the condition when teeth are not replaced. Such teeth are called monophyodont teeth and occur in *Sphenodon*. Such teeth are pointed, triangular, laterally compressed and arranged in two parallel rows.

Steapsin. Enzyme of pancreas which acts on fats.

Stegocephalia. Group of fossil amphibia having characters of both the fishes and amphibians.

Stellate Cells. Cells in cerebellar cortex which are present more superficial in location.

Stellate Neurons. A type of multipolar neurone which has several dendrities and one axon and branches of dendrites have been arranged in star-like manner.

Stellate Reticulum. Central mass of epidermal cells that is modified to form a connective tissue which separates outer and inner enamel epithelium of tooth bud.

Stenin. A protein present in brain and nerve ends. It has ATPase activity and some similarity with muscle myosin.

Stensen's Duct. Refers to the duct of parotid salivary glands (which are present in the region beneath external auditory meatus).

Sternebrae. In mammals sternum consists of a series of paired segmented ossifications which unite to form main body or mesosternum. These structure are known as sternabrae.

Sternum. Bone or bones of the ventral side of the thorax.

Stereospondylous Vertebrae. Vertebrae whose hypocentrum has become the main centrum and the pleurocentrum is reduced or completely lost. These vertebrae occur in extinct and present day amphibians.

Steriods. A naturally occurring group of chemicals allied to cholesterol. They include sex hormones, adrenal cortical hormones, bile salts etc. This term is also applicable to natural adrenal glucocorticoids like hydrocortisones and cortisone or synthetic analogues like prednisolone and prednisone.

Stigma. Sensitive pigmented eye spot which is found in certain protozoa *Euglena*.

Stimulus. A substance, action or other influence which on being applied with sufficient intensity to a host tissue brings about a response.

Stolon. A branch upon which colonial individuals may bud upright.

Stomach. Bag-like/sac-like division of alimentary canal in which there occurs storing and digestions of food.

Stomochord. A notochord like structure.

Stomodaeum. Portion of mouth cavity which is lined with ectoderm.

Stool. Means faeces including undigested food with mucous, bile pigments etc., which are passed out of alimentary canal by evacuation with bowels.

Stratified. Arranged in layers/strata one above the other e.g., stratified epithelium in the dermis of skin.

Stratum Granulosum. The cells in epidermis which form a layer of perfectly granular cells lying outside and above stratum spinosum. It is just below stratum lucidum.

Stratum Laxum. Refers to the outer layer of dermis in cartilagenous fihes nearly free of fibres.

Stratum Malpighi. Innermost layer of epidermis of skin. They are most living an active and continuously produce new cells by mitotic division. They are called *stratum germinativum*.

Stratum Spongiosum. Layer of cells of epidermis in mammals which are laying just above stratum malpighi.

Streptostylic. Refers to an autostylic type of jaw suspension in which the suspension is taking place at quadrate but quadrate in movable. It is found in snakes.

Striation. Refers to the formation of striae, stripes, lines etc.

Stria Vascularis. Vascular tissue over external wall of scala media which secretes endoymph.

Striped Muscles. They are also known as striated/voluntary/skeletal muscles and found in fascia. Each fibre is 5 cm long and µ100 wide having 100s of nuclei without cell boundaries. A fibre of striped muscle comprises of myofibrils each 2.4 cm long and 1 µ thick and embedded in sarcoplasm. Alternate bands of light (1-band) and dark (A-band) are present along the fibre. A-bond is crossed by central area called H-zone and each I-band is crossed by Z-disc. A-bands are composed of mysin and I-bands are composed of actin.

Strobilation. Budding in segments of sessile scyphozoan coelentrate and larvae which results in the formation of free swimming plate like ephyral larvae.

Stroma. The central connective tissue of ovary in mammals. In this portion the cells budded off from germinal epithelium are retained and surrounded by a group of cells that give rise the Graffian follicle. (It is in this structure the ovum gets matured).

Stratum Corneum. Refers to the outermost cornified layer of skin in amphibians and mammals.

Strychnine. A poisonous alkaloid ($C_{21}H_{22}N_2O_2$) which is able to block inhibitory synaptic transmission in vertebrate CNS.

Sturgeon. Common name for bony fish *Ascipenser*.

Stylopodium. Division of limbs that is able to form upper arm in fore limb and thigh in hind limb having skeletel support of humerus and femur respectively.

Subcardinal Veins. Veins in *Scoliodon* which run middorsally in abdominal cavity outside the peritoneum.

Subclavian. Under the collar bone (clavicle). It also refers to particular muscle, blood vessel or other structure.

Subclavius. Small muscle which produces upstroke of wing. It is also known as supracoracoideus or pectoralis minor.

Subcutaneous. A structure which is just below skin as subcutaneous connective tissue.

Subcutis. The loose, vascular and fatty layer of loose connective tissue and smooth muscle fibres as in fishes like *Scoliodon*.

Sublethal Gene. A lethal gene affecting the carrier between birth and attainment of reproductive age.

Sub-Lingual Glands. Salivary glands in mammals which are present just below tongue.

Submandibular. Refers to the salivary glands in mammals which open in the posterior part of lower jaw near the angles of jaws (its duct in man is known as Wharton's duct).

Submaxillary. Beneath the lower jaw.

Submucosa. It is usually thickest layer which is made up of dense elastic connective tissue-fibres having scattered fatty tissue. Fine capillaries of blood, lymph and nerve fibres traverse the submucosa.

Substrate. A substance which is acted upon by a ferment or an enzyme.

Sub-Unguis. The sole plate which is present below nail plate at the tip of digits in mammals like primates.

Succus Entericus. The internal juice which is secreted by Brunner's glands. It is capable of digesting any type of food into final form. It has sucrase, invertase, lipase, nucleosidase, elastase etc.

Sucrase. An enzyme hydrolysing sucrose to glucose and fructose. Also called invertase. Found mainly in the intestinal juice of vertebrates.

Sucrose. A disaccharide which is formed by union of glucose and fructose. It is found in cane sugar. Its chemical formula is $C_{12}H_{22}O_{11}$.

Sulcus. Means fissure or grooves.

Sulphydryl Group. Radical-SH attached to most enzymes. It forms active sites.

Summation. When a contracting muscle gets stimulated again during contraction phase, this second contraction gets added and the kymogrphic recordings exhibit M-like curves.

Suprachoroid. A thickened vascular portion of choroid which lies between choroid and sclerotic coats.

Suprarenal Gland. Tissue which corresponds to adrenal medulla. It is found as a series paired masses of cells which are extending from level of heart to hind end of coelom and surrounding sympathetic chain.

Suprascapula. Constituent of pectoral girdle in frog which is partly ossified and partly calcified cartilagenous.

Suppressor. A gene which prevents the expression of mutant gene present at a different genetic locus.

Surinam Toad. Common name for toad-like amphibian *Pipa-americana.*

Suspensory Ligament. Refers to the elastic suspensory structure lens. It is also called *zonule* and *composed of radially arranged fibres.*

Suture. Special fibrous joint which occurs exclusively is skull. It forms the tightest type of joint.

Swallowing. Means the passing of food (ingested one) from mouth to pharynx and then to oesophagus. It is helped by the movements of tongue.

Sweat Glands. The characteristic long tubular glands having ecrine or apocrine nature and lying deep in the dermis of mammals alone. Inner coiled portion of glands secretes a substance which is called sweat and is passed out of body through a duct. The glands help to maintain the temperature of body.

Swim Bladder. Also known as *air bladder.* It is characteristic of teleosts lying just below gut outside the coelom and dorsal to gut, with shiny white balls. It possesses hydrostatic function.

Symbiosis. Means living together/existence between two different species in which both are beneficial to each other e.g., presence of *Zoochlorallae* alga in the body of *Hydra* where the alga derives protection from the later while the oxygen produced by alga is taken up by the *Hydra. Opposite* to parasitism.

Sympathetic Nervous System. Refers to a portion of autonomic nervous system which is composed of a chain of ganglia on the both sides of vertebral column in thoracolumber region and sending fibres to all smooth muscle tissues such as heart, lung, alimentary canal etc.

Synalbumin. A protein having anti-insulin activity.

Synapse. A conjugation taking place between two directly interacting nerve cells where impulses in the presynaptic cells are able to influence activity of postsynaptic cell.

Synapsis. In meiotic cell division it refers to the pairing of homologous chromosomes during zygotene stage (they come in contact with each other so that the number of chromosomes in the dividing cell appears apparently reduced to half).

Synaptic Cleft. A space normally more than 200 A. Across this transmitting chemicals tend to move during neural transmission.

Synaptic Knobs. The structures at the tip of axons having synaptic vesicles. Also known as *terminal buttons*.

Synaptic Mediator. An agent that is able to mediate transmission at synapses between pregnaglionic and post-ganglionic sympathetic endings e.g. acetylcholine.

Synaptic Noise. Refers to the irregular changes in post-synaptic membrane potential which are produced by irregular subthreshold synaptic inputs.

Synaptic Transmitters. Substances like acetylcholine which facilitate the transmission of nerve impulse across synapses, e.g., acetylcholine.

Synpticulae. Means cross connections existing between two structures as in the gill slits of *Amphioxus* (a protochordate). They also refer to septa like structure found in mushroom coral *Fungia*.

Synaptic Vesicles. Refers to structures within synaptic knob which have chemicals. When released they either depolarise or hyperpolarise proximal dendrites during transmission of nerve impulse.

Synarthroses. Refer to imperfect joints which are immovable. They are known as immovable joints. They may be fibrous cartilagenous or even bony. They have been divided into several types like:

Fibrous. It is present between bones of skull. It is held edge to edge and separated by connective tissue that is able to form the sutures between them.

Cartilagenous. Two bones are firmly held and their slightest movement takes place by mean of cartilagenous disc present in between them e.g., public symphysis and one between two mammalian vertebrae.

Bony. In this mesenchyme between articulating bones is differentiated into bone e.g., mandibular symphysis of cat.

Gomphose. In this one of the articulating bones fits into a socket of other bone e.g., teeth in mammals.

Schindylose. In this one of the two bones is in form of a thin process which fits into a slit of other bone e.g., nasal process of premaxilla in the nasal bone of *Varanus*. *Compare* with diarthroses or movable joints.

Synchrony. Means co-ordinated movement of particular part/parts of the body at the same time, e.g., cilia of transverse row of a *Paramecium* undergoes beating in this manner in contrast to that of longitudinal row which under-goes metachronous beating.

Syncitium. Refers to undivided mass of cytoplasm having many nuclei as in the hypodermis of *Ascaris* (a nematode). It is a product of nuclear division which is not accompanied by the cytoplasmic division or cytokinesis.

Sydactyly. The webbed fingers or toes as in hind limbs of frog.

Syndrome. A group of symptoms and/or signs which together are able to produce a pattern or symptom complex typical of a particular disease.

Synergistic Muscle. Means a muscle which co-operates with another one. One partner in the body helps in cerebellar control.

Syngamy. Means the fusion of gametes forming a zygote. It is *opposite* to hologamy in which two individuals fuse throughout their body to form a zygote.

Synovial Cavity. Cavity which is present between two synovial membranes at articular surface between bones.

Synovial Fluid. A fluid filling in the synovial cavity (capsule) which forms a double walled structure binding articular caps. It has a large polysaccharide of hyaluronic acid and protein that give it a sticky and glairy consistency. It is secreted by synovial membranes and lubricates articular surfaces of bones.

Synsacrum. Means fusion of 3-4 sacral vertebrate as in mammals.

Synteny. Refers to the association of genes on one same chromosome. It is often used synonymously with linkage.

Syrinx. The voice box of birds.

System. A group of organs which are concerned together with same general function. In the bodies of individual there are a number of systems like digestive system, respiratory system, circulatory system, excretory system, nervous system and reproductive system.

Systemic Arch. A large artery/arteries in a vertebrate which corresponds to 4th embryonic aortic arches. It is able to carry blood from heart to dorsal aorta.

Systole. Means contraction phase of auricles, ventricles or contractile vacuole. In heart it implies contraction of auricles and the ventricle during which the blood passes from auricles to ventricle and then to outside of heart into main aches. In protozoa like *Amoeba, Euglena* and *paramecium* the contractile vacuole is able to discharge its contents outside the body during this phase. *Opposite* of diostole.

Systolic Pressure. The peak pressure which is reached during systole. Generally, it is equal to Age+100. Normal value is 120-140 mm of mercury (Also see *blood pressure*).

T

T_3(3, 5, 3'-Tri-Iodo Thyronine). One of the principal hormones which is secreted by thyroid gland.

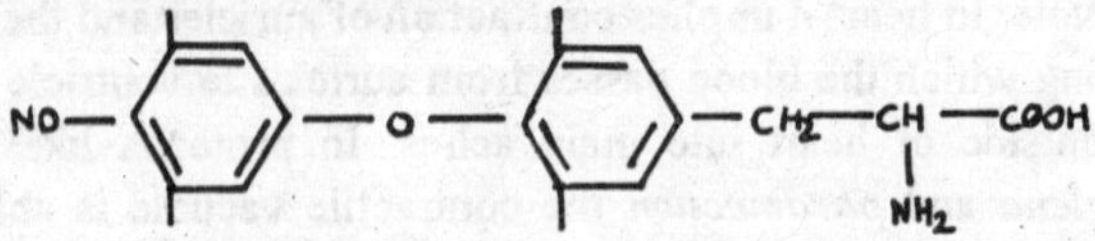

T_3 3, 5, 3'-Trio-iodo Thyronine

T_3-Reverse. The small amount of tri-iodo thyronine which is obtained by condensation of mon-iodotyrosine and di-iodotyrosine.

T_4 (3,5,3',5'-Tetraiodthyronine). It is also known as thyroxine. It is one of the principal hormone secreted by thyroid which is able to influence the body growth and metabolism.

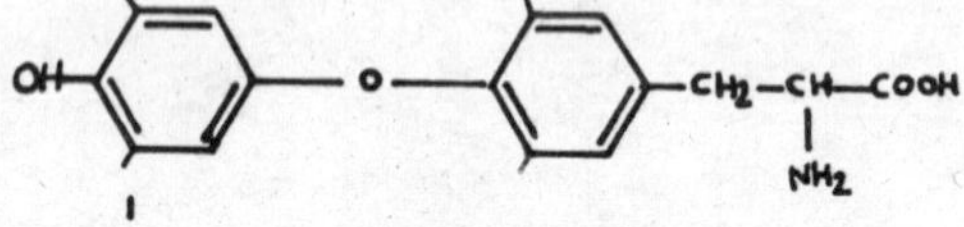

T_4 (3, 5, 3', 5' - Tetraiodo Thyronine)

T_3-Prop (3,5,3'. Triiodothyropropionic Acid). A thyroxine derivative in the body.

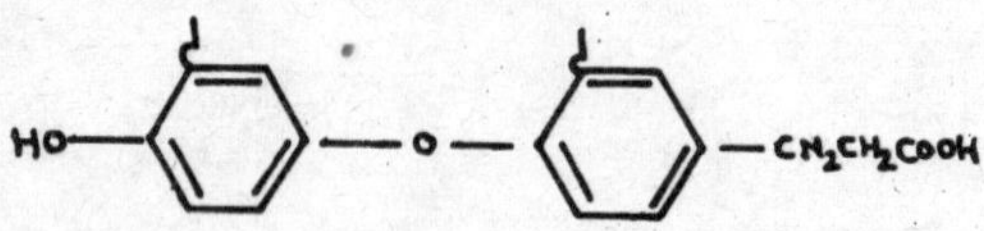

T3 - Prop (3, 5, 3' - Triodothyropropionic Acid).

T_4Prop (3, 5, 3'. 5'-Thyropropionic Acid). A thyroxine derivative in the body.

T-Cells Killer. Refers to T-lymphocytes (responsible for keeping cellular immunity) which are responsible for lysis of homografts and tumours independent of complement (Also refer to *Killer-T-cells*).

Tabes Disease. Disease of dorsal root. In this dorsal roots get interrupted during which all reflex contractions of bladder are ceased. The bladder tends to becomes distended, thin walled and hypotonic.

Tachycardia. An increase in heart rate above the normal level. It is opposite to bradycardia.

Tachypnea. The rapid, shallow breathing in animal.

Tractile. Term used for the sense of touch.

Tactile Disc. Pertaining to light touch receptor.

Tadpole. Refers to an aquatic tailed free swimming larval stage of frog and other animals such as *Herdmania*. The larva possesses gills, segmented body, well-developed alimentary canal for herbivorous alimentary canal for herbivorous diet. After undergoing a free swimming life for some period it metamorphosed into adult.

Tagma. Used for the body region such as thorax or abdomen.

Tagmosis. Means union of body segment into functional groups.

Tail. The postanal extension of body as seen in the all chordates. It may be lacking in adult condition in few cases like frog, man etc. but certainly present in embryonic stage. There are few non-chordates that may have some structure looking like a tail but it is not a true tail because anus is situated at its tip.

Tail Diphycercal. Refers to a tail in bony fish *Neo ceratodus* (a lung fish). It is perfectly symmetrical and primitive since ventral lobe of fin is not supported by hypurals but by separate radials.

Tail Feathers. Also known as *rectrices*. These constitute a type of contour feathers which are having unequal vane. However, they are arranged in a fan-like manner around the pygostyle and help to balance the body if the bird sits on the land and acts like rudder if the bird flies in the air.

Tail Gephyrocercal. Tail in some fishes in which tail fin of homocercal tail is secondarily simplified and tapering like niphycercal one.

Tail Heterocercal. Tail in fishes like *Scoliodon* which possesses small upper lobe (epichordal lobe) and large lower lobe (hypochordal lobe) and therefore, it is asymmetrical externally as well as internally.

Tail Homocercal. Tail is teleosts like *Labeo* which possesses equal hypochordal and epichordal lobe but end part of the tail (vertebral column) is bending upwards thereby making it asymmetrical internally.

Tapetum. Refers to the inner surface of choroid in certain animals, particularly fishes and cattle like cows, sheep, buffaloes etc. It is having platelets with light reflecting nitrogenous purine called guanine ($C_5H_5N_5O$) and hypoxanthine ($C_5H_4N_4O$) and situated behind retina. Light falling over platelets gets reflected back over the retain during night hours and thus visibility gets improved.

Target Cell. Cell preferentially binding and responding to specific hormones.

Tarsals. Bones present in ankle in the hind limb of vertebrates.

Tarsometatarsus. Bone of hind limb in birds which is formed by involving the fusion of dital tarsals with the metatarsals. It provides strength to support the weight of body when the bird lands. In male it has fighting spur that is helpful in fighting during sexual selection.

Tarsus. Distal end of leg of an insect. It is having one or more segments the distal most bearing claws. It also refers to ankle bone of vertebrates.

Taste. Chemoreception in organisms for near object which help to make discrimination between desirable and undesirable food.

Taste Buds. Receptors for taste which are in form minute barrelshaped structure and carried over small elevation in tongue called taste papillae. There are two kinds of columnar cells that constitute these structures—one which are centrally placed neurosensory cells and other that are peripheral none-sensory cells. They project over the surface as in mammals or project into small pit-like taste pores that are opening on the surface. They are supplied by, VII, IX and X cranial nerves.

Taste Papillae. Elevations in tongue which are able to carry taste buds. There are four types of these that are found in most vertebrates, (i) Filiform papillae—they are usually confined to anterior 2/3rd of tongue and rarely bear taste buds. In carnivores (dog, cat, lion etc.) they are covered with sharp horny caps which make to pull out minute pieces of flesh from bones, (ii) Fungiform papillae—these are less numerous than former and have few taste buds, (iii) Foliate papillae—they are leaf-like and present near the base of tongue on each side and have taste buds on their lateral sides, (iv) Circumvallate papillae— they are largest papillae of rounded form having slightly raised central areas and low trench-like margin. Taste buds are situated on lateral sides mostly in trench-like margin. Taste papillae are lacking in whales.

Taurocholates. Sodium and potassium salts of taurocholic acid. They are present in bile juice and prevent putrifaction of food in the alimentary canal.

Taxis. An orientational movement. For example, a taxis may be towards a desirable or away from a undesirable condition.

Taxonomy. Dealing with classification of organisms.

Tay-sach's Disease. A condition in which an autosomal recessive is inherited. In this an infant develops loss of muscular function, paralysis, blindness, deafness and other derrangements that usually cause death before the age of four years.

Tazon. Catagory or animal classification.

TBG (Thyroxine Binding Globulin). A globulin in the plasma having an clectrophoretic mobility between γ-1 and γ-2 globulins which binds thyroxine (the thyroid homone).

TBPA (Thyroxine-Binding Prealbumin). An albumin called preablumin in the plasma which tends to bind thyroxine.

TBV (Total Blood Volume). Means total blood volume in a man. It can be calculated by multiplying the plasma volume with

$\frac{100}{100 - \text{haematocrit}}$ e.g., if haematocrit is 30 and plasma volume is

3500 then total blood volume will be:

$$3500 \times \frac{100}{100 - 30} = 5000 \text{ ml.}$$

TBW (Total Body Water). Means the total water present in the body of an individual. Its percentage depends upon age and sex. In the following table total body water with relation to age and sex has been given.

Age (In years)	*Male*	*Female*
10-18	59%	57%
18-40	51%	51%
40-60	55%	47%
60-more	52%	46%

TCA-Cycle. Also known as Kreb's cycle/tricarboxylic acid cycle. It is a metabolic cycle in which there occurs a complete oxidation of acetyl coenzyme. The cycle gives rise to the regeneration of 36 ATP molecules, 6 CO_2 and $6H_2O$.

TEA (Tetra Ethyl Ammonium). A substance which is able to block some potassium channels in membrane. The chemical formula of substance is $(C_2H_5)_4N$.

Tear Glands. Lecrymal glands which are associated with eyes of land vertebrates. They act as the devices of keeping surface of eye-ball clean and moist.

Teat. The nipples present on the breast/mammary glands.

Tectonic Plates. Conditional earth plates which tend to move by the conveyor or belt-like motion of substrate.

Tectorial Membrane. A fine gelationous sheet which is situated on the organ of Corti in the ear in contact with cilia of hair crlls.

Tectocerebellar Tract. Afferent tract to cerebellum that transmit auditory and visual impulses via inferior and Superior colliculi. It gets distributed along folium, tuber, ansceriform lobule and enters cerebellum through superior peduncle.

Tectum Synoticum. A cartilage which connects two otic capsules above the brain.

Telencephalon. Fore brain of vertebrates.

Teleology. The scientific method of thinking in which a result is treated as a cause in explanation of natural phenomenon.

Telophase. Any of the final phases of mitosis in which the cells may undergo division or may not undergo division but daughter nuclei are always formed.

Teleoptiles. Refers to the final adult feathers in succession of feathers in a bird (compare with mesoptiles like prefilopulmes, preplumes, prepinnae).

Telereceptors. The distant receptors concerning with events like distance. Chemoreceptors concerned with smell may be regard telereceptors.

Telocentric. A chromosome in which centromere is so close to one end that one arm is tiny or inconspicuous.

Telodendria. The axor ends in a number of synaptic buttons having granules or vesicles in which synaptic transmitter secreted by nerve is being stored.

Telson. The term used for the terminal extension of last abdominal appendage of a crustacean such as cray-fish.

Teleostei. A group of higher bony fishes that also includes majority of modern living bony fishes like *Labeo Clarias, Exocoetus Anguilla* etc.

Teliolecithal. An egg having large amount of yolk. Also known as *mesolecithal* egg as found in *fishes* and *amphibia.*

Template. Mold or pattern for the formation of a duplicate as in gene duplication.

Temporal. The lateral areas of head above the zygomatic arch.

Temporal Fossae. In the temporal region of reptiles there are vacuities which are called *temporal fossae.* They are lacking in chelonians (turties, tortoises etc.) but present in many reptiles both extinct and present day living ones as well as some early mammals. The four types of the fossae are present in the different reptiles.

Anapsid. They are Characterised by absence of any fossa e.g. chelonia.

Synapsid. In these there is single fossa behind the orbit on each side and bounded by postorbital and squamosal above and jugal bones below. They are found in extinct reptiles and lower mammals.

Diapsid. Characterised by two fossae, upper supratemporal and lower infraorbital. The former is bounded by parietal and prefrontal above and squamosal and postorbital below, whereas later is bounded by squamosal and postorbital above and quadrate jugal and quadrate below. They are found mainly in dinosaurs and crocodilians and their modified forms are found in lizards and snakes.

Parapsid. They are same as synapsid but the squamosal and postorbital meet below the fossa. They are found in extinct reptiles.

Temporal Lobe. Lobe of cerebral hemisphere which lies in lower lateral area at temples having centres for speech and hearing.

Temporal Summation. Refers to the addition of activity in more than one synaptic knobs at the same time. Temporal summation takes place if repeated afferent stimuli result in new excitatory post-synaptic potentials before there occurs the decay of previous excitatory post-synaptic potentials.

Temporary Homeostatic Plug. If there occurs a damages to a blood vessel the endothelium is disrupted and an underlying layer of collagen gets exposed which attracts platelets that adhere to it and liberate serotonin and ADP. ADP in turn is able to attract other platelets and there occurs the formation of loose plug is aggregated platelets at the damaged part. Formation of this plug is not affected by clinically used does of heparin and dicumerol etc.

Tendons. Refers to the tough cord-like connective tissue extensions of body. They are present at the end of a muscle which attach it to other structure say bone. They consist of while fibres.

Teniae Coli. Fibres of external muscular layer of colon get collected into three longitudinal bands. These bands are shorter than the rest of colon. Therefore, the wall of colon forms outpocketings between then.

Tensors. The prepatagial muscle of birds namely tensor longus, tensor brevis, tensor accessorious and tensor posterius. They are ordinarily used for muscles that are useful for stretching any structure.

Tentacle. A flexible arm-like extension of body. It is present in many non-chordates like *Hydra*, *Obelia* etc. It helps in grasping and the movement. It is also used as offensive organ.

Tentculocyst. Refers to the sense organ present in some cnidarians like jelly-fishes. It lies around the margin of umbrella and his calcareous crystals and sensory cells. It makes the animal to balance the body during locomotion.

Tergum. The dorsal portion of the exoskeleton of any body segment in animals such as arthropods (for example cockroach). It is joined to ventral piece called sternum by lateral plates.

Terminal Buttons. Axons ending in postsynaptic knobs. Also called axon telodendria.

Terminal Cisternae. Refer to the portion of sarcoplasmic reticulum on both the sides of Z-line which is able to make close contract with T-tubules in striped muscle fibre. (Also see striped/skeletal/voluntary muscles).

Terminating Codon. A DNA base triplet which brings about the cessation of protein synthesis; the terminating codon is not able to code for any amino acid; e.g. UGA.

Terrapins. Members to reptilian group chelonia. They have characteristic angular rings on carapace. They are used for human consumptions. They are close relatives of turtles.

Terrestrial. The term is applicable to an organism that lives on the ground like cow, sheep, horse, etc., including the man, also a number of insects.

Territoriality. Refers to the behaviour of maintaining personal living space used for breeding purposes.

Tertiary Structure. The way in which a polypeptide chain is folded or bent to give rise the overall conformation of molecule.

Test. Shell over the body of many animals like *Herdamania* (where it consists of tunicin) *Arcella* (where it consists of pseudochitin), *polystomella* (where it is calcareous) and also in a number of other protozoan animals belonging to group testacea. It also refers to an experiment to substantiate or disprove an uncertain point.

Testes. The male reproductive gonad which produces the sperms and secretes the hormones essential for sexual and secondary sexual activities.

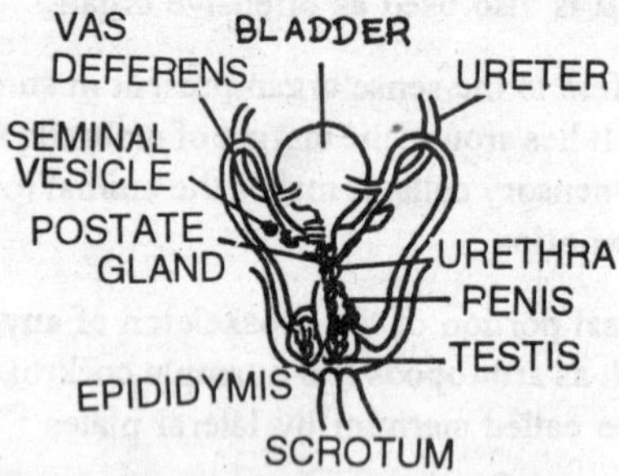

Fig. 3. Human male reproductive system

Testesterone. Hormone secreted by testes. It is responsible for developing secondary sexual characters of male such as beard, hair on the body, harsh voice, aggressiveness etc., alongwith proper development of reproductive organs, the proper functioning of reproductive organs in an individual. It is also to influence sexual behaviour and development of male accessory genetalia.

Testicles. Pertaining to testes.

Tetanus. Condition in which a muscle is subjected to permanent contraction and is never relaxed again. It is attributed to the fact that the muscle is subjected to repeated stimulations during its contraction phase for a very long duration in case of experimental muscles. In man it occurs due to a pathogenia disorder which is caused by bacteria Clostridium tetani present in the dust and manure. Patient of this disease develops spasm and rigidity of muscle and his body gets stretched into bow-shape leading finally into the death. ATS injections give rise to passive immunity for the disease.

Tetany. Refers to the condition of muscular hyperexcitability in which mild stimuli are able to produce cramps and spasm. It is common in persons having parathyroid deficiency and alkalosis. In infants it has been associated with gastrointestinal upset and rickets.

Tetany Hypocalcemic. A disease which occurs due to increased activity of motor nerve fibres. It is characterised by extensive spasms of skeletal muscles of extremities and larynx. It occurs due to fall in Ca^{2+} ions of the blood due to distributed calcium regulation by parathyroid. This disease is known as roussacus' sign.

Tetrac (3, 5, 3', 5',-Tetraidothyroacetic Acid). An acetic acid analogue which is formed by deamination of thyroxine and subsequent decarboxylation. The substance has been found to lower plasma cholesterol level to greater degree than it affects BMR.

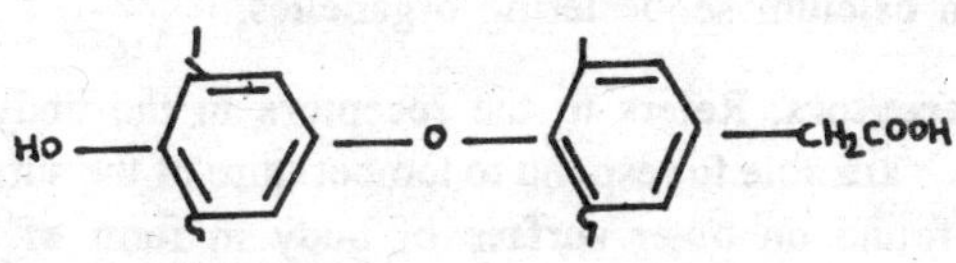

Tetrac (3,5, 3' 5'- Tetraiodothyro Acetic Acid)

Tetrapoda. The term used for a group of vertebrates that have four limbs i.e., one pair of anterior limbs in posterior region are called hind limbs. In the group all the vertebrates above the level of fishes i.e., amphibia, reptiles, birds and mammals are included.

TFN (Trifluoro Nitrophenyl). A chemical poison which affects specially the lampreys (*Pctromyzon*).

Thalamic Nuclei. The nuclei present in the thalamus and its pats (epithalamus, dorsal thalamus and ventral thalamus).

Thalamus. Portion of fore brain of vertebrate. It comprises of epithalamus, dorsal thalamus and ventral thalamus.

Thalassemia. An inherited blood disorder. In the disease, homozygote is suffering from a severe anaemia called *Thalassemia Major* and the heterozygote is suffering from milder form of anaemia called *Thalassemia minor*. As a general term it includes various anaemias due to effects of different genes.

Thebesian Valve. The tricuspid valve present at the right auriculoventricular opening of mammalian heart. It is able to control back-flow of blood from right ventricle to right auricle during ventricular systole.

Thebesian Vein. The term used for direct connection to the atria and ventricles from coronary veins in the heart of mammals.

Theca Interna. The cells of middle layer of stratum granulosum of the Graffian follicle in mammals. These cells serve to act primary source of estrogens (the female sex hormones).

Thecodont Teeth. A type of teeth lodged into definite theco or cup-like structures which are formed within jaw bones. These teeth occur exclusively in mammals.

Theophylline. Crystalline alkaloid present in tea. It inhibits phosphodiesterase thereby increasing c-AMP level. It further releases Ca^{2+} ions from calcium sequestering organelles.

Thermoreceptors. Refers to the receptors in the body of organisms which are able to respond to temperature of the surrounding. They are found on outer surface of body in form of nerve endings, specialised structures such as ampullae of Lorenzini (found in cartilagenous fishes like *Scoliodon*) or some other form.

Thermoregulation. The ability of an organism for regulating its body temperature. It involves morphological, physiological or behavioural adaptations by the individual which help it to maintain its body temperature within limits necessary for its survival. For example the human beings are having constant body temperature of 37°C. If there occurs an increase in the environmental temperature above this temperature, sweating by the body takes place which causes a heat loss and similarly a fall in the environmental temperature below this brings about shivering etc., so that the body's heat loss gets checked.

Thermotaxis. Means behavioural response of an organism to heat. If

Euglneae are placed in a tube and one end of tube is maintained warm (above 40°C) and the other maintained cold, they do not remain either towards the warm end or cold end but they move and settle some what midway where they get optimum temperature.

Thesocytes. Type of cells present in the mesoglea of sponges such as *Sycon.* They have amoeboid shape and a large nucleus. There are several rod like inclusions in the cytoplasm which might serve as reserve food materials. These cells function to store reserve food like liver in higher vertebrates.

Theta Rhythm. A pattern of regular waves in electroencephalogram which is having number of waves as 4-7 sec. This rhythm is generally common in case of children. A rhythm having less than 4 waves/sec. is known as delta-rhythm and more than this (180-30/sec) is known as β–rhythm.

Thiamine. Also known as vitamin B_1. It sevres as a co-factor in decarboxylation reactions. It occurs in skin, unrefined cereal grains and the deficiency of the substance gives rise to Beri-beri and neuritis.

Thick Filament. The muscle filament in striped muscle forming A-band and is composed of myosin.

Thin Filament. The filament in striped muscle composed of another protein called actin.

Thiouracil. A compound which is related to thiourea. It inhibits iodination of monoiodotyrosine and is thus able to block coupling thereby inhibiting reaction thyroid function.

Thirst. Refers to the feeling to take fluid. It generally gets stimulated by decrease in extra cellular fluid volume and processes like haemorrhage etc. It is under hypothalamic control (Also see *thirst centre*).

Thirst Centre. Sunfornical organ in diencephalon which is acting like a centre for stimulating thirst.

Thoracic Pump. During inspiration there occurs a fall in intrapleural pressure from -2.5 mm to -6 mm of mercury. This negative pressure transmits to great veins and to lesser extent to atria so that there occurs central venous pressure from about 6 mm of mercury, during expiration, to approximately 2 mm of mercury during quiet inspiration. The decrease in venous pressure during inspiration helps venous return. When diaphragm descends during inspiration there occurs an increase in intrabdominal pressure and this also squeezes blood towards heart because back-flow into large veins has been prevented by venous valves.

Thorax. The term used for the major division of an animal (vertebrate) which is just posterior the neck or head if the neck is absent. Also refers to that part of body of an arthropod situated between head and abdomen.

Thread Worms. Common name for nematodes like *Entrobius,* They are present in the intestine of man and other animals. They give more trouble to children and their presence is shown by itching tendency of host in the anal region. This itching occurs because the gravid female enters the lower part of alimentary canal of host and as soon as it comes in contact with air (possibly during itching by host) it bursts up to release its eggs (a device of autoinfection in the host).

Threonine. An essential hydroxyl substituted amino acid.

Threshold Potential. Potential which is large enough to produce the response e.g., action potential, muscle twitch.

Threshold Stimulus. The minimum strength of stimulation which is just sufficient to produce a detectable response or an all-or-none response.

Thrombasthenic Purpura. If there occurs the abnormal circulating platelet count, it may result in easy brusiability and subcutaneous bleeding which is known as thromb-asthenic purpura.

Thrombin. A substance which is formed from prothrombin during blood clothing. It converts fibrinogen into fibrin (which forms clot).

Thromboplastin. A substance which is released from blood platelets in the clotting sequence (see blood clotting/coagulation) when there occurs an injury to the tissue. This substance along with calcium ions induces the formation of prothrombinase or thrombokinase (that causes the conversion of prothrombin into thrombin).

Thrombocytes. Blood corpuscles, like platelets of mammals. They are found in the blood of other vertebrates. They are simple spindle shaped nucleated structures of the size almost of a lymphocyte and are produced by haemopoitic tissue. Also known as *spindle cells*.

Thrombocytopenia. Means a reduction in the number of platelets from their normal count.

Thrombocytopenic Purpura. When there occurs low platelet count. there occurs deficient clotting reaction and poor clotting reaction and poor constriction of ruptured blood vessels resulting in this syndrome. It is characterised by easy brusiability and multiple subscutaneous haemorrhage.

Thrombosis. A process of blood clot formation which takes place within the circulatory system. It is particularly seen in area having plaques which are formed as a result of arteriosclerosis. Its two major forms are known as clinically one coronary thrombosis leading to heart attack and other cerebral thrombosis leading to cerebral haemorrhage.

Throboxane. The most active cell regulaters which are formed from prostaglandin precursors. Its two types are known—thromboxane-A_2 and thromboxane - B_2(both being formed from endoperoxidase). Functions of these compounds are not understood completely.

Thymosin. A polypeptide which is extracted from thymic tissue and thymic hormone. It promotes maturation of T-lymphocytes in young condition and is able to maintain immunity reactions during that period. The substance has 108 amino acid residues and is rich in acidic amino acids.

Thymus. Two lobbed endocrine glands just shapped below thyroid of mammals. It is developed during childhood when it has defensive role in the body by secreting thymic hormone which is able to promote maturation of T-lymphocytes. In puberty it disappears as their secretion tends to hamper the development of reproductive function.

Thyrocalcitonin. Calcitonin is often known as thyrocalcitonin.

Thyroglossal Duct. A duct which marks the path of thyroid from tongue to neck (often persisting in the adulthood).

Thyroglobulin. A globulin binding tyrosine molecules in peptide linkage (from tyrosine are synthessed colloids by iodination and condensation). This glycoprotein is having 4 peptide chains and molecular weight of nearly 660000. It is synthesised by thyroid cells and secreted in the colloids. Hormone remains bound until it is secreted. After secretion colloid gets ingested and hormones are released into capillaries.

Thyrohyal. Posterior horns of hyoid apparatus that represent fourth branchials.

Thyroid Cartilage. A pair of cartilages present in the larynx of mammals. They enlarge to form a protuberance in the neck region, called Adam's apple, in adult man. Vocal cords are extending between these cartilage and arytenoids.

Thyroid Glands. The paired glands present in the neck region of vertebrates. They are endocrine in nature and they secrete thyroxine hormone. This hormone is responsible for body growth. Its activity is controlled by thyroid stimulating hormone which comes from anterior pituitary lobe. Hyposecretion by these glands results in myxodema and hyposecretion results in cretinism.

Thyroid Storm. Severe exacerebation of hyperthyroidism.

Thyrotropic Releasing Hormone. A hormone from hypothalamus that is able to influence anterior pituitary lobe to secrete thyroid stimulating hormone or thyrotropin.

Throtoxicosis. Means hyperthyrodism which is characterised by nervousness, weight loss, hyperphagia, heat intolerance, increased pulse (tachycardia), fine tremor of outstretched fingers, warm soft skin and high BMR.

Thyroxine. The hormone which is secreted by thyroid gland. It is involved in body growth and metabolism.

Tibia. Refers to the larger medial bone of vertebrate lower leg. Also refers to the part between femur and tarsus in insects like cockroach.

Tibiotarsus. In birds the distal end of tibia of hind limb gets fused with a proximal tarsal to form tibiotarsus bone. It give rise to (along with fibula) the skeleton of shank of the bird.

Tickle and Itch. Refers to the skin which is produced by low frequency stimulation of C-fibres. Tickle produced by very mild stimulation is regarded as a pleasurable sensation but the itch is make or less annoying.

Tidal Volume. Amount of air moving into lungs with each inspiration (or the amount of air moving out with each expiration). It is usually 0.5 litre in man and same in woman.

Tiedmann's Bodies. Refer to the spherical swellings (caeca or pouches) that are present on the inner wall of ring canal of water vascular system of echinoderms like star fist (*Pentaceros*). They are generally nine in number and have glandular epithelium. They are concerned with production of amoebocytes.

Tight Junction. Means the cell junction in which members of two cells get apposed and the outer layers of membranes get fused with each other.

Tissue. A group of cells that are similar in structure (with intercellular substances, if any) and carry out a specialised function. The study of tissues is called histology. The main types of tissues present in the body, of a vertebrate are epithelial tissue, which is able to form the various coverings in the body, connective tissue which connects various body parts by forming blood, muscles, bones etc., and the nervous tissue which forms a co-ordinating system among various body structures and is able to control their functioning.

Tissue Fluid. The extraceullar fluid constituting environment of the bony cells.

Tissue Macrophage. From bone marrow monocytes enter circulation but after 24 hours they enter the tissues to become tissue macrophages. All kinds of tissue macrophages including Kuffer's cells of live, alveolar macro-phages of lungs are coming from circulating monocytes.

T-Lymphocytes. Cellular immunity mediating bodies which are present throughout the body when they undergo encounter with the antigens on cells from other individual or antigen from tumour cells or viruses. When they get activated, they become enlarged, are dividing and releasing lymphokinin. Such lymphocytes are known as T-lymphocytes (Also known as *Killer- T-cells*).

Tone. Refers to the normal healthy state of tension in skeletal muscle. Also known as *tonus*.

Tongue. Mobile, muscular organ that is present in the mouth cavity. It is connected with speech, mastication, swallowing and taste. It is mostly joined posteriorly and free anteriorly but in some cases like frog it gets joined anteriorly and free posteriorly. In snakes and lizards it is bifid.

Tonic. A state of continuous muscular contraction as opposed to intermittent contraction. Also a medicine which is used to tone up the body after ailments etc.

Tonicity. The effective osmotic pressure of solution with respect to plasma. Those having similar osmotic pressure are termed as isotonic, those with greater pressure are termed as hypertonic and those with lesser osmotic pressure are said to be hypotonic.

Tooth. Structure/structures that are present on the jaws of mouth cavity in gnathostomate vertebrates (fishes, amphibia, reptiles, birds and mammals). They are mounted on jaw bones either on outer surface (Acrodont) or on inner surface (pleurodont). They are lodged into definite sockets of jaw bone (thecodont). They are all alike (homodont) or of different shapes (heterodont). They may be either permanent (monophyodont), replaced twice e (diphyodont) or replaced several times during the life of individual (polyphyodont). They find use mainly as structures for preventing escape of ingested food and also as masticatory structures in some individuals. The offensive and defensive function is also known in certain cases.

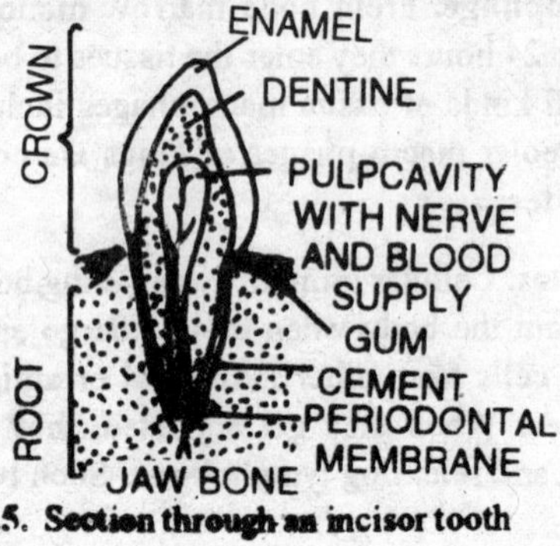

Fig.5. Section through an incisor tooth

Tornaria. The larva of enteropneust hemichordate such as Balanoglossus.

Tortoises. Mainly or entirely land dwelling forms that belong to group chelonia of reptiles. When compared with turtles and terrapins both of which are aquatic.

Toxemia. Refers to the generalised poisoning of the body which takes place by products of bacteria or damaged tissue.

Toxicity. Refers to the degree or quality of being poisonous.

Toxicology. Deals with the study of poisonous substances and their effect on the different parts of the organisms.

Toxin. A product of bacteria which is poisonous to other cells. When this substance is injected into animal or a man, it stimulates the production of an antibody (i.e., antitoxin) to it. Non-bacterial toxin is a poisonous nitrogenous compound which is released from diseased or injured tissue.

TPA (Tri Phosphopyridine Nucleodtide). One of the substances that accept hydrogen ions which are formed during dehydrogenation in biological oxidations. It is also known as *coenzyme-II* or *NADP* (according to new terminology).

TPNH. Reduced TPN.

T_pT. The tromyosin binding subunit of troponin. It occurs in association with action of thin filament.

Trabaculae. The inward processes or septa which are generally formed of connective tissue. They are found in the ventricles of heart in mammals. They are also known as *rods* or *bundles of fibres*.

Trachea. The wind pipe in vertebrates. It is the fibrocartila-genous tube which is lined with mucous membrane passing between larynx and bronchi. Its lumen has ciliated cells and mucous producing cells, the connective tissue supporting the walls has been supplied with blood vessels. The structure also occurs in insects where it gets connected to outside through spiracles. It undergoes division into several branches which are called bronchi, bronchioles etc., which form respiratory tree in vertebrates.

Tracheoles. Fine tracheae in close contact with tissue they are able to service.

Tractus Solitarius. Myelinated but slow-conducting taste fibres present in VII, IX and X-cranial nerves on each side under—go combination in medulla oblongata to form tractus solitarius.

Trail Endings. Refers to the nerve endings of gamma efferent fibres that are able to form extensive net-work primarily on nuclear chain fibres.

Train of Impulses. A rapid succession of action potentials which is propagated down a nerve fibre.

Trait. In genetics this term refers to alternate or variant forms of characteristics.

Transaminase. Enzyme catalysing transfer of NH_2 group from an amino acid to a keto acid. The reactions is known as transamination (one of the methods by which excess amino acids get metabolised in the body). The reaction in which amino group is transferred from alanine to pyruvic acids is an example of this process which has been catalysed by transaminase enzyme.

Transamination. In this process the NH_2 group from an amino acid gets transfer to a keto-acid. It is catalysed by specific enzyme called transaminase.

Trans Cellular Fluids. Refer to the fluids appearing in glandular secretions, CSF (cerebrospinal fluid), eye and other special fluids. These constitute extracellular fluids.

Transcortin. The globulin that is able to bid cortisol in the circulation. It is also known as corticosteroid—binding—globulin (CBG).

Transcription. Refers to the formation of RNA chain of complementary base sequence from informational base sequence of DNA. The process play significant role in protein synthesis and genetics.

Transducer. A mechanism translating energy or signals of one form into a different kind of energy or signals.

Transducer Molecule. A hypothesised intermediate molecule within cell membrane that is assumed to transmit a hormone initiated signal from the externally facing hormone receptor to the internally facing adenylate cyclase.

Transduction. Refers to the transfer of genetic material by means of a virus from one bacterium to other.

Transfer Reaction. A chemical reaction in which there occurs the removal of one portion of a molecule by a carrier.

Transferrin. The β_1 globulin that is able to transport iron which binds to it. It is also known as siderphilin. The amount of the substance is 130 μ g/dl in man and 110 μ g/dl in woman.

Transfusion. Means the introduction of fluid into tissue or into blood vessel. By blood transfusion it implies the transfer of blood into the vein. By intrauterine transfusion it means transfusion of blood to foetus which has been endangered by Rh compatibility (Red cells could be transfused directly into abdominal cavity of foetus on one or more occasions. This makes the induction of labour to be postponed until there is a time more favourable to foetal walfare).

Translation. Means the assembly of a polypeptide chain by using information in m-RNA for directing the amino acid sequence. The process plays significant role in protein synthesis.

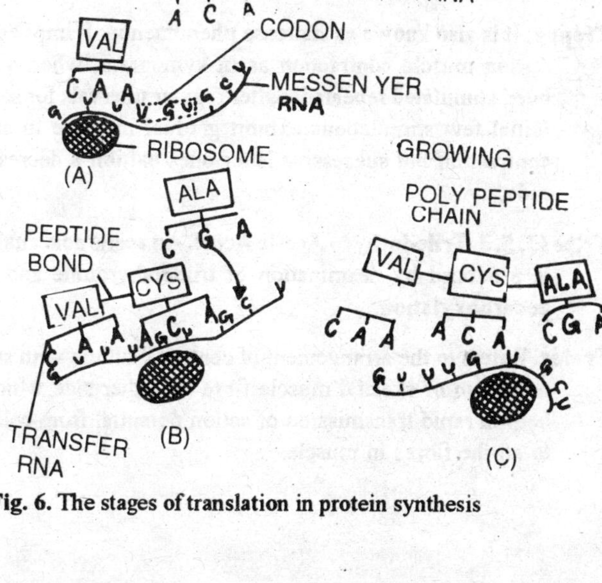

Fig. 6. The stages of translation in protein synthesis

Translocation. Chromosome aberration in which there occurs a shifting of a segment of chromosome to a new location in the same or completely different chromosome.

Transmission. Refers to the process of conducting wave/impulse from one place to other. The term is mostly restricted to conduction of nerve impulse.

Transmural. Means through the wall e.g., of a cyst, organ or a vessel.

Transmural Pressure. Equals to pressure on the one side of wall of blood vessel minus that on the other side.

Transphosphorylation. Means transfer of phosphate group between organic molecules by involving the inorganic phosphate stage.

Transexual. An individual who thinks himself/herself to be person of either sex. It is not regarded as homosexual because the latter person is able to identify his or her anatomical sex.

Transverse Process. One of the lateral projections in a vertabra of the vertebrates.

Trauma. Means bodily injury and also imotional shock.

Traumatic Shock. A shock which a body gets from an injury or emotional disturbances.

Treppe. It is also known as staircase phenomenon. It implies a condition during muscle contraction as in kymogram, when a muscle has been stimulated repeatedly after regular intervals for sometime, the initial few stimulations exhibit gradual increase in amplitude of contraction but successive latter ones exhibit a decreasing amplitude.

Triac (3, 5, 3-Trilodothyro Acetic Acid). An acetic acid analogue which is produced by deamination of triiodothyronine and subsequent decarboxylation.

Traids. Refers to the arrangement of central T-tubule with sarcoplasmic reticulum of skeletal muscle fibre on either side which serves to help in rapid transmission of action potential from cell membrane to all the fibres in muscle.

Tricuspid Valve. The trilobed valve that is present between right atrium and ventricle in mammalian heart. It is also known as the besian valve.

Trchocyst. One of the definite bodies lying in the ectoplasm. It is producing hair-like fibres in organisms like *Paramecium*. It is discharged on chemical, mechanical stimulation as well as at the time of death of the animal.

Trigeminal. The 5th cranial nerve is vertebrates. It is divided into three branches which supply the skin of face, tongue and the teeth.

Triglycerides. Esters of glycerol and 3 fatty acids. On hydrolysis they yield glycerol and 3 fatty acids, e.g., *deshi ghee* (a natural fat). They are formed in animals form carbohydrates.

Trilobites. Extinct marine arthoropods of Plaeozoic era.

Triiodothyronine. An iodine-bearing tyrosine derivative which is being synthesised in and secreted by thyroid gland; it raises cellular metabolism as does the thyroxine.

Triplet. Means the sequence of three purine or pyrimidine bases each of which represents a particular amino acid.

Triton X-100. A non-ionic detergent which finds use in cell biology to solubilise the lipids and certain cell proteins.

Trisomy. A chromosomal mutation which results an individual to have three copies instead of normal two copies (diploidy) of a given chromosome.

t-RNA. Small RNA molecule which could recognise a specific amino acid, transport it to ribosome and select its position properly in growing polypeptide chain. It is also known as *s-RNA* (souable-RNA).

Trochanter. Refers to the second segment of insect leg between coxa and femur. It also refers to the articulating surface which is found in the head of femur of vertebrates for attachment of thigh muscles. The trochanters are not present in the femur of frog but are present in reptiles (only one), birds (again one) and mammals (three).

Trochopore. Refers to the free swimming ciliated larva of annelids and molluscs which is having well—devloped alimentary canal, sense organs three ciliary bands. It is metamorphosed into adult.

Trophic Substances. Hypothetical chemical substances which are regarded to be releases from neuron terminals and to be regarded influencing the chemical and functional properties of the postsynaptic cell.

Trophoblast. The germinal layer from which chorion in mammals in developed.

Trophozoite. Refers to the sporozoite in the phase of life-cycle where its primary activity is to get nourishment and grow.

Tropism. Means orientation and locomotory responses, e.g., by photoropism it means orientation and locomotory response with respect to light, similarly, by heliotropism movements with respect to sun. Other such tropic movements include geotropism (with respect to gravity), thigmotropism (with respect to touch), hydrotropism (with respect to water) etc.

Tropomyosin. A long protein molecule present in the grooves of actin of thin filaments of striped muscles. It is inhibiting contraction of myosin cross-bridges with actin filaments.

Troponin. Means a complex of globular protein which is associated with actin and tropomyosin of thin (actin) filament of stripped mucle. It is having three subunits:

T_pC. It is a submit which bind Ca^{2+} molecules and brings about conformational change in troponin complex permiting the tropomyosin to move out of its blocking position on actin filament.

T_pI. It is an inhibitory subunit which keeps tropomyosin in its normal position thereby blocking sliding of actin filaments.

T_pT. It is tropomyosin binding subunit that is able to bind torpomyosin and action so as to make the latter to interfere with attachment of cross-bridges.

Truncus Arteriosus. Base of aorta as present in lower verte-brates like frog. In the lumen of this there is a spiral valve which protects the collapsing of its walls as well as distribution of blood through the organ.

Trypsin. Enzyme which promotes digestion of proteins in the vertebrates. It is present in the pancreatic juice of all the vertebrates.

Trypsinogen. Inactive form of typsin. It is secreted in this form so as to prevent autodigestion. The substance gets activated by enterokinase from success entericus.

Tryptophan. As essential aromatic amino acid in the body.

TSH (Thyroid Stimulating Hormone). A hormone which if secreted from pituitary gland. It is able to control activity of thyroid gland. It is secreted under the influence of thyrotropin releasing hormone from hypothalamus.

T/S Ratio. Defined as the ratio of thyroid to plasma or serum free jodine. it varies from 10-100. It becomes high if binding gets blocked by antithyroid compounds like thiouracil.

T-System. One of the structural elements of sarcotubular system. It consists of transverse tubules. It is continuous with membrane of muscle fibre. The space between two layers is an extension of extacellular space.

T-Tubules. Running around the perimeter of each myofibril at the level of Z-line membrane limited tubule called T-tubule (transverse tubule) is present. It is less than 0.1μ m in diameter and branches so that it becomes continuous with similar tubules surrounding neighbouring myofibrils at the same level.

Tube Feet. The locomotory organs present in the ambulacral grooves of star-fish and other echinoderms. They deal with water vascular system and consist of dorsal sac-like ampulla and ventral disc-like sucker. There are certain muscles over the ampulla and the sucker which regulate the contraction and expansion of these parts and in turn make the substratum or relax the hold of substratum. It is by undertaking such necessary contractions and expansion the animal carries out locomotion with these organs.

Tubule. A small tube-like structure which is present in the body. It may be of the following types:

Collecting tubule. It is associated with kidney and conveys urine to ureter.

Convoluted tubule. It is present in the anterior part of renal tubule.

Uriniferous tubule. It is syonymous with nephron.

Seminiferous tubule. It is associated with testes.

Tubulin. Action like 40 A^0 globular protein molecule which forms building black of microtubule.

Tufted Cells. Cells among dendrites of axons of receptors which are terminating in olfactory bulbs. Axons of tufted cells are passing posteriorly through the medial olfactory stria, the middleline in anterior commissure and finally crossing the opopsite olfactory bulb.

Tumour. A swelling, a mass of abnormal tissue which is similar to the normal tissue in structure but serves no useful function and grows at expense of body. Benign/simple/innocent tumours do not infiltrate adjacent tissue and do not likely to recover if removed. Malignant tumours infiltrate adjacent tissue and bring about metastasis.

Tunic. Test present in urochordates like *Hardmonia.* Due to this reason these animals are also known as tunicates.

Tunica. Refers to a lining membrane, a coat over an organ.

Tunica Adventitia. The coat/membrane present over an artery. Also known as *tunica externa.*

Tunica Albugenia. The touch protective connective tissue that covers the testes.

Tunica Interna. The lining of an artery. Also known as *tunica intima*

Tunica Media. The middle muscular coat of an artery.

Tanica Vaginals. The peritoneum of vaginal coelom which gets reflected over a part of testis after its descent into sacrotal sac in mammals. Increase in fiuid within this layer brings about hydrocoel.

Tunicate. A group of animals having test over the body e.g., *Herdmania* (an urochordate).

Tunicine. Refers to a polysaccharide which is identical to cellulose of plants. It forms test in case of urochordates such as *Herdmania* and others.

Turbinals. Thin scroll-like folded bones which are present in the olfactory part of nasal chamber of mammals. They are covered with highly vascular epithelium and consist of olfactory receptor cells, mucous cells, supporting cells etc. The presence of these structures makes the olfactory chamber to maintain constant temperature of the air that enters the respiratory tract.

Turbulence. A phenomenon in blood in which disorderly flow occurs when viscosity of blood is lower and is related to the diameter of vessel. Turbulent flow also produces sound whereas streamline flow in silent.

Turbulent Flow. Refers to the flow in which fluid moves in all the directions. It produces sound as opposite to silent/streamline flow.

Turgor. Cell distension which takes place due to its fluid contents.

Turner's Syndrome. The gonadial dysgensis which could be identified at birth. Individual of this syndrome are having chromosomes instead of 46 and one sex chromosome is lacking. Individuals are brought up as girls though they are genetically male. Such individuals have small female genetalia, scanty pubic hair, atrophic ovaries, webbed neck and valgus of elbow. They are generally sterile. The syndorome was discovered by American endocrinologist Hanry H. Turner in 1892.

Turle. Refers to the semi-aquatic species of reptilian group chelonia.

T-Wave. Portion of electrocardiogram which is associated with repolarisation of ventricles (usually relaxation).

Twitch Muscles. Most common striated vertebrate skeletal muscle type is usually pale in colour due to low myoglobin contents. It is having few mitochondria and its fibres are constructed of many clearly defined fibrils. These undergo contraction rapidly and obtain most of energy through araerobic metabolism.

Tymapanic Bone. A bone which is derived from angular bone of reptilian lower jaw (in mammals it is applied to petrous part of periotic bode).

Tympanic Bulla. The tympanic bone of mammals is closely applicable to a part of periotic boen and forms flask-shaped structure which is called tympanic bulla. The neck of this structure is forming external suditory meatus.

Tympanic Cavity. Air filled space/cavity in the middle ear that is communicating with pharynx through eustachian tube.

Tympanic Reflex. Sound transmission reduction taken place by contracting of tenson tympani and srapedial muscles pulling manubrium of malleus inward and foot plate of stapes: outwards initiated by loud sound is known as tympanic reflex.

Tympanum. Vibrating memebrane which takes part in sense of hearing. Also the hollow resonating chamber of syrinx in the birds.

Type-I Cells. Cells which are lining alveoli in man. They are flat having large cytoplasmic extensions and primary lining cells.

Type-II Cells. Cells which are lining alveoli of man. They are thicker and have numerous lamellar inclusion bodies and secrete surfactant.

Type 'J' Receptors. Juxtra capillary located receptors responding to pulmonary congestion and embolisation which produce shallow, rapid breathing.

Typhlosole. Refers to the median dorsal intestinal fold of mucous membrane in the intestine of enrthworm. It increases, the area of absorption for the intestine of the animal.

Tyramine. Toxic substance which is formed within colon by bacterial enzymes. It decarbxylates amino acids.

Tyrosine. An aromatic amino acid which takes part in the synthesis of catecholamines and thyroid hormones etc., in the body. In the body it is derived from an essential amino acid, phenylalanine.

U

Udder. The mamma of animals such as the cow and horse.

UDPGA (Uridine Diphospho-Glucuronic Acid.) Used for the basic source of glucuronic acid in cells.

UDP Glucuronyl Deficiency Type-II. Used for congential disease in which there occurs relative deficiency of glucuronyl transferase which catalyses conjugation of free bilirubin to glucuronic acid.

Ulna. The back one of the two bones of the lower front leg in 4-legged animals or the arm in man i.e., the one on the inside when the front of the hand is facing forward.

Ulnare. That one of small bones at the joint between arm or front leg and foot which is in line with and nearest to the ulna.

Utrafibril. Said of the smallest unit of chromosome which is visible under an electron microscope. It is 100 A thick and is having two DNA double helics which are separated by space of 25 A° across the associated protein. Smallest ultrafibril is single DNA having its associated protein. But the single DNA molecule is nearly roughly 20° A diameter. Therefore, two DNA molecules will have 40° A thickness and a gap of 25 A° between two DNA halices of ultrafibril will be having a total thickness of 100A°.

Ultrafiltration. In glomerulus as the blood passes from capillaries of afferent renal arteriole to efferent renal arteriole then due to pressur difference, certain substances from blood are fallen in Bowman's capsule and then diffuse to the neck of renal tubule through one celled thick wall of capsule. This filtration is termed as ultrafiltration.

Mobilical Cord. (1). Used for the cord-like connection between embryo or foetus of a mammal and the placenta. It consists of blood vessels and connective tissue.

(2). A similar cord in birds and reptiles connecting the embryo with the folk sac.

Umbilicus. (1) Used for the navel: a depression in the abdomen which is left by the separation of the umbilical cord of a mammal after birth.

(2) Of feathers a small hole at each end of the hollow stem of a quillfeather. At the proximal end there is a *inferior umbilicus*; at the distal end there is the *superior umbilicus*.

Umbo. Refers to the elevation near the hinge of bivalve shell observed in *Unio* around which the concentric lines of growth are added in the shell.

Umbraculum. Refers to a specially modified part of iris in camels and other mammals in form of pigmented fringe which works to shield retina from excessive glare.

Umbrellar Surfaces. Used for a *exumbrellar* and *subumbrellar* (upper and lower) surfaces of a medusa or jally-fish.

Unciform Bone. Used for describing a hook-shaped bone of the mammalian carpus which is formed by fusion of the fourth and fifth distal carpals; more particularly used of the human wrist but sometimes also of other mammals.

Unicinate Processes. Small processes which are projecting backwards from the ribs of birds and some reptiles, each overlapping the next posterior rib and thereby helping to strengthen the wall of the thorax.

Unconditioned Response. In classical conditioning, it is the response which naturally follows the unconditioned stimulus. It becomes conditioned response where there occurs the completion of conditioning.

Undulating Mambrane. A protoplasmic membrane which is used for locomotion or feeding in certain protozoa. In ciliated protozoa as in Paramecium it is used for feeding. In Trypanosoma, it is used for locomotion.

Undulipodium. Used for ship-like apparatus having 9 + 2 arrangement of inner flaments. It is a new term for cilium and flagellum.

Unguis. Refers to long plate of keratin which is found in claws, nails etc. It is sharply curved so as to enclose always smaller sole plate or subunguis between its edges.

Ungulate. Used for mammals with hoofs, e.g., horse, cow, sheep, pig etc. These animals get adapted for walking on hard land.

Unguiligrade. Describing walking or an adaptation for walking hoofs as observed in horses, sheep, cows, etc.

Unicellular Organisms. Organisms having only one cell. Some of these have been definitely classed in the animal kingdom (protozoa); others among plants as simple algae or fungi. The term *Protista* is often used to comprise both types.

Uniovular Twins. Monozygotic points.

Uniparous. Producing one offspring at a birth.

Unipolar (V) Leads. Refers to nine standard locations for exploring electrodes which are used for routine electrocardiography. Six leads are known as chest leads (precardial) and designated as V_1-V_6. There unipolar leads are V_R (right arm), V_L (left arm) and V_F (left foot). Augmented limb leads are generally used these days and designated as (aVR, aVL and aVF).

Unipolar Neurone. A nerve cell which is having two processes or fibres arising from the same point and appearing as one.

Unipotent Committed Stem Cells. Used for the cells in bone marrow which on stimulation become differentiated into one of the cell types occurring in red bone marrow monoblast, myeloblast, lymphoblast, proerythroblast and megakaryoblast).

Unit Mambrane. Used for describing sandwich-like profile of biological membrane which is observed under electron microscope. It is believed to represent binocular leaflet having hydrophobic central region between hydrophilic surface.

Universal Donor. Type 'O' individuals that are lacking agglutinogen and thus causing agglutination when their blood gets transfused to any other type of individual (belonging to any other blood group).

Universal Recepients. Persons of 'AB' type lack any agglutinins in their blood and their blood is not agglutinated when any other type of blood (of different blood group) gets transfused to them.

Urachus. Refers to stem-like structure which is connecting the bladder with umbilicus in foetus and is represented in postnatal life by fibrous cord which is situated between the apex of bladder and umbilicus called median umbilical ligament.

Uracil. A pyrimidine derivative, one of the four bases which together with sugars and phosphate groups, make up the molecule of RNA.

Urea. Co $(NH_2)_2$. A soluble nitrogenous waste substance which is excreted by many animals.

Urease. An enzme converting urea into ammonia and carbon dioxine; more common in plants than in animals.

Uremia. A clinical syndrome which occurs due to renal failure which in turn is resulting from either disease of kidneys themselves or disorders or disease elsewhere in the body which induces kidney dysfunction thereby retaining urea and other nitrogenous substances in blood. Fully developed syndrome is followed by nausea, vomiting, headache, hicough, weakness, dimness of vision, convulsions and coma.

Ureotelic Animals. Animals whose chief excretory product is urea, e.g. fish, Amphibia and mammals (*cf. Uricotelic*).

Ureter. A tube along which urine is flowing from the kidney to the bladder.

Urethra. A tube which is conducting urine from the bladder of a mammal to the exterior.

Uric Acid. $C_5H_4N_4O_3$. An important nitrogenous compound, the chief excretory product of birds, reptiles and insect (see *Uricotelic Animals*).

Uricase. An enzyme present in the liver and kidneys of many animals; its chief action is the conversion of uric acid to *allantoin* which is an important excretory product, particularly of herbivores.

Uricotelic Animals. Used for describing animals which excrete uric acid instead of usea, e.g. most reptiles, birds, insects, snails etc.

Uridine Diphospho-Galactose. A substance which is formed by reaction of galactose-6 phosphate with uridine diphosphoglucose. This is converted into uridine diphosphoglucose.

Uridine Diphospho Glucose. A subsance to which galactose (after it is subjected to phosphorylation) undergoes reaction during glycogen formation from galactose.

Urinary Bladder. Thin sac-like structure which is attached with cloaca in which urine formed by kidneys gets accumulated.

Urine. The waste liquid excreted by an animal, usually having nitrogenous constituents such as urea, acid or ammonium salts.

Uriniferous Tubules. The narrow tubes, often of great length and with several convolutions, conveying urine from the malpighian bodies of the vertebrate kidney to the ureters.

Urinogenital Duet. Refers to the common duct for the passage of excretory and genital products.

Urinogenital Sinus. A chamber or sac into which the urinogenital ducts open.

Urinogenital System. In vertebrates, excretory organs and reproductive organs work in close association since early development, thus both systems together are known as urinogenital system.

Urochordate. Group of protochordates in which notochord has been confined to the tail region during embryonic life and gets lost in adult stage with the loss of tail during metamorphosis e.g., Herdmania.

Urodaeum. Refers to the chamber/division of cloaca in reptiles and birds in which urinogenital organs open. It opens into procctodaeum.

Urodela. Amphibians with tail such as salimanders and newts.

Urogastrone. A substance having inhibitory effect on stomach secretion, found in urine.

Urohyal. One of the median bones in hyoid apparatus representing the basibranchial of first branchial arch.

Uropeltidae. Used for small burrowing snakes of Ceylon and South India, having the cranial bones solidly united and having the tail ending in a large shield.

Uropepsinogen. Refers to pepsinogen activity which is detected in plasma and urine.

Uroplatidae. Lizards from Madagascar that are resembling geckos except that the nasal bones are fused, the interclavicle small and the clavicles not dilated.

Uropods. Refers to the last pair of abdominal appendages of a crustacean like a lobster or crayfish; frequently flattened and forming part of the tail fan.

Uropygial Gland. The 'preen-gland' on the dorsal side of a bird's tail, secreting oil with which the bird makes its feathers waterprooff.

Uropygium. Refers to the smallconical tail-like profection in the birds. It has preen/uropygeal glands on its dorsal side.

Urosome. Refers to the hind-part of a crustacean such as *Cyclops* in which there is no true 'abdomen but a certain number of segments at the hind-end have been modified to form a 'tail'.

Urostyle. A bony rod which is formed from a number of fused vertebrae at the hind-end of the vertabral column in frogs and toads. Fused to the pelvic girdle it imparts greater rigidity to the latter when the frog is jumping.

US (Unconditioned Stimulus). In classical conditioning, the stimulus which naturally elicits the response in question.

Uterine Milk. A nutrient secretion from glands in the wall of the uterus, providing the first food for a developing mammalian embryo before the formation of the placenta.

UTP (Uridine Triphosphate). It is a phosphoylated derivative of macil and finds use in providing energy like ATP, CMP etc.

Utriculus. Refers to the portion of membranous labyrinth of internal ear; antains maculae having cells which on diplacement by oroliths send orientation to the brain. It is normally in a position which is superior to sacculus and has more importance.

Uterine Cycle. Comprises of cyclical changes in the structure and function of uterus with respect to the hormones which are associated with menstrual cycle. For example, shedding of endromitrium, development of uterine glands, secretory or progestational changes, formation of stratum functionale and preparation of next sloughing in case pregnancy has failed.

Uterine Villi. Structures resembling villi by which a mammalian placenta gets embeded in the wall of the uterus.

Uterus. The womb: a muscular enlargement of the two oviducts forming a paired or a single sac in which the embryo develops in female mammals. The two halves lead into a median vagina which is opening to the exterior by the vulva.

Uterus Masculinus. Refers to a sac-like structure dorsal to the bladder in the male mammal, homologous with the uterus of the female.

Utricle, Utriculus. A part of the membranous labyrinth of the ear which consists of a sac into which the three semi-circular canals open.

Uvea. A layer of pigmented cells which is covering the back of the iris in the eye of a vertebrate.

Uvula. A sensitive fleshy process which is hanging from the back of the soft plate in mammals.

V

Vacuity. Used for the space cavity in the body. The term is generally restricted in use to cavity in the temporal region of vertebrates specially reptiles. Also, see *temporal fossae.*

Vacuole. Refers to a small structure which is enclosing a cavity in cytoplasm. It may be having a fluid/liquid as in contractile vacuole or may have other substances like food (food vacule), excretory products (excretory vacuoles) secretory products (secretory vacuoles) etc.

Vagal. When pertaining to the vagus nerve (X-cranial nerve).

Vagina. (1) Refers to a passage leading from the uterus to the exterior in a female mammal.

(2) A similar dilatation of the oviduct in some invertebrate.

Vaginal Cycle. Includes cyclic changes which are taking place in vaginal wall with respect to different hormones of reproductive cycle. These changes in vaginal wall include cornification of vaginal wall's epithelium under the effect of estrogen and secretion of thick mucous and proliferation of epithelium which becomes infiltrated with WBCs under the effect of progesterone. The above mentioned changes have been reported in experimental rats and also reported to be taking place in buman being but not so clearly marked out.

Vaginal Smear. When a film of material is spread out on a glass slide for examination under microscope, the cells in vaginal lumen present in smear show conformational changes which are helpful in the identification of complication of vaginal cycle and reveal infection etc., which might be occurring in vaginal wall.

Vagotomy. Refers to the surgical division of vagus nerve which is done in connection with gastroenterostomy in the treatment of peptic ulcers or pyloroplasty.

Vagus. Used for the tenth cranial nerve of a vertebrate which is having numerous branches to the various viscera. The viscerai branches to the heart, alimentary canal etc. help for regulating unconscious auomatic activities. They form part of the parasympathetic system.

Vagus Ganglion. A ganglion from which branchiovisceral trunk of vagus nerve is arising.

Valine. An essential amino acid having substituted chains: Also called *neutral amino acid.*

Valve. (1) Any structure which is allowing the passage of a fluid in one direction only, as in the heart veins etc.

(2) In Mollusca, Brachiopoda and some Crustacea: one of the units which go to make up the shell.

Valve of Vieussens. A thin transverse band of white matter which is extending between the two anterior peduncles of the cetebellum immediately behind the corpora quadrigemina.

Vanadocytes. Green blood-cells having the element vanadium, fund in certains.

Vane (Vexillum). The flat distal part of a feather which consists of barbs and barbules linked together by small hooks.

Vanus. Used for describing the anal or posterior lobe of an insect's hand-wing: a fan-like expansion separated from the rest of the wing by a furrow.

Variation. Refers to the difference in structure or function which is shown by individuals of same species.

Variety. In taxonomic use it refers to a division of the species: a group of individuals loving single interbreeding population that are different in some minor respect from rest of species.

Vas Deferens. The main duct by which seminal fluid is passing from the testis to the exterior.

Vasa Efferentia. Refers to small tubules which are leading from the testis into the *vas deferens* (*q.v.*).

Vascular System. The blood system, having usually heart arteries, capillaries and veins.

Vaso-Constrictor. Causing constriction of blood-vessels, e.g. by the action of certain nerves or hormones.

Vaso-Dilator. Causing he enlargement of blood-vessels, e.g. by nerve or muscle action.

Vasomtor Centre. Refers to a centre in medulla oblongata which is controlling discharge rate in sympathetic vasoconstrictor nerves. A large diffused area in reticular formation from just below the obex to region of vestibular nuclei and floor of fourth ventricle to almost pyramids. Stimulation of rostral and lateral portions of this centre results in rise in blood pressure and techycardia whereas stimulation of a smaller portion around the obex results in a fall in the blood pressure (bradycardia).

Vasomotor Nerves. Nerves which are originating in the *medulla oblongata* and control the muscular walls of blood-vessels causing them to become dilated or constricted, thereby regulating the flow of blood to any part of the body.

Vasopressin. A hormone which is produced by the posterior lobe of the pituitary organ. It is stimulating the muscle fibres in the walls of arterioles and resulting in a rise in bloodpressure.

Vasa vasorium. Minute nutrient vessels of artery and vein walls.

Vector. An animal which is carrying parasites or pathogenic organisms to the host, e.g. a mosquito.

Vector Tissue. Refers to the specialised tissue in the bodies of certain leeches which is connecting particular regions of the skin to the ovisacs. When ocpulating, the sperm, in the form of a gelatinous spermatophore, gets deposited on the skin. It perforates this and makes its way along the vector tissue to the eggs.

Vegetive Hemisphere. Part of egg in which the rate of metabolism is lower than that of the animal pole; usually having yolk material.

Vegetative Nucleus. An alternative name for *megniculeus* (q.v.).

Vegetative Phase. The part of a life-cycle in which reproduction is asexual, e.g. by fission or by budding.

Vegetative Pole (Vegetal Pole). The part of an egg which is opposite to the nucleus and having most of the yolk. Eggs of this type are called *telolecithal* and are typical of chordates.

Veins. Vessels conveying blood from the ovarious parts of the body back to the heart.

Velar. Pertaining to or attached to the velum (*q.v*).

Velar Tentacles. Of *Amphioxus:* ciliated tentacles that are situated on the velum round the edges of the enterostome.

Verliger. A larval stage after trochophore in many molluscs like Pila.

Velum. A membranous band which is present between stomodaeum and pharynx of *Amphioxus*. It is bearing 12 backwardly directed tentacles along posterior border. Also known as *enterostome.*

Vena Bulbi. Refer to two small veins which are formed by union from net work of fine vessels arising from carotid artery and supplying wall of conus arteriosus. Vena bulbi posoterior is entering the anterior abdominal or hepatic portal vein, while vena bulbi anterior is joining left systemic.

Venae Cavae. Used for three large veins which are collecting deoxygenated blood from various parts and drain it into sinus venosus/right auricle. They are also named as superior and inferior venae cavae or pre-and post-cavals.

Vennule. Refers to a small sub-division of vein which is joined with ocapilllaries.

Ventral. Side/surface of body which is pertaining to belly; away from back opposite to dorsal.

Ventral Horn. Refers to the ventral portion of gray matter in vertebrate spinal cord in which mother nerve cell bodies have been situated

Ventral Root. Refers of a nerve trunk which is leaving spinal cord near its ventral surface having only motor neurons.

Ventricle. Refers to any of the small chambers in the anatomy of animals; term mainly refers to the chamber in the heart from which the blood is distributed. In brain the term refers to several large divisions of central space.

Ventricular Arrhytmias. Used for deviation from normal rhythm in ventricle. This may be attributed to development of ectopic focus in ventricle that causes extrasystole/premature beat which may disturb the normal cardiao rhythm.

Ventricular Fibrillation. Refers to a group of ectopic foci in the ventricle leading to extrasystole or premature beat and if focus discharges relatively at a rate higher than SA node it yields ventricular tachycardia and can also yield ventricular fibrillation.

Ventricular Systole. Refers to contraction of ventricles.

Ventricular Tachycardia. Refers to series of rapid, regular, ventricular *extrasytoles*. It is serious because during this cardiac output gets decreased and *ventriculat* fibrillation may be occasional complication.

Ventriculus. (1) A ventricle (see above).

(2) The gizzard of a bird.

(3) The stomach or mid-gut of an insect etc.

Venules. (1) Small veins of the blood system which are connecting capillaries with larger veins.

(2) The smaller 'veins' of an insect's wing.

Veratridine. A drug causing apnea, hypotension and bradycardia in experimental animals when it is injected into their body.

Vermiform Appeadix. Refers to a slender, tubular pouch-like projection from caecum of large intestine of some mammals.

Vermis. A worm-like structure which is forming the median part of the cerebellum in birds and mammals.

Vertebrae. Refers to the structural unit of vertebral columns of a vertebrate. Each of them principally is having a centrum (formed by replacement of notochord) and dorsal, ventral processes (that form neural spine), transverse processes, zygapophyses and hypocentrum.

Vertebraterial Canals. Refer to a pair of canals which are placed on either side of cervical vertebrae of birds and mammals specially, through which vertebral arteries are passing.

Vertebral Column. Refers to the 'backbone'. It is the most characteristic feature of vertebrates and consisting of a series of small bones of cartilages which are running the length of the body near the dorsal side, enclosing and protecting the spinal cord.

Vertebrarterial Canals. Refer to the holes for the passage of the vertebral arteries through the cervical vertebrae of mammals, birds and reptiles. Each hole is formed by fusing of a short double-headed rib to the side of the vertebra so that the canal is actually present between the vertebrae and the ribs but appears to go through the vertebrae.

Vertebrate. Animals having vertebral column e.g., fishes, amphibia, reptiles, birds and mammals.

Vertebro-sternal Ribs. True ribs attached to sternum through costal cartilages. Their number is 1-7.

Vertex. Refers to the dorsal region of an insect's head.

Vertigo. Refers to a sensation of rotation.

Vestibular Cerebellar Tract. Refers to the afferent tract in cerebellum which is transmitting vestibular inpulses from labyrinth directly and via vestibular nuclei. It is distributed mainly in fluccolondar lobe and enters the cerebellum through inferior peduncle.

Vestibule. Any passage or cavity leading into another, e.g.

(1) Protozoa: A depression in the ectoplasm leading to the 'mouth', as for in *Paramecium*.

(2) Mammals: The cavity leading from the vulva to the vagina in a female mammal.

Vestigial Organ. A small or imperfectly developed organ which has lost or changed its function during evolution, e.g., the vermiform appendix in Man.

Vexillum. The vane of a feather (*q.v.*).

Vibrissae. Stiff hair in or at the nostrils. Also a *whisker.*

Villi. Refer to the minute finger-like projections specially those present on the intestinal lining of vertebrates; increase an absorptive surface.

Villikinin. General and collective name for different hormones which are secreted from gastrointestinal tract such as VIP, GIP, bombesin, somatostatin etc. These hormones regulate intestinal activity.

Villi-placental. Trophoblastic villi lodged into crypts of uterine wall in case of placenta of mammals.

VIP. Refers to a hormone like substance which is isolated from intestinal wall having 28 amino acid residues and stimulates markedly the intestinal secretion. It is known as *vasoactive intestinal peptide.*

Vipers. Common name for poisonous snake *Ancistodor.* that is having rounded tail, large belly scales and distinct head bearing small scales. They may be having a pit between nostrils and eye. Generally found on higher altitudes.

Viruses. Refers to the non-cellular parasitic submicroscopic structures which are composed of protein shell and core of nucleic acid. Can pass even through microfilters and can be stored in bottles for long time without making any harm to them.

Viscers. Organs which are enclosed within great cavities in body specially the abdomen.

Visceral Arches. (1) Partitions or septa which are present between adjacent gill-slits of a fish.

(2) The bony or cartilagenous skeletal arches which are lying in these partitions.

Visceral Clefts. Refer to the openings which are leading from the pharynx to the exterior in chordates; in fish they form the gills-slits but in land animals the embryonic clefts usually become atrophied in the adult.

Visceral Fibres. Thinly myelinate nerve fibres which are running from the sympathetic ganglia to the principal viscera.

Visceral Sensory. Nerve fibres in the lower part of dorsal horn which are dealing with iternal reflexes.

Visceral Skeleton. Refers to frame work of jaws and gill arches in a vertebrate.

Visceral Smooth Muscles. See *unstripped muscle or smooth muscles.*

Visceral Motor. Refers to the nerve fibres which are composed of smaller motor units (cell bodies) and forming lateral horn of grey matter. They are concerned with internal reflexes.

Viscosity. Physical property of fluid that is determining the ease with which layers of fluid are moving fast to each other. Vision. Principal sensory modality of eye which makes the organism to see things around it. This takes place due to rods and cones present in the eye.

Visual Agnosia. Refers to the inability to see objects by a particular visual modality even through visual modality is intact. It occurs due to leisions in parietal lobe.

Visual Cortex. Refers to the outer most thin layer of grey matter which is present in the occipetal region of cerebrum; involved in the processing of visual information.

Visual Threshold. Minimal amount of light eliciting a sensation of light.

Vital Capacity. Refers to the total quantity of air that can be expelled from lungs. In man it is 3.44.0 × 10^3cc.

Vitalism. A theory which postulates that biological processes cannot adequately be explained by physical and chemical processes and laws.

Vitamins. Refer to any of the unrelated organic substances that do not yield energy but their presence is essential in diet. They are needed for normal growth and function and essential in minutest quantities. Two catagories to these substances are water soluble vitamins and fat soluble vitamins. Vitamin A, D, E and K are fat soluble whereas vitamin B and B-complex and C are water soluble.

Vitelline Mambrane. A thin membrane which is surrounding the egg/ yolk.

Vitreous Body. A name which is sometimes used to denote the crystalline cones in the compound eyes of arthropods.

Vitreous Humour. A transparent gelatinous substance which is filling the space behind the lens in the eye of a vertebrate.

Vivicum. Refers to all living things.

Viviparity. Refers to the ability of animals to give birth to living young one that is being developed from egg within the body of mother and is nourished with the help of a placenta.

Viviparous. An organism giving birth to living young one that is developed in the body of mother and is nourished by her own blood stream as most of autherian mammals.

VLDL (Very Low Density Lipoproteins). Refers to the substances which are principally formed in liver and in turn converted into intermediate low density lipoproteins during fat metabolism.

VMA (3-Methoxy-4Hydroxy Mandelic Acid). Refers to one of the degradation products formed by catabolism of epinephrine. 700 μ g of VMA is excreted per day.

Vocal Cords. Refers to a pair of membranes which are stretching between thè thyroid cartilage and the arytenoid cartilages in the larynx of a mammal. When these membranes vibrate, they produce vocal sounds which can be varied by altering their tension and position.

Vocal Sacs. Resonating facs on each side of the buccal cavity, e.g. in the males of certain frogs.

Volkmann's Canals. The canals which are connecting Haversian canals in the mammals and found below periosteum and in the centre of bon. Through these canals blood vessels reach interior of bone.

Volvent. Refers to small types of cnidarian nematocyst having an armed thread (with spines) used to capture prey.

Vomer. Refers to a median membrane bone which is present above the palatine bones and beneath the nasal cavity in the mammalian skull. It is homologous with the rostral bone of a bird and with the parasphenoid bone in lower vertebrates. The so-called 'vomer' of a frog is now considered to be a prevomer.

Vomiting. Used for describing ejection of stomach contents through mouth. Also regarded as sickness due to deficiency of vitamin or may be occurring due to infection or other disorders of gastrointestinal tract.

Vortical Sounds. During turbulent flow then the blood strikes an obstruction or passes through a narrow orifice at high speed swiling vorticals set up generate vortical sounds.

VIP (Ventricular Premature Beats). In ventricular arrhythias the extrasystoles originating in an ectopic ventricular focus have polonged QRS complexes due to slow speed of impulses from this focus through ventricular muscles to rest of ventricule. They are usually not capable of exciting the bundle of His and so retrograde conducting of auricles does not take place. P-wave is generally burried in QRS complexes of extrasystole. When normal impulses reach ventricles they are still in refractory stage following depolarisation from ectopic focus, however, a second succeeding impulse from SA-node produces normal beat. Hence ventricular extrasystoles have been followed by compensatory pause which does not take place in atrial extrasystole.

Vulva. Refers to the external opening of the vagina in female mammals.

v-Wave. Refers to a Portion of recording of jugular venous pressure which shows rise in atrial pressure before tricuspid valve is opened during diastole.

W

Wallace's Line. Refers to a line separating the Oriental from the Autralasian zoo-geographic region. It is running between the islands of Bali and Lombok, through the Macassar Straits, then south-east of the Philippine Islands.

Warm Blooded. Animal whose body temperature does not have tendency to fluctuate with external/environmental temperature. Also called *homeotherm*. Birds and mammals are examples of this catagory of animals.

Warning Colouration. Bright colours or markings on animals that are poisonous or otherwise dangerous, e.g. on many insects and snakes, giving them the advantage of being recognized and avoided by other carnivorous animals.

Warts. Refer to the rough outgrowths on skin as in toads. In human beings they may be present on palms, fingers etc.

Water Expulsive Vesicle. New name for contractile vacuole. An osmoregulatory apparatus which can secrete water from cytoplasm.

Water Pockets. Refer to a number of pouch-like diverticula from rumen of camel. They do not store water as believed but actually are the structures that contain food in the peocess of digestion.

Wax. A type of lipid which helps in the protection of organism. It is found in the epicuticle of many insects.

Wax Glands. Refer to the tubular glands which are found in passage of external auditory meatus. they produce an oily secretion that dries up into thin max-like flakes called wax of ear. The main function of this substance is to prevent dust entry into ear passage. These glands are also called *ceruminous glands*.

WBCs (White Blood Corpuscles). Colourless blood cells found in vertebrates. Also called *leucocytes*. (See *leucocytes*).

WEB. Thin membranous structure which is joining toes in certain animals, e.g., frog.

Webbed Feet. Refer to feet adapted for swimming by a web of skin present between the toes. This form of adaptation has taken place independently in a great many birds, in frogs and in such mammals as the duck-billed platypus (*Ornithrhymchus*), the otter and the beaver.

Weberian Ossicles. A chain of small bones which are probably detached portions of some of the anterior vartebrae and are connecting the air bladder with the ear capsule in fish of the group Ostariophysi. This group also includes many common fresh-water fish.

Wever's Law. A law which states that the perception of a small difference between two sensory stimuli has been found to depend on the ratio of that difference to their magnitudes but not on the absolute difference between them. In the perception of light by the human eye, for instance, if the brightness gets increased by 1% the difference is just noticable. For noise the ratio is about 1/3 for pressure on the skin it is usually between 1/10 and 1/30.

Weismannism. Refers to the doctrine of the continuity of the germ-plasm and the non-transmission of acquired characteristics. By this theory the reproductive cells are set apart at an early age from the rest of the body-cells and are not influenced by them. As the only connection between one generation and the next is through the germ-cells it means that characteristics acquired during life cannot be inherited. In the light of recent research on the action of physical and chemical factors influencing chromosomes, the rigid theory as defined by Weismann has been considerably modified.

Wenckebach Phenomenon. In a form of complete heart block the repeated sequence of beats occur in which PR-interval lengthens progressively until a ventricular beat gets dropped.

Whales. Common name for large sized aquatic mammals such as *Balenoptera indica* . They are having naked skin, paddle-like limbs and a very thick layer of fat in subdermis called blubber.

White Matter. Tissue of central nervous system having myelinated nerve fibres.

White Muscles. The muscles having mixture of three types of fibres (A, B, C).

White Ramus. A thin nerve (*ramus communicans*) consisting of white or medullated fibres connecting each spinal nerve with the corresponding lateral sympathetic ganglion. These fibres are called *preganglionic*, those which leave the ganglion in the reverse direction are non-medullated grey fibres and are called *post-ganglionic*.

White Reaction. When pointed object is made to draw lightly over the skin, the stroke lines would become paler. It is called white reaction and it appears in about 15 seconds.

Wild Type. Normal/standard form of a gene or an individual such as the form occurring typically in nature.

Wilson's Disease. Due to disturbance of copper metabolism there occurs an autosomal recessive physiological disorder. The body suffers with a severe copper anaemia and there occurs no urinary catecholamine excretion. The disease was discoverd by English dermatologist Sir J.E. Wilson (1809-1884).

Wings. Flattened membranous structures which are found in animals like insects and birds. They help in flying. The wings of birds are modified fore limbs and homologous with the fore limbs of other vertebrates which wings of insect are analogous structures.

Wish Bone. Used for describing fused clavicles and interclavicle forming a fork-shaped bone which is articulating between pectoral girdles and sternum and is acting like a spring between two halves of pectoral girdle.

Witch's Milk. The milk often found in newly born child.

Wolffian Duct. Refers to the mesonephric duct in vertebrates like reptiles, birds and mammals. Named in the honour of German anatomist Wolff.

Wolffan Vestige. refers to the posterior part of maesonephric kidney in reptiles lying obliquely at the posterior margin of testis.

Wolf-Parkinson-Syndrome. The accelerated AV conduction in some otherwise normal individuals who are prone to attacks of atrial arrhythmias.

Wolffan Ridges. Redges on an embryo from which the limbuds arise.

Wood-Pecker. Common name for *Brachypternus bengalensis* which is characterised by its habit of clinging to trunks park at intervals and searching for holes to pull out insects.

Workers. Among social insects such as been, ants wasps and termites, there are usually sterile females called workers which build the nest, fetch and store food and feed the other members of the colony; they far outnumber all the other castes. In bees the workers are developed from fertilized eggs, the larvae being given a diet of nectar and pollen. If given a more complete diet of 'royal jelly' which is partly predigested and concentrated, the same larvae may develop queen been.

Worms. Collective term for describing small rounded animals like tapeworms, pin-worms, hook-worms, thread-worms etc.

X

Xanthine. Describing a substance which is formed during conversion of urea to uric acid in the liver of uricotelic animals like pigeon. Chemically it is 2, 6-dioxypurine.

Xanthine Oxidase. An enzyme catalysing oxidation of xanthine to uric acid.

Xanthophyll. A yellow pigment which is associated with chlorophyll in plants and in the plant-like protozoa (Phytomastigina). It is an oxidation product of the orange coloured pigment carotin and is having the empirical formula $C_{40}H_{56}O_2$.

Xanthoplasts. Yellow chromatophores or protoplasmic structures having xanthophyll.

Xanthopterin. A yellow pigment which is derived from *pteridine*, occurring in the bodies of wasps and in the wings of certain butterflies.

Xantusilidae. Lizards of Central America and Cuba with short tongues, lacking osteoderms and lacking movable eyelids.

X-Chromosome. One of the sex chromosomes which is related to sex determination. Individuals having double dose (XX) are females while those having single dose and a Y-chromosome (XY) are male (in XY sex determination).

Xerophthalmia. Used for describing drying of retina and caused by deficiency of vitamin-A.

Xerostomia. Used for describing deficient salivation in some cases and such persons are having higher incidence of dental caries.

X-Gland. In some arthropods it is endocrine gland which is passing secretions along axons to sinus gland at the base of eye stalk.

X-Inactivation. Consideration and resultant inactivation of all or any X-chromosome in cell in excess of one.

Xiphisternum. Refers to the posterior and usually cartilagenous part of the sternum in many vertebrates.

Xiphoid Processes. There are posterolateral borders of sternum in birds which are expanded to form elongated flattened structures called xiphoid processes. Also called *xiphisternal processes*.

Xiphosura. Refers to class of chelicerate arthropods to which horse-shoe crab belongs.

X-linked Inheritance. Refers to the Inheritance of certain disorders like colourblindness and haemophilia etc., which occurs due to a mutant gene of X-chromosome. As males have XY genetic constitution, such a mutation affects them more readily (here Y-chromosome only concerns with sex) where as the females are rarely affected as they require a double dose of such chromosome (which is very rare).

XY-chromosomes. Refer to the chromosomes which are concerned with determination of sex. In some animals the males are having XY chromosomes and the female XX chromosomes. Certain other males are having XX chromosomes and female XY chromosomes. A few cases are known to have XO sex determination where males are having XO-chromosomes and females XX-chromosomes.

Y

Y-chromosome. Refers to a chromosome which is unpaired and is present only in one sex. In humans two type of sex chromosome called X and Y are known. Egg-cells always have an X-chromosome but sperm cells may have either an X or a Y. When an egg-cell gets fertilized by an X-sperm cell, a female offspring will be produced; when fertilized by a Y-sperm cell a male will be produced (see *Sex Chromosome*).

Yellow Bone Marrow. Inactive marrow which is infiltrated with fat.

Yellow Elastic Fibres. Elastic fibres which are composed of scleroprotein called elastin. It is insoluble upon boiling but dissolves in proteolytic enzyme solutions. It occurs in ligaments between vertebrae in the wall of large arteries and in areolar tissues.

Yellow Elastic Tissue. Connective tissue having a network of yellow fibres composed of *elastin.* It occurs nearly pure in ligaments and mixed with white fibres in areolar tissue.

Yellow Enzymes (Flavoproteins). Yellow substances which are formed from proteins combined with riboflavin nucleotides (vitamin B_2). A number of these are known which apparently act as hydrogen carriers during cell respiration.

Yellow Spot (Macula Lutea). Refers to that part of the retina which is most sensitive to colour and which has numerous cones and very few if any rods. It is found in the eyes of humans and higher apes and is in the centre of the retina immediately opposite the lens. Clear perception of colour therefore will be only possible when the eyes look straight at an object and the image falls on this part.

Yolk. The food store of an egg which consists largely of proteins and fats. Normally it is formed within the egg-cell and is contained by the vitalline membrane; sometimes, however, as in the

Platyhelminthes, it is sccreted by separate *vitalline glands*.

Yolk Plug. Refers to a mass of yolk which protrudes through and partly blocks the blastopore in the developing embryo of such animals as the frog.

Yolk-sac. A sac having yolk which is communicating directly with the gut in the embryos of birds, reptiles and some fish. In mammalian embryos there is a similar sac which, however, is small and does not contain yolk. It absorbs nutritive secretions from the uterus but ceases to be functional when the placenta has been formed.

Yolk Stalk. A stalk-like structure which connects yolk sac with intestine of embryo in case of *Scoliodon*.

Z

Z-Disc. Refers to a thin membrane, also called *Krause's mambrane*, traversing the fibrils of striped muscle at right angles and dividing each fibril into *sarcomeres*.

Zebra. Common name for striped horse *Equus Zebra.*

Zeis Glands. Refer to the sebaceous glands which are associated with follicies of eye-lashes. Their secretion helps to keep eye-lashes smooth and oily.

Zero Nerve. Terminal nerve which is emerging out from ventral side of cerebrum through the neuropore and passing into olfactory mucous membrane. It is sensory nerve and is bearing a ganglion mid-I way from its origin.

Zero Order Kinetics. Kinetics in which rate of reaction does not depend upon concentration of any of its reactants. This would take place if enzyme concentration was the limiting factor.

Zeugpopodium. Also formed as antebrachium forming fore limb and shank in hind limb. Skeleton support in former is provided by radius and ulna bones and in the later by tibia and fibula bones.

Zona Glomerulosa. Refers to the zone of adrenal gland's cortex consisting of whole of cells which are continuous with columns of cells that form zona fasciculata. Enzymatic mechanisms for aldosterone biosynathesis is confined to zona glomerulosa.

Zona Fasciculate. Adrenal cortex is divided in three zones the middle of which is known as zona fasciculata. The columns have been separated by venous sinuses.

Zona Pellucida. Vitalline membrane in mammals which is often called zona pellucida.

Zona Radiata. Refers to the striated primary membrane in mammalian egg which is separated by follicular cells.

Zona Reticularis. Refers to the innermost zone of adrenal cortex where the cellular columns get interlaced in a network.

Zonula Occludens. Refer to tight junctions existing between epithelial cells, generally having a ring-shaped configuration and helping to occlude transepithelial extracellar passage.

Zonule of Zinn. Fibres which are holding crystalline lens of eye in place.

Zoo-Geographic Regions. Regions of the earth which are often separated by some natural barrier such as a mountain range or an ocean, each having its own characteristic fauna. Nine such regions have been recognized, viz;

1. *Palaearctic* (Northern Europe and Asia),
2. *Nearctic* (North America, Greenland etc.),
3. *Ethiopian* (Africa and Arabia),
4. *Madagascan*,
5. *Neotropical* (Mixico, Central and Sough America),
6. *Oriental* (India, Burma and S.E. Asia west of Wallace's Line),
7. *Australasian* (Australia, New Guinea and islands S.E. of Wallace's Line),
8. *New Zealand* (including neighbouring islands),
9. *Polynesian* (numerous Pacific islands of volcanic origin having no apparent connection with the continents). See also under *Wallace's Line*.

Zoogeography. Branch of zoology which is dealing with the study of geographical distribution of animals.

Zooid. One of the members of hydroid or siphonophore colony. In a limited sense it refers to a particular kind of individual as a hydranth or gonagium. Term is also used for describing a subordinate individual that is formed by transverse fission of animal such as *Planaria*.

Zoology. Branch of science which is dealing with study of animal life.

Zooplankton. Refers to the floating or drifting animals that constitute a part of aquatic ecosystem.

Zygapophyses. Refer to one of the four projections (two anterior and two posterior) extending from neural arch of a vertebra. Those of posterior pair articulates with anterior pair of vertebrae next behind.

Zygantra. Refer to the pair of articular surfaces which are present at the posterior part of vertebrae in snakes.

Zygodactylus. Used for describing foot in birds like parrot in which first and four digits are directed backwords and third and second directed forwards for providing a grip on the branch of a tree.

Zygomatic Arch (Zygoma). Refers to an arch which is forming the cheek-bone of a mammal, consisting of a backward process of the jugal bone joined to a forward process of the squamosal bone.

Zygomatic Process. A process of squamosal which is taking part in the formation of zygomatic arch.

Zygonema. Used for describing a stage during meiosis when the pairing of homologous chromosomes occurs. Also called *synapsis*.

Zygosphenes. Articular surfaces of the anterior ends of vertebrae of snakes.

Zygote. A fertilized ovum: a cell which is formed by the union of the male and female gamete.

Zygotene. Refers to a stage in meiosis at which the chromosomes come together in homologous pairs. The two of a pair are usually having the same shape and size but one is of paternal and the other of maternal origin.

Zygote Nucleus. Refers to the nucleus which is formed by the fusion of male and female nuclei when two gametes combine to form a zygote.

Zymogen Cells. Cells in stomach mucosa that are secreting pepsin.

Zymogen Granules. In the salivary glands, these are the secretory granules having salivary envymes. These get discharged from aciner cells into ducts.

Zymoprotein. An alternative name for an enzyme (*q.v.*)

Zwitterion. A molecule which is carrying both negativlity and positivity or negatively or positively ionised or ionisable sites.